Annals of Mathematics Studies

Number 120

Simple Algebras, Base Change, and the Advanced Theory of the Trace Formula

by

James Arthur and Laurent Clozel

PRINCETON UNIVERSITY PRESS

PRINCETON, NEW JERSEY

1989

Clothbound editions of Princeton University Press books are printed on acid-free paper, and binding materials are chosen for strength and durability. Paperbacks, while satisfactory for personal collections, are not usually suitable for library rebinding

Printed in the United States of America
by Princeton University Press, 41 William Street
Princeton, New Jersey

Library of Congress Cataloging-in-Publication Data

Arthur, James, 1944-
Simple algebras, base change, and the advanced theory of the trace formula / by James Arthur and Laurent Clozel.
p. cm. – (Annals of mathematics studies ; no. 120)
Bibliography: p.
ISBN 0-691-08517-X : ISBN 0-691-08518-8 (pbk.)
1. Representations of groups. 2. Trace formulas. 3. Automorphic forms. I. Clozel, Laurent, 1953- . II. Title. III. Series.
QA171.A78 1988
512'.2–dc19 88-22560
CIP

ISBN 0-691-08517-X (cl.)
ISBN 0-691-08518-8 (pbk.)

Contents

Introduction

The general theory of automorphic forms is in some ways still young. It is expected eventually to play a fundamental unifying role in a wide array of arithmetic questions. Much of this can be summarized as Langlands' functoriality principle. For two reductive groups G and G' over a number field F, and a map ${}^LG' \to {}^LG$ between their L-groups, there should be an associated correspondence between their automorphic representations. The functoriality principle is very deep, and will not be resolved for a long time.

There is an important special case of functoriality which seems to be more accessible. It is, roughly speaking, the case that ${}^LG'$ is the group of fixed points of an automorphism of ${}^LG$. In order that it be uniquely determined by its L-group, assume that G' is quasi-split. Then G' is called a (twisted) endoscopic group for G. Endoscopic groups were introduced by Langlands and Shelstad to deal with problems that arose originally in connection with Shimura varieties. Besides being a substantial case of the general question, a proper understanding of functoriality for endoscopic groups would be significant in its own right. It would impose an internal structure on the automorphic representations of G, namely a partition into "L-packets", which would be a prerequisite to understanding the nature of the general functoriality correspondence. However, the problem of functoriality for endoscopic groups appears accessible only in comparison with the general case. There are still a number of serious difficulties to be overcome.

When the endoscopic group G' equals GL(2), Jacquet and Langlands [25], and Langlands [30(e)], solved the problem by using the trace formula for GL(2). In general, it will be necessary to deal simultaneously with a number of endoscopic groups G', namely the ones associated to those automorphisms of ${}^LG^0$ which differ by an inner automorphism. One would hope to compare a (twisted) trace formula for G with some combination of trace formulas for the relevant groups G'. There now exists a (twisted) trace formula for general groups. The last few years have also seen progress on other questions, motivated by a comparison of trace formulas. The purpose of this book is to test these methods on the simplest case of general rank. We shall assume that G' equals the general linear group $\mathrm{GL}(n)$. A special feature of this case is that there is essentially only one endoscopic group to be considered.

There are two basic examples. In the first case, G is the multiplicative group of a central simple algebra. Then G' is the endoscopic group associated to the trivial automorphism of ${}^L G^0 \cong \mathrm{GL}(n, \mathbf{C})$. This is the problem of inner twistings of $\mathrm{GL}(n)$. In the second case, G is attached to the general linear group of a cyclic extension E of degree ℓ over F. In order to have uniform notation, it will be convenient to write $G^0 = R_{E/F}(\mathrm{GL}(n))$ for the underlying group in this case, while reserving the symbol G for the component $G^0 \rtimes \theta$ in a semidirect product. The trace formula attached to G is then just the twisted trace formula of G^0, relative to the automorphism θ associated to a generator of $\mathrm{Gal}(E/F)$. In this second case, the identity component of the L-group of G^0 is isomorphic to ℓ copies of $\mathrm{GL}(n, \mathbf{C})$, and G' comes from the diagonal image of $\mathrm{GL}(n, \mathbf{C})$, the fixed point set of the permutation automorphism. This is the problem of cyclic base change for $\mathrm{GL}(n)$. In both cases we shall compare the trace formula of G with that of G'. For each term in the trace formula of G, we shall construct a companion term from the trace formula of G'. One of our main results (Theorems A and B of Chapter 2) is that these two sets of terms are equal. This means, more or less, that there is a term by term identification of the trace formulas of G and G'.

A key constituent in the trace formula of G comes from the right convolution of a function $f \in C_c^\infty(G(\mathbf{A}))$ on the subspace of $L^2(G^0(F)\backslash G^0(\mathbf{A})^1)$ which decomposes discretely. However, this is only one of several such collections of terms, which are parametrized by Levi components M in G. Together, they form the "discrete part" of the trace formula

(1)

$$I_{\mathrm{disc},t}(f) = I^G_{\mathrm{disc},t}(f) =$$

$$\sum_M \|W_0^M\| \, \|W_0^G\|^{-1} \sum_{s \in W(\mathfrak{a}_M)_{\mathrm{reg}}} \|\det(s-1)_{\mathfrak{a}_M^G}\|^{-1} \operatorname{tr}(M(s,0)\rho_{P,t}(0,f)),$$

in which $\rho_{P,t}$ is a representation induced from the discrete spectrum of M, and $M(s,0)$ is an intertwining operator. (See §2.9 for a fuller description of the notation, and, in particular, the role of the real number t.) Theorem B of Chapter 2 implies an identity between the discrete parts of the trace formulas of G and G'. We shall describe this more precisely.

Let S be a finite set of valuations of F, which contains all the Archimedean and ramified places. For each $v \in S$, let f_v be a fixed function in $C_c^\infty(G(F_v))$. We then define a variable function

$$f = \prod_v f_v$$

in $C_c^\infty(G(\mathbf{A}))$ by choosing functions $\{f_v : v \notin S\}$ which are spherical (i.e. bi-invariant under the maximal compact subgroup of $G^0(F_v)$). For each valuation v not in S, the Satake transform provides a canonical map $f_v \to f'_v$ from the spherical functions on $G(F_v)$ to the spherical functions on $G'(F_v)$. Our results imply that there are fixed functions $f'_v \in C_c^\infty(G'(F_v))$ for the valuations v in S, with the property that if

$$f' = \prod_v f'_v,$$

then

$$I^G_{\text{disc},t}(f) = I^{G'}_{\text{disc},t}(f'). \tag{2}$$

Given the explicit nature (1) of the distribution $I^G_{\text{disc},t}$, and the fact that the spherical functions $\{f_v : v \notin S\}$ may be chosen at will, we can see that the identity (2) will impose a strong relation between the automorphic representations of G and G'. In particular, we shall use it to establish global base change for $\mathrm{GL}(n)$.

Chapter 1 is devoted to the correspondence $f_v \to f'_v$. We shall also establish a dual correspondence between the tempered representations of $G(F_v)$ and $G'(F_v)$. For central simple algebras, the local correspondences have been established by Deligne, Kazhdan and Vignéras [15]. We can therefore confine ourselves to the case of base change. The correspondence is defined by comparing orbital integrals. For a given f_v, we shall show that there exists a function $f'_v \in C_c^\infty(G'(F_v))$ whose orbital integrals match those of f_v under the image of the norm map from $G(F_v)$ to $G'(F_v)$. At the p-adic places we shall do this in §1.3 by an argument of descent, which reduces the problem to the known case of a central simple algebra.

The main new aspect of Chapter 1 is the proof in §1.4 that the matching of orbital integrals is compatible with the canonical map of spherical functions. The proof is in two steps. We first define a space of "regular spherical functions"; if one represents a spherical function as a finite Laurent series, they are defined by the condition that certain singular exponents do not occur. For these regular functions, the required identities of orbital integrals can be proved inductively by simple representation-theoretic arguments.

An argument of density using the version of the trace formula due to Deligne and Kazhdan then shows that the identities hold for all spherical functions. This argument relies in an essential way on a result of Kottwitz, which proves the identities of orbital integrals for units in the Hecke algebra. Once the comparison theory of spherical functions has been established, it will be easy to obtain the local correspondence of tempered representations (§1.5). It takes the familiar form of a lifting from the representations of $G'(F_v)$ to the representations of $G^0(F_v)$ that are fixed by θ. We shall also prove identities between local L-functions and ε-factors related by lifting (§1.6). For the Archimedean places, the local lifting of representations is already known ([32], [11(a)]). We shall establish the matching of orbital integrals, as well as a Paley–Wiener theorem, in §1.7.

In Chapter 2 we shall compare two trace formulas. The trace formula

$$\begin{aligned}&\sum_M |W_0^M|\,|W_0^G|^{-1} \sum_{\gamma\in(M(F))_{M,S}} a^M(S,\gamma)I_M(\gamma,f)\\ &= \sum_t\sum_M |W_0^M|\,|W_0^G|^{-1}\int_{\Pi(M,t)} a^M(\pi)I_M(\pi,f)\end{aligned} \tag{3}$$

for G will be matched with a formula

$$\begin{aligned}&\sum_M |W_0^M|\,|W_0^G|^{-1} \sum_{\gamma\in(M(F))_{M,S}} a^{M,\mathcal{E}}(S,\gamma)I_M^{\mathcal{E}}(\gamma,f) =\\ &\sum_t\sum_M |W_0^M|\,|W_0^G|^{-1}\int_{\Pi^{\mathcal{E}}(M,t)} a^{M,\mathcal{E}}(\pi)I_M^{\mathcal{E}}(\pi,f)\,d\pi\end{aligned} \tag{3$^{\mathcal{E}}$}$$

obtained by pulling back the trace formula from G' to G. Theorem A establishes an identification of the geometric terms on the left-hand sides of the two formulas, while Theorem B gives parallel identities for the spectral terms on the right. (It is the identity of global spectral terms $a^{G,\mathcal{E}}(\pi)$ and $a^G(\pi)$ which gives the equation (2), and leads to the global correspondence of automorphic represntations.) The two theorems will be proved together by means of an induction argument. We shall assume that all the identities hold for groups of strictly lower dimension. This hypothesis will actually be needed in §2.12 to construct the right-hand side of $(3)^{\mathcal{E}}$. It will also give us considerable scope for various descent arguments. These arguments lead to the identity of $a^{M,\mathcal{E}}(\gamma)$ and $a^M(\gamma)$ in most cases (§2.5), of $a^{M,\mathcal{E}}(\pi)$ and $a^M(\pi)$ in most cases (§2.9), and of $I^{M,\mathcal{E}}(\pi,f)$ and $I^M(\pi,f)$ in all cases (§2.10). They also provide partial information relating $I^{M,\mathcal{E}}(\gamma,f)$ and $I^M(\gamma,f)$ (§2.5, §2.6, §2.7). However, some intractible terms remain in the end, and these must be handled by different methods. In §2.13 and

§2.14, we shall show that for suitable f,

$$\gamma \to I_M^{\mathcal{E}}(\gamma, f) - I_M(\gamma, f), \qquad \gamma \in M(F_S),$$

is the orbital integral in γ of a function on $M(F_S)$. This allows us to apply the trace formula for M. We obtain a relation between the spectral sides of (3), of $(3)^{\mathcal{E}}$, and of the trace formula for M. By comparing the resulting distributions at both the Archimedean and discrete places, we are then able to deduce vanishing properties for the individual terms (§2.15, §2.16). We shall finally complete the induction argument, and the proofs of the two theorems, in §2.17.

As an application of the identity (2), we shall establish base change for GL(n) in Chapter 3. For GL(2), the complete spectral decomposition of the space of automorphic forms is known, and this makes it possible to compare very explicitly the discrete spectra of $\mathrm{GL}(2, \mathbf{A}_F)$ and $\mathrm{GL}(2, \mathbf{A}_E)$. Such explicit information is not available for $n > 3$. If it were, and in particular, if there was a strong enough version of multiplicity one, we would have no trouble deducing all the results on base change directly from the formula (2). We must instead restrict the category of automorphic representations considered to those that are "induced from cuspidal", a natural notion coming from the theory of Eisenstein series. To prove that the lifting exists, and preserves this special kind of automorphic forms, we use (2) in combination with the very precise results obtained by Jacquet and Shalika about the analytic behavior of L-functions associated to pairs of automorphic representations.

Assume that E/F is a cyclic extension of number fields, of prime degree ℓ, with Galois group

$$\{1, \sigma, \sigma^2, \ldots, \sigma^{\ell-1}\}.$$

Given the local lifting, we may define the global lifting as follows. Let $\pi = \bigotimes_v \pi_v$ be an automorphic representation of $\mathrm{GL}(n, \mathbf{A}_F)$, a tensor product over all places v of F; let $\Pi = \bigotimes_w \Pi_w$ be an automorphic representation of $\mathrm{GL}(n, \mathbf{A}_E)$, w denoting a place of E. We say that Π is a (strong) base change lift of π if, for any $w|v$, Π_w lifts π_v. Our main result is Theorem 3.5.2, and applies to representations *induced from cuspidal.* Let π, Π stand for such representations of $\mathrm{GL}(n, \mathbf{A}_F)$, $\mathrm{GL}(n, \mathbf{A}_E)$. We prove that

(i) If Π is σ-stable, – i.e., Π is equivalent to $\Pi \circ \sigma$ – it is a base change lift of finitely many π.

(ii) Conversely, given π, there is a unique σ-stable Π lifting π.

In fact, our results are more explicit. In particular, assume that Π is a *cuspidal* representation of $\mathrm{GL}(n, \mathbf{A}_E)$. We show that

(iii) If $\Pi \cong \Pi^\sigma$, there are exactly ℓ representations π lifted by Π. They are all twists of one of them by powers of the class field character associated to E/F.

(iv) Assume $\Pi \not\cong \Pi^\sigma$. Then the data $(\Pi, \Pi^\sigma, \ldots, \Pi^{\sigma^{\ell-1}})$ define, through the theory of Eisenstein series, an automorphic representation of $\mathrm{GL}(n\ell, \mathbf{A}_E)$. This representation is σ-stable and lifts exactly one cuspidal representation π of $\mathrm{GL}(n\ell, \mathbf{A}_F)$.

Taken together, (iii) and (iv) imply, as shown in §3.6, the existence of automorphic induction, a functor sending automorphic representations of $\mathrm{GL}(n, \mathbf{A}_E)$ to those of $\mathrm{GL}(n\ell, \mathbf{A}_F)$. In particular, this theorem contains, for $n = 1$, Kazhdan's result about the map which sends idèle class group characters of E to cuspidal representations of $\mathrm{GL}(\ell, \mathbf{A}_F)$; this in turn generalized the version given by Labesse and Langlands of the classical construction by Hecke, Maaβ, Weil and Jacquet–Langlands of the forms on GL(2) associated to characters of a quadratic extension.

In §3.7 we apply these theorems to problems related to representations of Galois groups. In particular, we prove the existence of the cuspidal automorphic representation associated to an irreducible representation of a nilpotent Galois group. However, Artin's conjecture is already known for nilpotent groups. Indeed, cyclic or solvable base change alone does not give any new cases of the Artin conjecture. (See §3.7.) Recall that Langlands' application of base change to the Artin conjecture for GL(2) already required another tool, either the lifting from GL(2) to GL(3), or the Deligne–Serre characterization of holomorphic forms of weight 1.

Finally, we observe that our results lead to an interesting property of the representations obtained by (solvable) automorphic induction from Abelian characters. The principle of functoriality implies a multiplicative structure on the set of automorphic representations. If π_n and π_m are automorphic representations of $\mathrm{GL}(n, \mathbf{A}_F)$ and $\mathrm{GL}(m, \mathbf{A}_F)$, there should exist an associated automorphic representation $\pi_n \boxtimes \pi_m$ of $\mathrm{GL}(nm, \mathbf{A}_F)$. If π_n comes from an Abelian character by solvable induction, we can show that this product exists for arbitrary π_m.

The base change problem has an interesting history. For GL(2) and quadratic extensions, it was first studied by Doi and Naganuma in connection with modular curves ([17(a)], [17(b)]). They relied on Weil's converse to Hecke theory, as did Jacquet [24(a)] in further work. Saito [34] introduced the use of a twisted form of the trace formula, and treated certain examples of Hilbert modular forms in cyclic extensions of prime degree of totally real fields. His method was cast by Shintani [39] and Langlands [30(e)] into the mold of automorphic forms on adèle groups. They proved the existence of a lifting from automorphic forms on $\mathrm{GL}(2, \mathbf{A}_F)$ to automorphic forms on $\mathrm{GL}(2, \mathbf{A}_E)$, E/F again being a cyclic extension of prime degree. Shintani also introduced a local notion of lifting: this makes it possible to obtain analogous results for an extension E/F of p-adic fields. The case of GL(3) was later considered by Flicker [19]. We refer the reader to the beginning of Langlands' book [30(e), §1-3] for a more complete introduction of the base change problem and its history, as well as the famous applications to Artin's conjecture in dimension 2.

We would like to thank Robert Langlands for his encouragement while this work was in progress. We are especially indebted to him for suggesting that we exploit the cancellation of singularities, a technique that comes in at a crucial stage in §2.14. We would also like to thank Hervé Jacquet and Robert Kottwitz for useful discussions. This work has been supported in part by NSERC Grant A3483 (J. A.) and a Sloan Fellowship, as well as NSF Grant DMS-8600003 (L. C.).

NOTATIONAL CONVENTIONS: The notation of the introduction will prevail in Chapter 2. In Chapters 1 and 3, which are concerned mainly with base change, we will use a more classical notation. Here we will write G or G_n for the general linear group $\mathrm{GL}(n)$.

If F is a nonArchimedean local field, $\mathcal{O}_F$ will denote the ring of integers.

We shall index our results by both chapter and paragraph. However, we shall omit the numbers of the chapters when referring to theorems, formulas, paragraphs etc., of a current chapter.

Simple Algebras, Base Change,

and the

Advanced Theory of the Trace Formula

CHAPTER 1

Local Results

1. The norm map and the geometry of σ-conjugacy

For this section E/F is a cyclic extension of order ℓ of fields of characteristic 0; we denote by Σ the Galois group, by σ a generator of Σ. We do not assume that ℓ is prime.

As we have agreed, G will stand for $\mathrm{GL}(n)$ throughout Chapters 1 and 3. Recall that $g, h \in G(E)$ are called *σ-conjugate* if $g = x^{-1}hx^{\sigma}$ for an $x \in G(E)$.

If $x \in G(E)$, we will write Nx for the element $xx^{\sigma}\cdots x^{\sigma^{\ell-1}} \in G(E)$; it is called the *norm* of x.

LEMMA 1.1:

(i) If $x \in G(E)$, Nx is conjugate in $G(E)$ to an element y of $G(F)$; y is uniquely defined modulo conjugation in $G(F)$.

(ii) If Nx and Ny are conjugate in $G(E)$, then x and y are σ-conjugate.

Otherwise stated, the norm map is an injection from the set of σ-conjugacy classes in $G(E)$ into the set of conjugacy classes in $G(F)$. We will write $\mathcal{N}x$ for the conjugacy class in $G(F)$ so obtained.

Proof. ([30(e)]). Part (i). Let $p_1(X)|p_2(X)|\cdots|p_r(X)$ be the elementary divisors of the matrix Nx; thus $p_i(X) \in E[X]$. We have $(Nx)^{\sigma} = x^{-1}(Nx)x$. This shows that in fact $p_i(x) \in F[X]$, so the conjugacy class of Nx is defined over F.

For Part (ii), we will need the following construction. Let $u = Nx$. By (i) we may assume that $u \in G(F)$. Let G_u be the centralizer of u, an F-group; it is the set of invertible elements of $\mathfrak{g}_u$, where $\mathfrak{g} = M_n = \mathrm{Lie}(\mathrm{GL}(n))$.

Let $G_{x,\sigma}(F)$ be the σ-centralizer of x: it is the set of all $g \in G(E)$ such that $g^{-1}xg^{\sigma} = x$; it is the set of F-points of a group over F, which we denote by $G_{x,\sigma}$. It is easy to check that $G_{x,\sigma}(F) \subset G_u(E)$; moreover, the F-structure on $G_{x,\sigma}$ is defined by $z \mapsto xa^{\sigma}x^{-1}$. In other terms, $G_{x,\sigma}$ is an inner form (in fact an E/F-form) of G_u, the cocycle being given by $c_{\sigma} = \mathrm{Ad}(x) \circ \sigma$.

The same construction applies to the Lie algebra (which is also naturally an associative matrix algebra): we define $G_{x,\sigma}$ and G_u in the same manner, and G_u is an E/F-form of $G_{x,\sigma}$. Hilbert's Theorem 90 (cf. [35, Exercise 2, p. 160]) then gives

$$H^1(\Sigma, G_{x,\sigma}(E)) = 0.$$

But then an easy cocycle computation gives (ii). ∎

We will say that $x \in G(E)$ is *σ-semi-simple* if the class $\mathcal{N}x$ is semi-simple. In that case, of course, G_u is a semi-simple algebra, isomorphic to a product $\prod_{i=1}^{r} M_{n_i}(F_i)$ where F_i/F are field extensions; G_u is isomorphic to $\prod \mathrm{GL}(n_i, F_i)$ seen as an F-group, and $G_{x,\sigma}$ is an inner form of this group which defines a product of central simple algebras.

Assume now that F is a *global* field. We will need to extend the definition of the local norms to the places of F which are not inert in E. This is easy and we do not give details. Assume for example that v is a place of F which splits in E. Then $E \otimes F_v = F_v \oplus \cdots \oplus F_v$ (ℓ factors), Σ acting by cyclic permutations; we set $N(g_1, \dots g_\ell) = (g_1, \dots g_\ell)(g_2, \dots g_1) \cdots (g_\ell, g_1 \cdots g_{\ell-1}) = (g_1 g_2 \cdots g_\ell, g_2 \cdots g_1, \dots, g_\ell g_1 \cdots g_{\ell-1})$. It is conjugate in $G(E_v)$ to an element of the form $(h, h, \dots h) \in G(F_v)$. The general case is an obvious composite of the split case and the inert case.

LEMMA 1.2: *Assume F is a global field. Then, if $u \in G(F)$, $u = Nx$ has a solution in $G(E)$ if and only if it has a solution in $G(E_v)$ for any place v of F.*

Proof. Only the "if" part need be proved. We will first treat the case of a semi-simple u. We may write u as a diagonal matrix

$$\begin{pmatrix} u_1 & & & & & & & & \\ & \ddots & & & & & & & \\ & & u_1 & & & & & & \\ & & & u_2 & & & & & \\ & & & & \ddots & & & & \\ & & & & & u_2 & & & \\ & & & & & & \ddots & & \\ & & & & & & & u_k & \\ & & & & & & & & \ddots & \\ & & & & & & & & & u_k \end{pmatrix}$$

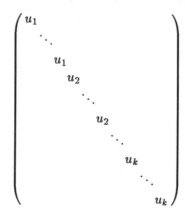

where u_i generates a field extension F_i/F of degree m_i, embedded in $\mathrm{GL}(m_i, F)$. The centralizer $\mathfrak{g}_u$ is then (if $u_1 \neq u_2 \neq \cdots \neq u_k$) a product of matrix algebras $M_{k_i}(F_i) \subset M_{k_i m_i}(F)$, where u_i appears k_i times. It is easy to see that the problem actually takes place in $\prod_i M_{k_i m_i}(F)$; thus we may assume that u has only one eigenvalue, say $u_1 \in F_1^*$. We set $k = k_1$, $m = m_1$.

Let us first assume that $F_1 = F$. The hypothesis is that for any place v, $u = Nx_v$, $x_v \in \mathrm{GL}(k, E_v)$. Taking determinants, we have $\det u = N(\det x_v) \in NE_v^*$. Thus $u_1^k \in NE^*$ since in F^*, an element which is a local norm everywhere is a global norm. We now use the following lemma:

LEMMA 1.2.1: (J.-J. Sansuc). *Let k, ℓ be two integers and E/F an extension of local or global fields of characteristic 0, cyclic of degree ℓ. Assume $x \in F^*$ is such that $x^k \in N_{E/F}E^*$. Then there exists an étale algebra F_0/F of degree k such that $x \in N_{E_0/F_0}E_0^*$, where E_0 is the cyclic étale algebra $E \otimes F_0$ over F_0.*

Proof. Assume first that F is a p-adic field. Then any field extension F_0/F of degree k has the requested property. Indeed, there is a commutative diagram (Serre [35, p. 201])

$$\begin{array}{ccccccc}
F^*/N_{E/F}E^* = \hat{H}^0(\Sigma, E^*) & \xrightarrow{\approx} & H^2(\Sigma, E^*) & \hookrightarrow & \mathrm{Br}\, F & = & \mathbf{Q}/\mathbf{Z} \\
\downarrow \qquad\qquad \downarrow & & \downarrow & & \downarrow & & \downarrow x^k \\
F_0^*/N_{E_0/F_0}E_0^* = \hat{H}^0(\Sigma, E_0^*) & \to & H^2(\Sigma, E_0^*) & \hookrightarrow & \mathrm{Br}\, F_0 & = & \mathbf{Q}/\mathbf{Z}
\end{array}$$

where Br denotes the Brauer group; whence a square

$$\begin{array}{ccc}
F^*/N_{E/F}E^* & = & \mathbf{Z}/\ell\mathbf{Z} \\
\downarrow & & \downarrow x^k \\
F_0^*/N_{E_0/F_0}E_0^* & \hookrightarrow & \mathbf{Z}/\ell\mathbf{Z}
\end{array}$$

which implies the result.

If now $F = \mathbf{R}$, the only nontrivial case is when $E/F = C/\mathbf{R}$ and k is even. It suffices to take $F_0 = \mathbf{C}^{k/2}$.

Now assume F is a number field. We may choose a finite set S of places of F such that, if $v \notin S$, x is a local norm in the extension $(E \otimes F_v)/F_v$. For every finite place $v \in S$, set $n_v = k$; if v is infinite, set $n_v = 1$ if k is odd or if k is even and $F_v = \mathbf{C}$, and $n_v = 2$ otherwise. We now quote the following theorem (Artin–Tate [3, p. 105]):

THEOREM 1.2.2: *Assume F is a number field, S a finite set of places of F and $(n_v)_{v\in S}$ integers such that $n_v = 1$ or 2 $(F_v \cong \mathbf{R})$ and $n_v = 1 (F_v \cong \mathbf{C})$. There exists a cyclic extension F_0/F, of degree $n = l.c.m.(n_v)$ such that, for $v \in S$, the extension $F_{0,v}/F_v$ is a field extension, cyclic of degree n_v.*

We apply the theorem, with the n_v fixed before. By the p-adic and Archimedean cases, we see that for $v \in S$, $x \in N_{E_{0,w}/F_{0,w}}(E_{0,w}^*)$ for any place w of F_0 above v. If $v \notin S$, x is a local norm from $E \otimes F_v$ and *a fortiori* from $E_0 \otimes F_v$. Therefore x (considered as an element of F_0) is a local norm everywhere and thus a global norm. This finishes the proof of Lemma 1.2.1. ∎

We can now prove Lemma 1.2 for u scalar $(F_1 = F)$. Since $u_1^k \in NE^*$, Lemma 1.2.1 ensures the existence of a field F_0/F ($\cdots$ in fact cyclic) of degree k such that $u_1 \in N_{E\otimes F_0/F_0}(x)$ for $x \in (E \otimes F_0)^*$. We can embed the extension F_0 of F into $M_k(F)$; then $E \otimes F_0$ is embedded into $M_k(E)$, and this yields an element of $M_k(E)$ whose norm equals u.

Now let us treat the general case of an element

$$u = \begin{pmatrix} u_1 & & \\ & \ddots & \\ & & u_1 \end{pmatrix} \qquad (k \text{ copies}),$$

where $u_1 \in F_1^*$ embedded into $\mathrm{GL}(m, F)$ and u_1 generates F_1. The centralizer of u in $M_{km}(F)$ is then isomorphic to $M_k(F_1)$ as an F-algebra.

Assume u is a local norm at the place v of F: $u = N_{E_v/F_v} x_v$, $x_v \in \mathrm{GL}(mk, E_v)$, where $E_v = E \otimes F_v$.Then, since $u = u^\sigma$:

$$x_v u = x_v u^\sigma = x_v x_v^\sigma \cdots x_v^{\sigma^{\ell-1}} x_v = u x_v.$$

Thus x_v lies in $M_k(F_1 \otimes E_v)$. If u is a local norm everywhere, we see that it is a local norm in the F-algebra $M_k(F_1)$. Applying the case already proved of the lemma to $M_k(F_1)$ and the extension $F \otimes F_1/F_1$ (this may not be a field but the extension to cyclic étale algebras is obvious), we see that there is an $x \in M_k(E \otimes F_1)$ such that $N_{E/F} x = u$.

This solves the problem in the semi-simple case.

We now treat the general case. Assume $u \in \mathrm{GL}(n, F)$ is a local norm everywhere. Let $u = sn$, s semi-simple, n unipotent, be its Jordan decomposition.

Notice first that the norm map may be defined by considering the non-connected group H over F defined by $H = (\mathrm{Res}_{E/F}\mathrm{GL}(n)) \rtimes \Sigma$, where Σ acts on $\mathrm{Res}_{E/F}\mathrm{GL}(n)$ by F-automorphisms via its action as a Galois group: in particular its action on $\mathrm{GL}(n,E) = (\mathrm{Res}_{E/F}\mathrm{GL}(n))(F)$ is its Galois action on $\mathrm{GL}(n,E)$. The norm map is just the ℓ-th power in $H(F)$; more precisely,

$$(g,\sigma)^\ell = (Ng,1)$$

for $g \in \mathrm{GL}(n,E)$. The group H is linear, and the Jordan decomposition is available in it.

Assume now that u is a local norm at the place v of F. We then have

$$(u,1) = (g,\sigma)^\ell \qquad g \in \mathrm{GL}(n,E_v). \tag{1.1}$$

Using Jordan decomposition in $H(E_v)$ we write:

$$(g,\sigma) = (s_1,\sigma)(n_1,1) = (n_1,1)(s_1,\sigma) \tag{1.2}$$

with $n_1 \in \mathrm{GL}(n,E_v)$ unipotent, $s_1 \in \mathrm{GL}(n,E_v)$ such that (s_1,σ) is semi-simple. Taking ℓ-th powers in H, we see that this last condition is equivalent to Ns_1 being semi-simple. On the "connected component" $\mathrm{GL}(n,E)$ of $H(E)$, equations (1.1) and (1.2) translate as

$$s = Ns_1, \quad n = n_1^\ell \tag{1.3}$$

$$s_1 n_1^\sigma = n_1 s_1. \tag{1.4}$$

Since $n \in \mathrm{GL}(n,F)$ we see first, taking logarithms, that $n_1 \in \mathrm{GL}(n_1,F)$. We will write $X = \log n \in M_n(F)$. By (1.4) we see now that s_1 commutes with X.

We now finish the proof of Lemma 1.2 under the assumption that

$$s = \begin{pmatrix} x & & \\ & \ddots & \\ & & x \end{pmatrix}$$

is a scalar matrix in $\mathrm{GL}(n,F)$. Let

$$V_0 = F^n = \mathrm{Ker}\,X^k \supset V_1 = \mathrm{Ker}\,X^{k-1} \supset \cdots \supset V_i = \mathrm{Ker}\,X^{k-i} \supset \cdots \supset \{0\}$$

be the flag associated to the nilpotent matrix X. As s_1 commutes with X, it acts on V_i/V_{i+1} as an endomorphism s_1^i defined over E_v. Clearly this graded action commutes with the taking of norms, so we see that $N_{E_v/F_v}(s_1^i)$ is the diagonal matrix with entries equal to x; the norm is taken, of course, in $\mathrm{GL}(a_i)$ where $a_i = \dim V_i/V_{i+1}$. Taking determinants, we conclude that x^{a_i} is a norm from E_v to F_v.

If $a = g.c.d.(a_i)$, this implies that x^a is a norm from E_v to F_v; as this applies to all places of F, we see that $x^a \in F^*$ is a norm from E^*. By the proof of the lemma in the semi-simple case, we see that the diagonal matrix

$$\begin{pmatrix} x & & \\ & \ddots & \\ & & x \end{pmatrix}$$

(a entries) is the norm of an element $t \in \mathrm{GL}(a, E)$. By the Jordan canonical form, n_1 may be written as a matrix with square blocks of dimension a, equal to 0 or 1. Such a matrix commutes with the element $s_2 \in \mathrm{GL}(n, E)$ having diagonal blocks $(t \cdots t)$ and all its σ-conjugates; clearly $N(s_2 n_1) = u$.

Finally, in the general case, equation (1.4) shows that n_1 preserves the decomposition of n-space according to the eigenvalues of s. Thus, as in the semi-simple case, we may first assume that s has only one eigenvalue $u_1 \in F_1^*$. Just as in the semi-simple case, we see that the problem actually takes place in the algebra $M_k(F_1)$ where s is identified to a semi-simple diagonalized element. This reduces to the previously treated case. ∎

We now study more especially the case of elliptic regular elements. (Recall that $u \in \mathrm{GL}(n, F)$ is regular elliptic if its eigenvalues generate an extension of F of degree n.)

LEMMA 1.3: *Assume $u \in G(F)$ is elliptic regular; let $F_1 \cong \mathfrak{g}_u(F)$ be the field generated by u.*

(i) The equation $u = Nx$ has a solution if and only if $u \in N_{L/F_1} L^*$, where $L = E \otimes F_1$; the norm from $E \otimes F_1$ to F_1 is defined by the structure of F-algebra on F_1.

(ii) In particular, if F_1 is F-isomorphic to E, $u = Nx$ has a solution.

Proof. If $u = Nx$, x commutes to u, hence $x \in \mathfrak{g}_u(F)$ which is isomorphic to L as an F-algebra. Thus $u \in N_{L/F_1} L^*$, and the converse statement in (i) is clear also. If $F_1 \cong E$, then F_1 splits over E and the norm map is onto; this proves (ii). ∎

LEMMA 1.4: *Assume E/F is a cyclic extension of local fields, and $u \in G(F)$ is elliptic regular. Then u is a norm if and only if* $\det u \in N_{E/F} E^*$.

Proof. If u is a norm, we see, taking determinants, that *det* u is one. Conversely, assume that $\det u$ is a norm. We will rely on Lemma 1.3(i). Let us first assume that F_1 and E are linearly disjoint. Let L be their compositum:

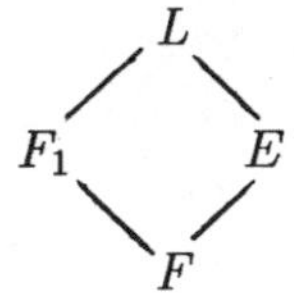

The extension L/F_1 is Abelian, of Galois group Σ. If K is a local field, denote by j_K the local reciprocity map: $K^* \to \mathrm{Gal}(K_{ab}/K)$, where K_{ab} is an Abelian closure of K. Then $u \in N_{L/F_1}L^*$ if and only if $j_{F_1}(u) = 1$ on L. On the other hand, $\det u \in F^*$ is naturally identified with $N_{F_1/F}(u)$. Then $\det u \in N_{E/F}E^*$ if and only if $j_F(\det u) = 1$ on E. By the compatibility of the local reciprocity maps (Serre [35, p. 178]):

$$j_F(N_{F_1/F}u) = c_{F_1/F}\, j_{F_1}(u)$$

where $c_{F_1/F}$ is the canonical map: $\mathrm{Gal}(F_{1,ab}/F_1) \to \mathrm{Gal}(F_{ab}/F)$. Assume then that $\det u$ is a norm, i.e., $j_F(N_{F_1/F}u) = 1$ on E. Then, by the isomorphism $\mathrm{Gal}(E/F) \underset{\sim}{\to} \mathrm{Gal}(L/F_1)$, we see that $j_{F_1}(u) = 1$ on L, which shows that u is a norm.

We now treat the general case. Let F'/F be a maximum subextension of E such that F' and F_1 are linearly disjoint. By the transitivity of Abelian norms, we have $\det u \in N_{F'/F}(F')^*$. Thus u is the norm of an element v in $(F_1 \otimes F')^*$, by the case already proved. But in the extension E/F', the field $F_1 \otimes F'$ splits totally: in particular, every element in $(F_1 \otimes F')^*$ is equal to $N_{E/F'}x$, for $x \in (F_1 \otimes E)^*$. By composition of (Abelian) norms, we see that u is a norm. ∎

We note that Lemma 1.4 is equivalent to the following assertion in Galois cohomology. Let $\mathbf{F}^*$ be the multiplicative group regarded as an F-torus. Let $\mathbf{F}_1^*$ be the F-torus canonically associated to F_1: thus $\mathbf{F}_1^*(E) = (F_1 \otimes E)^*$ for E an F-algebra. The norm map $N_{F_1/F}$ sends $\mathbf{F}_1^*$ to $\mathbf{F}^*$. Let, for $\mathbf{T}$ an F-torus, $\hat{H}^i(\Sigma, \mathbf{T}(E))$ denote the i-th Tate cohomology group of $\Sigma = \mathrm{Gal}(E/F)$ in $\mathbf{T}(E)$. Then:

Lemma 1.5: *The norm is an injection:*

$$\hat{H}^0(\Sigma, \mathbf{F}_1^*(E)) \xrightarrow[N_{F_1/F}]{} \hat{H}^0(\Sigma, \mathbf{F}^*(E)).$$

We finish with a last definition. We will say that (in the local or global case) $x \in G(E)$ is *σ-regular* if its norm is a regular semi-simple conjugacy class in $G(F)$.

2. Harmonic analysis on the non-connected group

2.1. In this section, unless otherwise stated, E/F is a cyclic extension of local non-archimedean fields of characteristic 0. Let $\Sigma = \langle\sigma\rangle$ be the Galois group. Let ℓ be the order of Σ.

We will denote by $\tilde{G}(E)$ the semi-direct product $G(E) \rtimes \Sigma$, the Galois group acting by the action on $G(E)$ defined by the F-structure. In an obvious way, this can be seen as the group of F-points of a non-connected linear algebraic group H defined over F.†

The standard theory of admissible representations of (connected) reductive p-adic groups extends to such groups – cf. [11(e)]. We will be mostly interested in the following type of admissible representations of $\tilde{G}(E)$. Assume that Π is an irreducible, admissible representation of $G(E)$ on a space $\mathcal{V}$. We say that Π is *σ-stable* if it is equivalent to the representation Π^σ defined by

$$\Pi^\sigma(g) = \Pi(\sigma g), \qquad g \in G(E).$$

By definition, there is then a nonzero intertwining operator $I_\sigma : \mathcal{V} \to \mathcal{V}$ between Π and Π^σ. By Schur's lemma, I_σ^ℓ, which intertwines Π and itself, must be scalar. We may first normalize I_σ by assuming $I_\sigma^\ell = 1$. This defines I_σ up to an ℓ-th root of unity.

We want to make a canonical choice of I_σ. This will rely on Whittaker models.

Assume first that the representation Π is *generic* (cf. [27(a), §1]). By definition, there is a linear form $\lambda \neq 0$ on $\mathcal{V}$ such that

$$\lambda(\pi(n)v) = \theta(n)\lambda(v), \qquad v \in \mathcal{V}$$

for any $n = \begin{pmatrix} 1 & x_1 & & * \\ & \ddots & \ddots & \\ & & \ddots & x_{n-1} \\ & 0 & & 1 \end{pmatrix}$ in the upper unipotent group;

θ is the character $n \mapsto \psi\ (\mathrm{tr}_{E/F}(x_1 + \cdots + x_{n-1}))$, where ψ is a non-trivial additive character of F. Moreover, the space of such functionals has dimension one. In that case, we normalize I_σ by requiring that ${}^tI_\sigma\lambda = \lambda$ for the dual action on $\mathcal{V}^*$.

†Note the slight abuse of notation: $\tilde{G}(E)$ *is not* the set of E-points of an algebraic group over E.

In general, by the Langlands classification for p-adic groups ([40(a)], [9]), Π can be realized as the unique submodule of an induced representation

$$\mathbf{R} = \mathrm{ind}_{M(E)N(E)}^{G(E)}(\Pi_M \otimes 1).$$

Here MN is a parabolic subgroup of G, that we may take to be defined over F; Π_M, a representation of $M(E)$, is essentially tempered, and in particular generic ([24(b)], [4]). By uniqueness of the Langlands classification, Π_M must be σ-stable. If I_σ^M is the normalized intertwining operator on Π_M, we define I_σ^G, acting on the space of $\mathbf{R}$, by induction (cf. [11(b), §6.2]). The restriction of I_σ^G to Π is then the normalized intertwining operator on Π.

To check that the definition of I_σ is independent of the choices involved, it is enough to check the following lemma, which follows easily from the transitivity properties of Whittaker vectors; we omit the proof.

LEMMA 2.1: *Assume Π_M is an irreducible, σ-stable, generic representation of a Levi subgroup $M(E)$. Assume $\Pi_G = \mathrm{ind}(\Pi_M)$ is irreducible and generic. Then the normalized intertwining operator on Π_G coincides with the operator induced from the normalized intertwining operator on Π_M.*

For any irreducible, σ-stable Π, we now define the canonical extension of Π to $\tilde{G}(E)$ by setting

$$\Pi(g \rtimes \sigma^i) = \Pi(g)\, I_\sigma^i.$$

This is an irreducible, admissible representation of $\tilde{G}(E)$. We define the *twisted character* of the representation Π as the distribution on $G(E)$ whose value on $\phi \in C_c^\infty(G(E))$ is given by

$$\Theta_{\Pi,\sigma}(\phi) = \mathrm{trace}(\Pi(\phi)\, I_\sigma).$$

Thus the twisted character is actually the trace of the canonical extension of Π on the component $G(E) \rtimes \sigma$ of $\tilde{G}(E)$.

PROPOSITION 2.2: *The twisted character $\Theta_{\Pi,\sigma}$ is given by a locally integrable function, locally constant in the neighborhood of σ-regular elements.*

Proof. This results from Theorem 1 of [11(e)]. The theorem says that the representation Π extended to $\tilde{G}(E)$ has a locally integrable character. We only have to check that the regular elements of $G(E) \rtimes \sigma$, as defined in [11(e)], are just the elements $g \rtimes \sigma$ where $g \in G(E)$ is σ-regular.

By definition ([11(e), §1]), $g \rtimes \sigma$ is regular if $D_{\tilde{G}}(g \rtimes \sigma) \neq 0$; here $D_{\tilde{G}}$ is given by

$$\det(T - \mathrm{Ad}(g) \circ \sigma + 1) = T^n D_{\tilde{G}}(g \times \sigma) + \text{ terms of higher degree.}$$

By Lemma 1 of [11(e)], this implies that $g \times \sigma$ is semi-simple in $H(F)$; so $(g \rtimes \sigma)^\ell = (Ng, 1)$ is semi-simple, which implies that g is σ-semi-simple. But then the eigenspace, for the eigenvalue 1, of $\mathrm{Ad}\, g \circ \sigma$, is just the Lie algebra of the set of F-points of the σ-centralizer of g: since over E this group is isomorphic to the centralizer of Ng, we see that Ng must be regular, so g is σ-regular. Conversely, if g is σ-regular, it is easy to check by the same argument that $g \rtimes \sigma$ is regular in $H(F)$. ∎

We will need next the analog, for twisted characters, of a result of Casselman relating characters and Jacquet modules. We briefly recall Casselman's theorem. If $g \in G(F)$, there is a canonical way to associate to g an F-parabolic subgroup $P_g = M_g N_g$ of G; $P_g(F)$ is the set of points contracted by $\mathrm{Ad}\, g$ ([10(c)]). Assume $g \in G(F)$ is regular. Then $g \in M_g(F)$; the Jacquet module Π_{N_g} associated to an admissible representation Π ([10(a),(b)]) is a representation of $M_g(F)$ and

$$\mathrm{trace}\, \Pi(g) = \mathrm{trace}\, \Pi_{N_g}(g).$$

Now assume that Π is a representation of $G(E)$. If $\Pi \cong \Pi \circ \sigma$ and $P = MN$ is defined over F, the operator I_σ acts on the Jacquet module Π_N; we denote again by I_σ the operator so defined. We will write $\mathrm{trace}(\Pi(g)\, I_\sigma)$ for the character $\Theta_{\Pi,\sigma}(g)$.

PROPOSITION 2.3: *Assume $\Pi \cong \Pi \circ \sigma$. Assume $g \in G(E)$; we assume that $Ng = h \in G(F)$. Let $P_h = M_h N_h$ the associated F-parabolic subgroup. Then, if Ng is regular:*

$$\mathrm{trace}(\Pi(g)\, I_\sigma) = \mathrm{trace}(\Pi_{N_h}(g)\, I_\sigma).$$

The proof is an easy paraphrase of Casselman's ([10(c)]), and is omitted. ∎

2.2. The next result of harmonic analysis that we will need is the analogue of a theorem of Kazhdan on the approximation of orbital integrals by characters. Before proving this, however, we need to extend to the twisted case the form of the trace formula due to Deligne and Kazhdan.

We first state the non-twisted version, in the form in which we will use it in later proofs.

Assume that E/F is now a cyclic extension of number fields. We write $\mathbf{A}$ for the adèles of F, $\mathbf{A}_E$ for the adèles of E. Let Z be the diagonal subgroup of G; let $Z_1 = N(\mathbf{A}_E) \subset Z(\mathbf{A}) \cong \mathbf{A}^*$. We fix a unitary character χ of Z_1, such that $\chi = 1$ on $Z_1 \cap F^*$.

Let us choose two finite places v_1, v_2 of F which split completely in E. Let f be a smooth function on $G(\mathbf{A})$ such that:

(1) $f(zg) = \chi(z)^{-1} f(g)$ for $z \in Z_1$.

(2) f is a tensor product of local functions $f = \prod f_v$; at almost all places, $f_v(zk) = \chi(z^{-1})$ if $z \in Z_1 \cap F_v^*$, and $k \in G(\mathcal{O}_v)$; f_v is zero on all other elements.

(3) f_{v_1} is a coefficient of a supercuspidal representation of $G(F_{v_1})$.

(4) f_{v_2} is supported on the set of regular elliptic elements of $G(F_{v_2})$ whose image in $PGL(n, F_{v_2})$ is regular.

Let r be the representation of $G(\mathbf{A})$, by right translations, on $L^2(G(F)Z_1\backslash G(\mathbf{A}), \chi)$, the space of L^2 functions on $G(F)\backslash G(\mathbf{A})$ which transform by χ under Z_1. Let r_{cusp} the subrepresentation on the space of cusp forms.

If $\gamma \in G(F_v)$, let, for $f_v \in C_c^\infty(G(F_v))$

$$\Phi_{f_v}(\gamma) = \int_{G_\gamma(F_v)\backslash G(F_v)} f(g^{-1}\gamma g)\frac{dg}{dt}$$

denote the corresponding orbital integral, for some choices of dg and dt.

LEMMA 2.4: (Deligne–Kazhdan, cf. [15], [21]). *If f is as above, the operator $r(f)$ sends the space of L^2 automorphic forms in the space of cusp forms; moreover*

$$\text{(I)} \qquad \operatorname{trace} r_{\text{cusp}}(f) = \sum_{\{\gamma\}} \operatorname{vol}(G_\gamma(F)Z_1\backslash G_\gamma(\mathbf{A}))\Phi_f(\gamma).$$

The sum ranges over the set of regular elliptic conjugacy classes in $G(F)$; $\Phi_f(\gamma)$ is the orbital integral

$$\Phi_f(\gamma) = \prod_v \Phi_{f_v}(\gamma).$$

The trace is taken for a measure $dg_{\mathbf{A}} = \prod dg_v$ on $G(\mathbf{A})$; that measure enters in the local orbital integrals, and the volume of $G_\gamma(F)Z_1\backslash G_\gamma(\mathbf{A})$ is computed for the product of the local measures on the tori $G_\gamma(F_v)$ figuring in the local orbital integrals.

Let now χ_E be a unitary character of $Z(\mathbf{A}_E)$, trivial on $Z(E)$. We consider the representation of $G(\mathbf{A}_E)$ on $L^2(G(E)\,Z(\mathbf{A}_E)\backslash G(\mathbf{A}_E), \chi_e)$. This space carries a natural action of Σ. Write I_σ for the operator associated to σ. Assume now that the function ϕ on $G(\mathbf{A}_E)$ satisfies conditions analogous to (1) and (2) above, and moreover:

(3') On $G(E_{v_1}) \cong G(E_{w_1}) \times G(E_{w_2}) \times \cdots \times G(E_{w_\ell})$ (ℓ factors), we have $\phi_{v_1} = (\phi_{w_1}, \dots \phi_{w_\ell})$ where each ϕ_{w_1} is a coefficient of the same supercuspidal representation π of $G(E_{w_i}) \cong G(F_{v_1})$.

(4') Let $\phi_{v_2} = (\phi_{w_1}, \dots \phi_{w_\ell})$ be the analogous decomposition at v_2 ($w_1, \dots w_\ell$ are not the same as in (3)). Let $\Omega_i = \mathrm{Supp}(\phi_{w_i})$, where $\mathrm{Supp}(f)$ denotes the support of f.

Then $\Omega_1 \Omega_2 \cdots \Omega_\ell$ is contained in the set of elements of $G(F_{v_2})$ with regular elliptic image in $PGL(n, F_{v_2})$.

Let again r denote the right representation, r_{cusp} its cuspidal part.

LEMMA 2.5: *Under those assumptions, the operator $r(\phi)$ sends L^2 automorphic forms into cusp forms, and*

$$\text{(II)} \qquad \mathrm{trace}(r_{\mathrm{cusp}}(\phi)\, I_\sigma) = \sum_{\{\delta\}} \mathrm{vol}(G_{\delta,\sigma}(F) Z(\mathbf{A}_E) \backslash G_{\delta,\sigma}(\mathbf{A}_E)) \Phi_{\phi,\sigma}(\delta).$$

Here $\{\delta\}$ runs over the σ-conjugacy classes of elements of $G(E)$ with elliptic regular norms. The group $G_{\delta,\sigma}$ is the σ-centralizer of δ, an F-torus.

$$\Phi_{\phi,\sigma}(\delta) = \prod_v \int_{G_{\delta,\sigma}(F_v) \backslash G(E_v)} \phi_v(g^{-1} \delta_v g^\sigma) \frac{dg}{dt}$$

is the product (over the places of F) of the local twisted orbital integrals. Normalization of measures is as in Lemma 2.4.

We sketch the proof, following Henniart's article [21]. First, the image of $r(\phi)$ is in the space of cusp forms, by Lemma 2.4. It is clear that I_σ preserves the cusp forms. Thus $r(\phi)\, I_\sigma$ is trace-class – since the cuspidal part of $r(\phi)$ is – and

$$\mathrm{trace}(r_{\mathrm{cusp}}(\phi)\, I_\sigma) = \mathrm{trace}(r(\phi)\, I_\sigma).$$

As in [21, §4.9] we obtain the trace by integrating along the diagonal the kernel associated to $r(\phi) I_\sigma$, whence

$$\mathrm{trace}(r(\phi)\, I_\sigma) = \int_{G(E)Z(\mathbf{A}_E)\backslash G(\mathbf{A}_E)} \left\{ \sum_{\gamma \in Z(E) \backslash G(E)} \phi(g^{-1} \gamma g^\sigma) \right\} dg.$$

At the place v_2, we have $E_{v_2} \cong F_{v_2} \oplus \cdots \oplus F_{v_2}$ (ℓ factors), the Galois group acting by cyclic permutations.

Let $\gamma \in G(E)$; the image of γ in the completion E_{v_2} is of the form $(\gamma_1, \gamma_2, \dots, \gamma_\ell)$, $\gamma_i \in G(F_{v_2})$. If $g = (g_1, \dots, g_\ell) \in G(E_{v_2})$, we have

$$g^{-1} \gamma g^\sigma = (g_1^{-1} \gamma_1 g_2, g_2^{-1} \gamma_2 g_3, \dots, g_\ell^{-1} \gamma_\ell g_1).$$

Assume then that $\phi_{v_2}(g^{-1}\gamma g^\sigma) \neq 0$. By assumption 4', we have $g_i^{-1}\gamma g_{i+1}^\sigma \in \Omega_i$, whence, taking the product:

$$g_1^{-1}(\gamma_1\gamma_2\ldots\gamma_\ell)g_1 \in \Omega_1\Omega_2\ldots,\Omega_\ell.$$

This shows that $N\gamma$ is regular elliptic at v_2, and *a fortiori* as a global element. Therefore, in the sum appearing in the expression of trace$(r(\phi)I_\sigma)$, only elements with regular elliptic norms appear.

LEMMA 2.6: *The function*

$$F(g) = \sum_{\substack{\gamma\in Z(E)\backslash G(E)\\ N\gamma \text{ elliptic regular}}} |\phi(g^{-1}\gamma g^\sigma)|$$

is compactly supported on $G(E)\,Z(\mathbf{A}_E)\backslash G(\mathbf{A}_E)$.

Proof. First of all, the sum is finite, uniformly for g in a compact set. Moreover, $\phi(g^{-1}\gamma g^\sigma) \neq 0$ implies $g^{-1}\gamma g^\sigma \in C = \mathrm{Supp}(\phi)$, whence $\gamma \in gCg^{-\sigma}$. Taking norms, we get

$$N\gamma \in g(CC^\sigma\cdots C^{\sigma^{\ell-1}})g^{-1},$$

with $CC^\sigma\cdots C^{\sigma^{\ell-1}}$ compact. Henniart ([21, Appendice 3]) shows that the set of g satisfying this condition for some elliptic regular δ (in lieu of $N\gamma$) is compact modulo $G(E)\,Z(\mathbf{A}_E)$. This proves the lemma. ∎

This shows that in the expression for the trace, we may now permute sum and integral; the usual manipulation then yields Lemma 2.5. (Note that no indices appear in the term multiplying an orbital integral, because we have assumed that the images of the relevant elements *in the projective group* are regular.) ∎

Let us denote by

$$\Phi_{\phi,\sigma}(\delta) = \int_{G_{\delta,\sigma}(F)\backslash G(E)} \phi(g^{-1}\delta g^\sigma)\frac{dg}{dt}$$

the local twisted orbital integral; thus E/F is an extension of local fields, and $G_{\delta,\sigma}$ is the twisted centralizer. (The convergence of this integral will be checked in §3. If δ is σ-semi-simple, in particular, the orbit is closed in $G(E)$ so convergence is obvious.)

PROPOSITION 2.7: *Assume that E/F is an extension of local fields (Archimedean or not), and that $\delta \in G(E)$ is σ-regular. Then, if $\phi \in C_c^\infty(G(E))$ is such that*

$$\Theta_{\Pi,\sigma}(\phi) = 0$$

for any tempered, σ-stable representation Π of $G(E)$, the orbital integral $\Phi_{\phi,\sigma}(\delta)$ vanishes.

Proof. First, an argument of descent reduces to the case of δ having elliptic norm. Assume $N\delta$ is not elliptic: then $\delta \in M(E)$ for a Levi subgroup M of G defined over F. Let $P = MN$ be an associated parabolic subgroup. Let $K_E = G(\mathcal{O}_E)$. Then, if

$$\bar{\phi}(g) = \int_{K_E} \phi(kgk^{-1})\, dk,$$

a standard descent formula gives ([40(b), 29(a)])

$$\Phi_{\phi,\sigma}(\delta) = |\Delta_{G/M}(N\delta)|^{-1}\Phi^M_{\bar{\phi}^{(P)},\sigma}(\delta).$$

Here the orbital integral on the right is taken in $M(E)$;

$$|\Delta_{G/M}| = |(D_{G/M})|^{\frac{1}{2}}$$

is a certain discriminant – see §4; and $\bar{\phi}^{(P)}$ is the constant term of $\bar{\phi}$, defined by:

$$\psi^{(P)}(m) = \delta_P(m)^{\frac{1}{2}} \int_{N(E)} \psi(mn)\, dn$$

(cf. eg. [29(a), §5]). On the other hand, if the representation Π is induced from a σ-stable representation Π_M of $M(E)$, and the intertwining operators correspond, an easy extension of a theorem of Harish–Chandra [20(d)] yields:

$$\Theta_{\Pi,\sigma}(\phi) = \Theta_{\Pi_M,\sigma}(\bar{\phi}^{(P)}).$$

Therefore, if we assume the proposition for M, we see that the twisted orbital integral of δ must vanish.

Assume now $N\delta$ is elliptic regular. We choose a global field k, and an extension k' of k, such that at the place v_0 the extension k'_{v_0}/k_{v_0} is isomorphic to E/F. We then apply Lemma 2.5. We choose first the supercuspidal representation Π of $G(k_{v_1})$. On elliptic elements close enough to 1, its character is then equal to the formal degree, and hence $\neq 0$. The twisted orbital integral of $\phi_{v_1} = (\phi_{w_1}, \dots \phi_{w_\ell})$ is the orbital integral of $\phi_{w_1} * \cdots * \phi_{w_\ell}$; for correct choices of the functions, it will be a non-zero multiple of the character of Π.

Therefore, if δ^* is an element of $\mathrm{GL}(n, k')$ approximating δ at v_0, we may assume, taking δ^* close to 1 at v_1, that the twisted orbital integral of ϕ_{v_1} does not vanish at δ^*. We may also assume, using finite approximation, that δ^* has elliptic norm (with regular image in $PGL(n)$) at the place

v_2. We may then choose the other functions ϕ_v in such a way that only one orbital integral appears in the right-hand side of (II), and that it is of the form $c\Phi_{\phi_{v_0},\sigma}(\delta^*_{v_0})$ with $c \neq 0$. (This is possible because, for GL(n), we may separate global (twisted) orbits using only local conditions: indeed, (twisted) conjugacy classes of semi-simple elements are parametrized by the coefficients of the characteristic polynomial, and obviously these are known as soon as they are known at a local place.) We then have an identity:

$$\mathrm{trace}(r_{\mathrm{cusp}}(\phi_{v_0} \otimes \phi^{v_0})\, I_\sigma) = c\Phi_{\phi_{v_0},\sigma}(\delta^*_{v_0}).$$

However, only generic representations occur in the left-hand side [27(a)]; by the assumption on ϕ_{v_0}, then, it vanishes, which shows that $\Phi_{\phi_{v_0},\sigma}(\delta^*_{v_0}) = 0$. We will see (§3) that the twisted orbital integral is smooth on the σ-regular set. Since $\delta^*_{v_0}$ can be made close to δ, this proves the proposition. ∎

2.3. Finally, we will close this section by studying the representations which play, in the twisted case, the role of the discrete series. We will say that a representation of $G(E)$ is *σ-discrete* if it is tempered, σ-stable, and cannot be induced from a (tempered) σ-stable representation of a standard Levi subgroup.

LEMMA 2.8: *Assume the representation Π is σ-discrete. Then there exist $m|n$, and a discrete series representation Π_1 of* GL(m, E) *such that $\Pi_1^{\sigma^r} \cong \Pi_1$ and $\Pi_1^{\sigma^i} \not\cong \Pi_1$ $\left(1 \leq i < r = \frac{n}{m}\right)$, and such that Π is induced from the representation $(\Pi_1, \Pi_1^\sigma, \dots \Pi_1^{\sigma^{r-1}})$ of the Levi subgroup of type $(m, \dots, m)$. Conversely, any such representation is σ-discrete.*

Proof. Since Π is tempered, it is induced from a representation $(\Pi_1, \Pi_2, \dots, \Pi_k)$, Π_i discrete, of a Levi subgroup of type $(m_1, \dots m_k)$. Since $\Pi^\sigma \cong \Pi$, we must have, by the standard classification results, $(\Pi_1^\sigma, \dots \Pi_k^\sigma) = (\Pi_1, \dots \Pi_k)$ up to permutation. Since Π is not induced from a σ-stable representation of a Levi subgroup, this permutation must be transitive on $(1, \dots k)$. Thus Π is induced from $(\Pi_1, \dots, \Pi_1^{\sigma^{k-1}})$ with $\Pi_1^{\sigma^k} \cong \Pi_1$. Finally, k must be minimal for this property, otherwise Π would again be properly induced. The converse is plain. ∎

We now state a Paley–Wiener theorem for σ-stable representations, due to Rogawski [33(c)]. It extends to the twisted case the Paley–Wiener theorem of Bernstein, Deligne and Kazhdan [6]. We formulate it in a way slightly different from [33(c)], since the normalized intertwining operators I_σ are available to us. Recall from [6] that the set Irr$(G(E))$ of irreducible admissible representations of $G(E)$ has a natural decomposition into "components" associated to cuspidal representations of Levi subgroups. Moreover, if M is a Levi subgroup (defined over F), and Π_M an admissible representation of $M(E)$ of finite length, we may twist Π_M by an unramified one-dimensional character of $M(E)$. Let $\mathrm{Irr}_\sigma(G(E))$ be the set of σ-stable representations in $\mathrm{Irr}(G(E))$.

PROPOSITION 2.9: (Rogawski). *Assume λ is an additive functional, with values in $\mathbf{C}$, on the Grothendieck group of σ-stable representations of $G(E)$ of finite length. Assume*

(i) $\lambda : \mathrm{Irr}_\sigma(G(E)) \to \mathbf{C}$ is supported on a finite number of components.

(ii) For any proper Levi subgroup M/F, and σ-stable Π_M of finite length,

$$\chi \mapsto \lambda\big(\mathrm{ind}_{N(E)N(E)}^{G(E)} \Pi_m \otimes \chi\big)$$

is a regular function of the unramified character χ. Then there is $\phi \in C_c^\infty(G(E))$ such that

$$\lambda(\Pi) = \mathrm{trace}(\Pi(\phi)\, I_\sigma), \quad \Pi \in \mathrm{Irr}_\sigma(G(E)).$$

COROLLARY 2.10: *Assume that Π is a σ-discrete representation of $G(E)$, with central character X. Then there is a function ϕ on $G(E)$, compactly supported modulo the center $Z(E)$, and such that $\phi(zg) = X(z)^{-1}\phi(g)$, with the following properties:*

(i) $\mathrm{trace}(\Pi(\phi)\, I_\sigma) = 1$.

(ii) $\mathrm{trace}(T(\phi)\, I_\sigma) = 0$ *for any tempered, σ-stable $T \neq \Pi$ with central character X.*

$$\left(\text{Note that} \quad T(\phi) = \int_{Z(E)\backslash G(E)} \phi(g) T(g) dg \quad \text{is well-defined}\right).$$

Such a function ϕ will be called a *pseudo-coefficient* of $\Pi \rtimes \sigma$.

A proof of this can be given, using the Langlands classification, by the same method as for Proposition 1 of [11(d)], as soon as Proposition 2.9 is known. ∎

Finally, we will need the following results:

LEMMA 2.11: *Assume that Π is irreducible, generic and σ-stable. Assume $\Theta_{\Pi,\sigma}$ is not identically 0 on the set of σ-elliptic elements. Then Π is σ-discrete modulo torsion by a character.*

Proof. By a result of Zelevinsky [42, Theorem 9.7], Π is a full induced representation from an essentially square-integrable representation of a Levi subgroup. Inducing by stages, we may write

$$\Pi = \mathrm{ind}_{M(E)N(E)}^{G(E)}(\Pi_M \otimes 1)$$

where Π_M is essentially tempered and dominant in the sense of the Langlands classification (cf. [9, XI.2.9]). Then $\Pi^\sigma = \mathrm{ind}(\Pi_M^\sigma \otimes 1)$. By the uniqueness of the Langlands classification, this implies that Π_M is σ-stable. If M is proper, the formulas for induced twisted characters ([11(b), Proposition 6]) show that $\Theta_{\Pi,\sigma} = 0$ on the σ-elliptic set. Therefore $M = G$, and Π is essentially tempered. It is then easy to see (cf. Lemma 6.4) that, up to torsion by an Abelian character, Π is induced from a σ-discrete representation. Again, the condition on the character implies that Π is itself σ-discrete. ∎

The following lemma is a sort of dual of Lemma 2.11:

LEMMA 2.12: *Assume that* Π *is* σ*-discrete.*

(i) $\Theta_{\Pi,\sigma} \not\equiv 0$ *on the* σ*-elliptic set.*

(ii) If $\phi \in C_c^\infty(G(E), X)$ *is a pseudo-coefficient of* Π*, the twisted orbital integrals of* ϕ *are 0 on non-*σ*-elliptic elements and do not vanish identically on the* σ*-elliptic set.*

Proof. We may assume the central character X equal to 1 and work on $PGL(n)$. If ϕ is a pseudo-coefficient of Π, the descent argument used in the proof of Proposition 2.7 shows that $\Phi_{\phi,\sigma}(\gamma) = 0$ for $N\gamma$ non-elliptic: indeed, $\mathrm{trace}(\Pi(\phi)\, I_\sigma) = 0$ for any σ-stable Π properly induced from σ-stable. But now, the identity

$$1 = \mathrm{trace}(\Pi(\phi)\, I_\sigma) = \int_{G_{\sigma-\mathrm{ell}}} \Theta_{\Pi,\sigma}(g)\phi(g)\, dg$$

implies both (i) and the rest of (ii). ∎

3. Transfer of orbital integrals of smooth functions

In this section E, F are non-Archimedean fields of characteristic 0; E is cyclic over F; σ, Σ are as above. We want to compare orbital integrals of functions in $C_c^\infty(G(F))$ and twisted orbital integrals of functions in $C_c^\infty(G(E))$. We fix, once and for all, Haar measures dg on $G(F)$ and dg_E on $G(E)$.

Assume $\delta \in G(E)$ has a regular norm $\gamma \in G(F)$. As explained in §1, $G_{\delta,\sigma}$ is an inner form of G_γ; since G_γ is a torus, these two groups are therefore isomorphic over F. In fact if $\gamma \in G(F)$, one has $G_{\delta,\sigma}(F) = G_\gamma(F)$, a canonical isomorphism.

In particular, in the expression of the twisted orbital integral

$$\Phi_{\phi,\sigma}(\delta) = \int\limits_{G_{\delta,\sigma}(F)\backslash G(E)} \phi(g^{-1}\delta g^\sigma)\frac{dg_E}{dt}$$

and the orbital integral

$$\Phi_f(\gamma) = \int\limits_{G_\gamma(F)\backslash G(F)} f(g^{-1}\gamma g)\frac{dg}{dt},$$

we will always assume that the choice of measures dt is the same on $G_\gamma(F)$ and $G_{\delta,\sigma}(F)$.

We will prove:

PROPOSITION 3.1:

(i) Assume $\phi \in C_c^\infty(G(E))$. Then there exists $f \in C_c^\infty(G(F))$ such that, for regular $\gamma \in G(F)$:

$$(*) \qquad \Phi_f(\gamma) = \begin{cases} 0 & \text{if } \gamma \text{ is not a norm} \\ \Phi_{\phi,\sigma}(\delta) & \text{if } \gamma = N\delta, \quad \delta \in G(E). \end{cases}$$

(ii) Conversely, given $f \in C_c^\infty(G(F))$ satisfying $(*)$, there exists $\phi \in C_c^\infty(G(E))$ such that

$$\Phi_{\phi,\sigma}(\delta) = \Phi_f(N\gamma) \quad \textit{for} \quad \delta \in G(E).$$

As usual, the study of orbital integrals begins with a compactness lemma:

LEMMA 3.2: *Let $\delta \in G(E)$; assume $N\delta = \gamma$ is a semi-simple element of $G(F)$. Let $M = G_{\delta,\sigma}$.*

Assume that $\delta \in T(E)$, where T is a maximal torus of G over F. Then there is a neighborhood V of 1 in $T(F)$ with the following property:

For any compact $\Omega \subset G(E)$, there is a compact set $\omega \subset M(F)\backslash G(E)$ such that, for $g \in G(E)$:

$$g^{-1}V\delta g^{\sigma} \cap \Omega \neq \emptyset \quad \textit{implies} \quad M(F)g \in \omega.$$

Note that by I, §1, the group M is reductive, an inner form of G_{γ}.

We also remark that, by the properties of the norm map, σ-regular elements are always σ-conjugate to elements of $T(E)$ for some maximal torus T over F. Clearly the twisted orbital integrals of an element $\delta \in T(E)$ depend only on its class in $T(E)^{1-\sigma}\backslash T(E)$ where $T(E)^{1-\sigma} = \{tt^{-\sigma} | t \in T(E)\}$. There is an exact sequence

$$1 \to T(E)^{1-\sigma} \to T(E) \xrightarrow{N} T(F)$$

since, by Hilbert's Theorem 90, $H^1(\Sigma, T(E)) = 1$ for any torus of $\mathrm{GL}(n)$. Here $N = N_{E/F}$, the norm map. Consequently, since the map $T_F \to T_F$ given by $t \mapsto t^{\ell}$ is an isomorphism in a neighborhood of 1, we see that $T(F)$ gives, near 1, a parametrization of $T(E)^{1-\sigma}\backslash T(E)$ or of the σ-conjugacy classes in $T(E)$.

Proof of Lemma 3.2 (cf. Shelstad, [38(b), Theorem 4.2.1]). We first reduce to the case that γ is central. Assume that $g^{-1}t\delta g^{\sigma} \in \Omega$, for some $t \in T(F)$. Taking norms, we have $g^{-1}t^{\ell}N\delta g \in \Omega_1 = \Omega\Omega^{\sigma}\dots\Omega^{\sigma^{\ell-1}}$. By the usual version of this lemma ([20(c), p. 52]), there is a neighborhood U of $N\delta$ in $T(E)$ such that, if $x \in U$, and $g^{-1}xg \in \Omega_1$, this implies $g \in M(E)\omega_1$, with ω_1 compact in $G(E)$. So if V_1, a neighborhood of 1 in $T(F)$, satisfies $V_1^{\ell} \cdot N\delta \subset U$, we have

$$g^{-1}V_1\gamma g^{\sigma} \cap \Omega \neq \emptyset$$

implies $g^{-1}Ug \cap \Omega_1 \neq \emptyset$, whence $g \in M(E)\omega_1$; writing $g = mx$, $x \in \omega_1$, $m \in M(E)$, we then have $x^{-1}m^{-1}V_1\gamma m^{\sigma}x^{\sigma} \cap \Omega \neq \emptyset$, whence $m^{-1}V_1\gamma m^{\sigma} \cap \Omega_2 \neq \emptyset$, where

$$\Omega_2 = M(E) \cap \{x\Omega x^{-\sigma} \mid x \in \omega_1\}$$

is compact in $M(E)$. But then, assuming the lemma in M (where γ is central) we see that for suitable V_1, the relation $m^{-1}V_1\gamma m^{\sigma} \cap \Omega_2 \neq \emptyset$ implies $M(F)m \in \omega_M$, $\omega_M \subset M(F)\backslash M(E)$; then $g = mx \in \omega_M\omega_1 \subset M(F)\backslash G(E)$. ∎

We now assume that γ is central.

LEMMA 3.3: *Assume $\Omega_1 \subset G(E)$ is compact. Then there is a neighborhood V_1 of 1 in $T(F)$, and Ω_2 compact in $G(E)$ such that the conditions $g \in G(E)$, $t \in V_1$, $gt^{\ell}g^{-1} \in \Omega_1$ imply $gtg^{-1} \in \Omega_2$.*

Proof. Let $\mathfrak{g}_E = \text{Lie } G(E)$. According to Harish–Chandra [20(g), p.330], there is an open, closed, invariant set $\mathfrak{g}_E^0$ of $\mathfrak{g}_E$ such that

(i) $\exp : \mathfrak{g}_E^0 \to G(E)$ is defined, and a diffeomorphism,

(ii) $\mathcal{O}_E \mathfrak{g}_E^0 = \mathfrak{g}_E^0$,

(iii) $\exp(Adx X) = x \exp X x^{-1}$, $x \in G(E)$, $X \in \mathfrak{g}_E$.

Let $\mathfrak{g}_E^1 = \ell \mathfrak{g}_E^0 \subset \mathfrak{g}_E^0$. Take $V_1 = T(F) \cap G_E^1$, where $G_E^1 = \exp(\mathfrak{g}_E^1)$. Assume $t \in V_1 : t = \exp H$, $H \in \mathfrak{g}_E^1$. Then, if $gt^\ell g^{-1} \in \Omega_1$, we have $\exp(Adg(\ell H)) \in \Omega_1 \cap G_E^1$ whence $Adg(H) \in \frac{1}{\ell} \exp^{-1}(\Omega_1 \cap G_E^1)$, a compact set. This implies the lemma. ∎

We may now prove Lemma 3.2. Let Ω be as in the lemma, and set $\Omega_1 = \gamma^{-1}\Omega\Omega^\sigma \ldots \Omega^{\sigma^{\ell-1}}$; define V_1 by Lemma 3.3. If $t \in V_1$ and $g^{-1}t\delta g^\sigma \in \Omega$, we have $g^{-1}t^\ell g \in \Omega_1$, whence by Lemma 3.3, $g^{-1}tg \in \Omega_2$, or $g^{-1}t^{-1}g \in \Omega_2^{-1}$, which implies $g^{-1}\delta g^\sigma = (g^{-1}t^{-1}g)(g^{-1}t\delta g^\sigma) \in \Omega_2^{-1}\Omega$. However, the map

$$M(F)\backslash G(E) \to G(E)$$

given by

$$\dot{g} \mapsto g^{-1}\delta g^\sigma$$

is proper since the orbit of the σ-semi-simple element δ by σ-conjugation is closed: it is just the neutral component of the orbit of $(\delta, \sigma) \in G(E) \times \Sigma$, and the orbit of the semi-simple element (δ, σ) is closed in $G(E) \times \Sigma$ by an easy extension of Borel's results [8(a), III.g]. So the condition $g^{-1}\delta g^\sigma \in \Omega_2^{-1}\Omega$ implies that g remains in a compact set modulo $M(F)$.

We now return to the proof of Proposition 3.1 (i). Obviously the assertion is local (in the space of conjugacy classes); it is easily reduced to the following (we assume the function ϕ given):

LEMMA 3.4: *Assume $\delta_0 \in G(E)$ has semi-simple norm. Let $\{T_1, \ldots T_r\}$ be the F-maximal tori of G, up to conjugacy, such that $\delta_0 \in T(E)$. Then we can choose neighborhoods $V_1, \ldots, V_r$ of 1 in $T_1(F), \ldots, T_r(F)$, and a function $f \in C_c^\infty(G(F))$ such that for regular γ,*

$$\Phi_f(\gamma) = \begin{cases} 0 & \gamma \text{ not a norm} \\ \Phi_{\phi,\sigma}(\delta) & \gamma = N\delta, \ \delta \in \delta_0 V_i. \end{cases}$$

Proof. (cf. Rogawski [33(a)]). Set $M = G_{\delta_0,\sigma}$. For all i, let $V_i \subset T_i(F)$ satisfy the conditions of Lemma 3.2 with $\Omega = \text{Supp}(\phi)$. Then $g^{-1}V_i\delta_0 g^\sigma \cap \Omega = \emptyset$ if $M(F)g \notin \omega = \bigcup \omega_i$, where ω_i is defined by Ω and V_i. Thus, if $\delta \in V_i\delta_0$ is σ-regular, the function

$$g \mapsto \int\limits_{T_i(F)\backslash M(F)} \phi(g^{-1}m^{-1}\delta m^\sigma g^\sigma)\frac{dm}{dt}$$

vanishes for $M(f)g \notin \omega$.

We choose $\alpha \in C_c^\infty(G(E))$ such that $\bar{\alpha}(g) = \int_{M(F)} \alpha(mg)dm$ is equal to 1 if $M(F)g \in \omega$, and to 0 otherwise. Define

$$\psi(m) = \int_{G(E)} \alpha(g)\phi(g^{-1}m^{-1}\delta_0 g^\sigma)dg.$$

The function ψ is in $C_c^\infty(M(F))$: it is obviously smooth since α has compact support, and if $\psi(m) \neq 0$, there is $g \in \mathrm{Supp}(\alpha)$ such that $\phi(g^{-1}m^{-1}\delta_0 g^\sigma) \neq 0$, whence $m^{-1}\delta_0 \in \Omega_1\Omega\Omega_1^{-\sigma}$, where $\Omega_1 = \mathrm{Supp}(\alpha)$, $\Omega = \mathrm{Supp}(\phi)$: so m must be in a compact set.

For σ-regular $\delta \in V_i\delta_0$, we have:

$$\int\limits_{T_i(F)\backslash G(E)} \phi(g^{-1}\delta g^\sigma)\frac{dg}{dt}$$

$$= \int\limits_{M(F)\backslash G(E)} \bar{\alpha}(g)\frac{dg}{dm} \int\limits_{T_i(F)\backslash M(F)} \phi(g^{-1}m^{-1}\delta m^\sigma g^\sigma)\frac{dm}{dt}$$

$$= \int\limits_{M(F)\backslash G(E)} \left\{\int_{M(F)} \alpha(m_1 g)dm_1\right\}\frac{dg}{dm} \int\limits_{T_i(F)\backslash M(F)} \phi(g^{-1}m^{-1}\delta m^\sigma g^\sigma)\frac{dm}{dt}$$

$$= \int\limits_{M(F)\backslash G(E)} \frac{dg}{dm}\left\{\int_{M(F)} \int\limits_{T_i(F)\backslash M(F)} \alpha(m_1 g)\phi(g^{-1}m^{-1}\delta m^\sigma g^\sigma)\frac{dm}{dt}dm_1\right\}.$$

By making the change of variable $g \mapsto m_1 g$, and grouping the integrals along $M(F)$, this can be written:

$$\int_{G(E)} \alpha(g)dg \int\limits_{T_i(F)\backslash M(F)} \phi(g^{-1}m^{-1}\delta m^\sigma g^\sigma)\frac{dm}{dt}.$$

Writing now $\delta = t\delta_0$, $t \in V_i$, whence $m^{-1}\delta m^\sigma = m^{-1}t\delta_0 m^\sigma = m^{-1}tm\delta_0$, we have:

$$\int\limits_{T_i(F)\backslash G(E)} \phi(g^{-1}t\delta_0 g^\sigma)\frac{dg}{dt} = \int_{G(E)} \alpha(g)dg \int\limits_{T_i(F)\backslash M(F)} \phi(g^{-1}m^{-1}tm\delta_0 g^\sigma)\frac{dm}{dt}$$

$$= \int\limits_{T_i(F)\backslash M(F)} \psi(m^{-1}tm)\frac{dm}{dt}.$$

In other terms,

$$\Phi_{\phi,\sigma}(t\delta_0) = \Phi_{\psi}^{M(F)}(t)$$

where the integral on the right is taken in $M(F)$ and non-twisted.

As recalled in §1, the centralizer G_{γ_0} of γ_0 is isomorphic to a product of groups of the form

$$\mathrm{Res}_{F_i/F}(\mathrm{GL}(m_i, F_i))$$

where F_i is a field extension and Res denotes restriction of scalars. Since M is an inner form of G_{γ_0}, it is a product of groups of the form

$$\mathrm{Res}_{F_i/F}(\mathrm{GL}(r_i, D_i))$$

with D_i a division algebra on F_i of degree d_i^2, and $d_i r_i = m_i$.

By a theorem of Deligne, Kazhdan and Vignéras [15, Theorem B2c] and Rogawski [33(b)], we may then associate to ψ a function f_1 on $G_{\gamma_0}(F)$ having the following property. (We write $\tilde{M}$ for the group G_{γ_0}. If $m \in M(F)$, denote by $\tilde{m}$ the conjugacy class in $\tilde{M}(F)$ corresponding to m by the Skolem–Noether theorem; if $\tilde{m} \in \tilde{M}(F)$ is not so obtained, we say that it does not originate in $M(F)$.) For regular $\tilde{\delta} \in \tilde{M}(F)$,

$$\Phi_{f_1}^{\tilde{M}(F)}(\tilde{\delta}) = \begin{cases} 0 & \text{if } \tilde{\delta} \text{ does not originate from } M(F) \\ \Phi_{\psi}^{M(F)}(\delta) & \text{if } \tilde{\delta} \text{ corresponds to } \delta. \end{cases}$$

If we combine this with the equation relating ψ and ϕ, we get

$$\Phi_{\phi,\sigma}(t\delta_0) = \Phi_{f_1}^{\tilde{M}(F)}(\tilde{t}), \quad t \in V_i.$$

We want to compare the orbital integral of ϕ at $t\delta_0$, however, with the orbital integral of f at $N(t\delta_0) = t^\ell\gamma_0$. Since γ_0 is central in $M(F)$ or $\tilde{M}(F)$, and the map $m \mapsto m^\ell$ is a conjugacy-preserving diffeomorphism from an invariant neighborhood of 1 in $\tilde{M}(F)$ onto its image (cf. Lemma 3.3), we may find $f_2 \in C_c^\infty(\tilde{M}(F))$ such that

$$\Phi_{f_1}^{\tilde{M}(F)}(\tilde{t}) = \Phi_{f_2}^{\tilde{M}(F)}(t^\ell\gamma_0).$$

(We have used the obvious relation between the correspondence $m \mapsto \tilde{m}$ and the norm map.)

Since $\tilde{M}(F)$ is the centralizer of γ_0, we may find a function $f \in C_c^\infty(G(F))$ such that, for t close to 1:

$$\Phi_f(t^\ell\gamma_0) = \Phi_{f_2}^{\tilde{M}(F)}(t^\ell\gamma_0).$$

This results from the fact that the germs on G all come from germs on $\tilde{M}$ (cf. [41(a), §2.6]) and the independence of germs (cf. 15, Proposition 2b]).

Tracing back all comparisons, we have

$$\Phi_{\phi,\sigma}(t\delta_0) = \Phi_f(t^\ell \gamma_0).$$

If $t \in G_{\delta_0,\sigma}(F)$, we have $t^{-1}\delta_0 t^\sigma = \delta_0$ whence

$$N(\delta_0 t) = \delta_0 t \delta_0^\sigma t^\sigma \dots \delta_0^{\sigma^{\ell-1}} t^{\sigma^{\ell-1}} = t^\ell \gamma_0,$$

as an easy computation shows. This shows that any semi-simple element $u\gamma_0 \in \tilde{M}(F)$, close to γ_0, which comes from M is a norm. In particular, Φ_f vanishes near γ_0 on elements which are not in the image of the norm map. This finishes the proof of Lemma 3.4. ∎

To prove part (ii) of Proposition 3.1, we just have to reverse the construction. We start with f on $G(F)$ satisfying the vanishing conditions; it is enough to construct ϕ on $G(E)$ in an invariant neighborhood of any $\delta_0 \in G(E)$. We keep the previous notations, thus $\tilde{M}$ is the centralizer of γ_0 and M the σ-centralizer of δ_0.

Given f, we may construct the function f_2 on $\tilde{M}(F)$, and then, by the converse of the theorem of Deligne–Kazhdan–Vignéras and Rogawski, the function ψ on $M(F)$ associated to $f_1(\tilde{m}) = f_2(\tilde{m}^\ell \gamma_0)$ on $\tilde{M}(F)$. An easy extension to the twisted case of the results of Vignéras ([41(a), §2.5]) shows that ψ can be lifted to a function ϕ on $G(E)$, the σ-orbital integrals of which then correspond to the orbital integrals of f.

Implicit in this argument has been the fact that, if f satisfies the vanishing conditions, the function f_1 satisfies the vanishing conditions involved in the comparison between $M(F)$ and $\tilde{M}(F)$. That is implied by the following result:

LEMMA 3.5: *Assume $\gamma_0 = N\delta_0 \in T(F)$, for T a maximal F-torus of G. Assume that $\gamma_0 t$ is a regular norm for some t arbitrarily close to 1 in $T(F)$. Then T is $G(\overline{F})$-conjugate to an F-torus T^1 in M.*

Proof. Assume $\gamma_0 t = (N\delta_0)t = Ng$, $g \in G(E)$. Then g commutes with $\gamma_0 t$, so $g \in T(E)$.

Now let t_n be a sequence such that $t_n \to 1$; write $\gamma_0 t_n = N(g_n)$, $g_n \in T(E)$. Since the norm: $T(E) \to T(F)$ is a local fibration by $T(E)^{1-\sigma}$, we may assume that the sequence (g_n) converges: we then get $g_0 \in T(E)$ with $Ng_0 = \gamma_0$. Since $Ng_0 = N\delta_0$, g_0 is σ-conjugate to δ_0. So there is a σ-conjugate of δ_0 such that $\mathrm{Ad}(g_0)\sigma t = t$ for $t \in T(F)$, i.e., such that T embeds in $G_{g_0,\sigma}$ over F. This implies that a $G(\overline{F})$-conjugate of T embeds in $G_{\delta_0,\sigma}$. ∎

With this the proof of Proposition 3.1 is complete. ∎

We will say that two functions ϕ and f verifying the relations of Proposition 3.1 are *associated*. We now want to compare the orbital integrals of associated functions on singular elements. We start with the semi-simple ones. We recall a construction of Kottwitz [29(b)]. Let $\delta \in G(E)$ have semi-simple norm. The group $G_{\delta,\sigma}$ is then a product of multiplicative groups of central simple algebras. Let $e(\delta)$ be the sign $e(G_{\delta,\sigma})$ defined in [29(b)]: thus $e(\delta) = (-1)^{r(G_\gamma)-r(G_{\delta,\sigma})}$ where $\gamma = N\delta$ and $r(H)$ denotes the F-rank of an F-group H.

LEMMA 3.6: *Assume $\phi \in C_c^\infty(G(E))$, $f \in C_c^\infty(G(F))$ are associated. Then, if $\gamma \in G(F)$ is semi-simple,*

$$\Phi_f(\gamma) = \begin{cases} 0 & \text{if } \gamma \text{ is not a norm} \\ e(\delta)\Phi_{\phi,\sigma}(\delta) & \text{if } \gamma = N\delta. \end{cases}$$

Here the measures defining the orbital integrals are as follows. Recall that we have fixed dg_E, dg on $G(E)$ and $G(F)$. We have to fix the measures on the groups $G_{\delta,\sigma}(F)$ and $G_\gamma(F)$. These groups are products of multiplicative groups of simple algebras, and we choose compatible measures on them as in [33(b), §3].

Proof of Lemma 3.6. We first prove the vanishing part. Assume $\gamma \in G(F)$ is not a norm. It is enough to show that if $\gamma \in T(F)$, T being a maximal torus, then t is not a norm, for regular $t \in T(F)$ close enough to γ_0. (This will imply that for f associated to ϕ, the regular orbital integrals of f close to γ vanish: by standard theory of Shalika germs, we conclude that $\Phi_f(\gamma) = 0$.) Assume then that $t_n = N(x_n)$ for a sequence of regular $t_n \to \gamma$. Then $x_n \in T(E)$; since the norm map $T(E) \to T(F)$ has compact fibers, we may extract a convergent subsequence, which implies that γ is a norm.

To prove the identity of orbital integrals in Lemma 3.6, we just have to retrace the proof of Lemma 3.4. Recall that given ϕ, we had constructed a function ψ on $G_{\delta,\sigma}(F)$, then f_2 on $G_{\gamma_0}(F)$ and f on $G(F)$; they are related by

$$\Phi_{\phi,\sigma}(t\delta) = \Phi_\psi(t) = \Phi_{f_2}(t^\ell\gamma) = \Phi_f(t^\ell\gamma),$$

t being an element in $G_{\delta,\sigma}(F)$ such that $t\delta$ is σ-regular.

In the computation of Φ_ψ and Φ_{f_2}, we have used measures on $G_{\delta,\sigma}(F)$ and $G_\gamma(F)$, that we take to be compatible measures. We have, by construction of ψ:

$$\begin{aligned}\psi(1) &= \int_{G(E)} \alpha(g)\phi(g^{-1}\delta g^{\sigma})dg \\ &= \int_{M(F)\backslash G(E)} \left\{\int_{M(F)} \alpha(mg)\phi(g^{-1}\delta g^{\sigma})dm\right\}d\dot{g} \\ &= \Phi_{\phi,\sigma}(\delta),\end{aligned}$$

the twisted orbital integral being computed by means of dg/dm, where dm is the measure on $G_{\delta,\sigma}(F)$.

Analogously, the usual computation on the non-twisted side shows that

$$f_2(\gamma) = \Phi_f(\gamma),$$

the orbital integral being computed by means of $dg/d\tilde{m}$, where $d\tilde{m}$ is the measure on $G_\gamma(F)$. We conclude by quoting the following result:

LEMMA 3.7: (Rogawski [33(b), Lemma 3.3]). *Assume $M(F)$ is an inner form of $\tilde{M}(F) = \mathrm{GL}(m, F)$. Choose associated measures dm on $M(F)$ and $d\tilde{m}$ on $\tilde{M}(F)$. Assume the functions f on $M(F)$ and $\tilde{f}$ on $\tilde{M}(F)$ are such that*

$$\Phi_{\tilde{f}}(\tilde{t}) = \begin{cases} 0 & \tilde{t} \text{ not from } M(F) \\ \Phi_f(t) & \tilde{t} \text{ associated to } t \in M(F). \end{cases}$$

(If $t, \tilde{t}$ are associated, we normalize the measures on the corresponding orbits by taking the same measure on the corresponding torus.) Then

$$f(1) = e(M)\tilde{f}(1).$$

We apply this to ψ and the translate of f_2 by the central element γ; this implies that $f_2(\gamma) = e(\delta)\psi(1)$, whence Lemma 3.6. ∎

Finally, we will now describe the (twisted) orbital integrals on all elements, by reducing them to semi-simple orbital integrals. This has been known to a number of people; we rely on unpublished notes of Kottwitz.

We first treat the non-twisted case. Let $\gamma = su = us$ be the Jordan decomposition of $\gamma \in G(F)$, with s semi-simple and u unipotent. Let $A = u - 1$, a nilpotent matrix. We consider the flag of subspaces of F^n:

$$V_0 = F^n \supset V_1 = AV_0 \supset \cdots \supset V_i = A^i V_0 \supset \cdots \supset V_k = \{0\}.$$

Let $P = P_u$ be the parabolic subgroup stabilizing this flag. If $g \in G(F)$ commutes with u, it is clear that g leaves the flag invariant, whence

$$G(F)_u \subset P(F).$$

Let $N(F)$ be the unipotent radical of $P(F)$, $\mathfrak{n}$, $\mathfrak{p}$ the corresponding Lie algebras.

LEMMA 3.8:

$$\mathrm{ad}(A)\mathfrak{p} = \mathfrak{n}.$$

The proof is easily supplied by using a matrix representation of P and noticing that A gives a surjective map: $V_i/V_{i+1} \to V_{i+1}/V_{i+2}$. (Alternatively, note that a statement equivalent by duality to Lemma 3.8 is proved by Howe [23, Lemma 2(b)].) ∎

Since s commutes with u, $s \in P(F)$; since it is semi-simple, we may choose a Levi component M of P over F such that $s \in M(F)$. We have $P(F)/N(F) \cong M(F)$. Let $X \subset P$ be the inverse image in P of the orbit of s in M. Then X is a smooth and irreducible subvariety of P and we have an isomorphism:

$$M_s(F)\backslash M(F) \times N(F) \xrightarrow{\approx} X(F)$$
$$(m,n) \mapsto m^{-1}smn.$$

The element γ lies in X; let X_0 be the orbit of γ by conjugation under $P: X_0 \subset X$.

LEMMA 3.9: $X_0(F) = P_\gamma(F)\backslash P(F)$ *is open and dense in* $X(F)$.

Proof. We compute the tangent space to X_0 at γ. The differential at 1 of the map

$$P(F) \to P(F),$$
$$p \mapsto p^{-1}\gamma p,$$

is equal to $\mathrm{Ad}(\gamma) - 1$.

Thus we must check that $\mathfrak{m} + \mathfrak{n} = (\mathrm{Ad}(su) - 1)\mathfrak{p} + \mathfrak{m}_s$, where $\mathfrak{m}_s$ is the Lie algebra of $M_s(F)$.

The element $\gamma = su$ acts on $\mathfrak{p}$ by the adjoint action, and $\mathrm{Ad}\,\gamma = (\mathrm{Ad}\,s)(\mathrm{Ad}\,u)$ is its Jordan decomposition; therefore

$$(\mathrm{Ad}(su) - 1)\mathfrak{p} = \sum_{\lambda \neq 1} \mathfrak{p}_\lambda \oplus \mathrm{Im}((Adu - 1) : \mathfrak{p}_1 \to \mathfrak{p}_1)$$

where $\mathfrak{p}_\lambda$ denotes the λ-eigenspace of s in $\mathfrak{p}$. (Here we work on an algebraic closure of F.) In particular $(\mathrm{Ad}(su) - 1)\mathfrak{p}$ contains the image of $Adu - 1$ acting on $\mathfrak{p}$. We show that this contains $\mathfrak{n}$. We must show that, if $Y \in \mathfrak{n}$, $Y = uZu^{-1} - Z$, for some $Z \in \mathfrak{p}$. Since (for matrix multiplication) $\mathfrak{n}N(F) = \mathfrak{n}$, it is equivalent to showing that $Y = uZ - Zu$, or $Y = [A, Z]$ where $A = u - 1$. This is true by Lemma 3.8.

Therefore $(\mathrm{Ad}(su)-1)\mathfrak{p}$ contains $\mathfrak{n}$; it also contains $\sum_{\lambda\neq 1}\mathfrak{m}_\lambda$. This proves that $\mathfrak{m}+\mathfrak{n}=\mathfrak{m}_s+(\mathrm{Ad}(su-1)\mathfrak{p}$.

This implies that the F-map $P_\gamma\backslash P\to X$ given by conjugating γ is open. By [8(a), Proposition 6.6], we see that $X_0\subset X$ is open; since X is connected, $X_0(F)$ is open and dense in $X(F)$. Moreover, since $G_u\subset P$, we have $P_\gamma=G_\gamma$ and therefore $H^1(F,P_\gamma)=0$ (cf. Proof of Lemma 1.1). Hence $X_0(F)=P_\gamma(F)\backslash P(F)$. ∎

We now choose left-invariant Haar measures dn, dm, dm_s, dg, dg_γ on $N(F), M(F), M_s(F), G(F), G_\gamma(F)$. Then $dx=\frac{dm}{dm_s}\times dn$ defines a measure on $X(F)$. Then, under conjugation by $P(F)$, dx is relatively invariant with factor δ_P; by restriction we get a measure dx_0 on $X_0(F)$ with the same property: thus $dx_0=dg_\gamma\backslash dp$ for some left-invariant Haar measure dp on $P(F)$. Since the measure of $X(F)-X_0(F)$ is null, we have

$$\int_{G_\gamma(F)\backslash P(F)} f(p^{-1}\gamma p)\frac{dp}{dg_\gamma}=\int_{M_s(F)\backslash M(F)}\int_{N(F)} f(m^{-1}smn)dn\frac{dm}{dm_s}.$$

Combining this with the formula

$$\int_{G(F)} f(g)dg=\int_{K_F}\int_{P(F)} f(pk)dkdp,$$

with $K_F=G(\mathcal{O}_F)$, we have proved:

PROPOSITION 3.10: *For any $f\in C_c^\infty(G(F))$,*

$$\int_{G_\gamma(F)\backslash G(F)} f(g^{-1}\gamma g)\frac{dg}{dg_\gamma}$$
$$=\int_{M_s(F)\backslash M(F)\times N(F)\times K_F} f(k^{-1}m^{-1}smnk)dk\,dn\frac{dm}{dm_s}.$$

In particular, if $f(k^{-1}gk)=f(g)$, we have

$$\Phi_f(sn)=\delta_P(s)^{-\frac{1}{2}}\Phi^M_{f^{(P)}}(s).$$

Here $f^{(P)}$ is the constant term of f, defined after Proposition 2.7. The measures are normalized according to: $\frac{dp}{dg_\gamma}=dx_0=\frac{dm}{dm_s}\times dn$.

We now sketch the proof of the corresponding twisted result. Let $\delta\in G(E)$. Write the Jordan decomposition of $\delta\rtimes\sigma$:

$$(\delta,\sigma)=(s,\sigma)(n,1)=(n,1)(s,\sigma),$$

with (s,σ) semi-simple in $G(E) \rtimes \Sigma$, n unipotent in $G(E)$. Thus

$$\begin{aligned}\delta &= sn^\sigma = ns,\\ N\delta &= \delta\delta^\sigma \dots \delta^{\sigma^{\ell-1}}\\ &= sn^\sigma s^\sigma n^{\sigma^2} \dots s^{\sigma^{\ell-1}} n\\ &= (Ns)n^\ell, \text{ as an easy computation shows.}\end{aligned}$$

We will assume, as we may up to σ-conjugation, that $\gamma = N\delta \in G(F)$. We write $\gamma = tu$, $t = Ns$, $u = n^\ell$. Since u is F-rational, we see that $n = u^{1/\ell}$ is also. Therefore $sn = ns$.

Let P be the F-parabolic subgroup defined, as above, by u (or n). We have $s \in G_u(E) \subset P(E)$.

Since $t = Ns$ is semi-simple in $G_u(F)$, there is a Levi subgroup M_1 of G_u, over F, such that $t \in M_1(F)$. We have $G_u \cong M_1 \times N_1$ as F-group, where N_1 is the unipotent radical of G_u. Since $t = Ns$ and $s \in G_u(F)$, this implies that there is an element $s_1 \in M_1(E)$ such that $t = Ns_1$. Then $N(s_1n) = tu$ since s_1 commutes with n and u. This implies that s_1n and $\delta = sn$ are σ-conjugate: so up to σ-conjugacy, we may assume that $s \in M_1(E)$ with M_1 a Levi subgroup of G_u over F, and *a fortiori* $s \in M(E)$, M being a Levi subgroup of P over F. We have $P(E)/N(E) \cong M(E)$.

Now we define Y, an F-variety in $\mathrm{Res}_{E/F} P$, as the inverse image of the σ-orbit of s in $M(E)$. Then

$$\begin{aligned}M_{s,\sigma}(F)\backslash M(E) \times N(E) &\overset{\rightarrow}{\approx} Y(F)\\ (m,n) &\mapsto m^{-1}sm^\sigma n.\end{aligned}$$

Let Y_0 be the orbit of δ under the twisted action of $P(E)$: thus $Y_0 \subset Y$.

LEMMA 3.11: *$Y_0(F) = P_{\delta,\sigma}(F)\backslash P(E)$ is open and dense in $Y(F)$.*

Proof. The arguments in the proof of Lemma 3.9 also show that we have $Y_0(F) = P_{\delta,\sigma}(F)\backslash P(E)$. It is enough to show that the tangent space at δ to Y_0 is the tangent space to Y. Since $Y_0 \subset Y$, we just count dimensions. Let X_0 be the variety associated to γ, as in Lemma 3.9. We denote by $\dim(V)$ the F-dimension of an F-manifold V. Then

$$\begin{aligned}\dim Y_0(F) &= \dim P(E) - \dim P_{\delta,\sigma}(F)\\ \dim X_0(F) &= \dim P(F) - \dim P_\gamma(F);\end{aligned}$$

since $P_{\delta,\sigma}$ is a form of P_γ:

$$\dim Y_0(F) - \dim X_0(F) = \dim P(E) - \dim P(F).$$

On the other hand, by Lemma 3.9:

$$\dim X_0(F) = \dim N(F) + \dim(M_t(F)\backslash M(F)).$$

From the two last equations, using the decomposition $P = MN$, and recalling that M_t is an F-form of $M_{s,\sigma}$, one easily derives:

$$\begin{aligned}\dim Y_0(F) &= \dim P(E) - \dim M_{s,\sigma}(F)\\ &= \dim N(E) + \dim(M_{s,\sigma}(F)\backslash M(E)).\end{aligned}$$

This proves the lemma. ∎

Proceeding as in the proof of Proposition 3.10, we then obtain, writing K_E for $G(\mathcal{O}_E)$:

PROPOSITION 3.12:
(i) If $\phi \in C_c^\infty(G(E))$, we have

$$\int_{G_{\delta,\sigma}(F)\backslash G(E)} \phi(g^{-1}\delta g^\sigma)\frac{dg}{dh}$$

$$= \int_{M_{s,\sigma}(F)\backslash M(E)\times N(E)\times K_E} \phi(k^{-1}m^{-1}sm^\sigma nk^\sigma)dk\, dn\frac{dm}{dm_s}.$$

In particular, if $\phi(k^{-1}gk^\sigma) \equiv \phi(g)$ for $k \in K_E$:

$$\Phi_{\phi,\sigma}(sn) = \delta_P(Ns)^{-\frac{1}{2}}\Phi^M_{\phi^{(P)},\sigma}(s).$$

(ii) In particular, the twisted orbital integral converges.

Here δ_P is the module of $P(F)$; we have $\delta_P(Ns) = \delta_{P(E)}(s)$, $\delta_{P(E)}$ the module of $P(E)$. The measures are related by $\frac{dp}{dh} = dy_0 = \frac{dm}{dm_s} \times dn$.

COROLLARY3.13: *Assume ϕ on $G(E)$, f on $G(F)$ are associated. Then, $\gamma = s_\gamma n_\gamma$, $\delta = s_\delta n_\delta$ being the Jordan decompositions:*
(i) $\Phi_f(\gamma) = 0$ if s_γ is not a norm,
(ii) $\Phi_f(\gamma) = e(s_\delta)\Phi_\phi(\delta)$ if $\gamma = N\delta$.

This is clear: we may replace f, ϕ by their averages under $K(F)$ or $K(E)$ conjugation, and then use Proposition 3.10 and Proposition 3.12. (The measures must be normalized so as to make Lemma 3.6 correct for the semi-simple orbital integrals.) ∎

4. Orbital integrals of Hecke functions

Let E/F be an *unramified* extension of local non-Archimedean fields. Let $\mathcal{H}_F$ be the Hecke algebra of compactly supported functions on $G(F)$, bi-invariant by $K_F = G(\mathcal{O}_F)$; let $\mathcal{H}_E$ be the analogue for E. In this section, we show that if $\phi \in \mathcal{H}_E$ and $f = b\phi$ is its base change image in $\mathcal{H}_F$, ϕ and f are associated in the sense of §3. This is the so-called "Fundamental Lemma" for Hecke functions.

If G is any unramified group, Kottwitz has shown that the analogous result holds for stable orbital integrals of the *units* of the Hecke algebras. When the stabilization of the twisted trace formula is understood, the method presented here should extend to prove base change for stable orbital integrals in the unramified situation. That is why we have worked in more generality than is required for $\mathrm{GL}(n)$.

4.1. A subspace of the Hecke algebra

Let G be any connected split reductive group over F. Then G is a Chevalley group, and we will take it to be defined over $\mathbf{Z}$.

We write $\mathcal{H}$ for the Hecke algebra of $G(F)$ with respect to the maximal compact subgroup $K = G(\mathcal{O}_F)$.

Let H be a maximal split torus in G; let $W = W(G, H)$ be the Weyl group. We fix a minimal parabolic subgroup $P_0 = HN_0$. All these groups are taken over $\mathbf{Z}$. Let $K_H = H(\mathcal{O}_F)$. We write $\mathcal{H}_H$ for the Hecke algebra $C_c^\infty(H(F), K_H)$. The Satake isomorphism

$$S : \mathcal{H} \to (\mathcal{H}_H)^W$$

associates to f the function

$$Sf(h) = \delta_{P_0}(h)^{\frac{1}{2}} \int_{N_0} f(hn)dn.$$

Here as below, we will, when convenient, write X instead of $X(F)$ for the F-points of a group X.

More generally, if $P = MN$ is a parabolic subgroup of G, we have the constant term along P (cf. §2)

$$f^{(P)}(m) = \delta_P(m)^{\frac{1}{2}} \int_N f(mn)dn \quad (m \in M).$$

For relevant facts about these notions see [20(d)], [40(b)]. Thus $Sf = f^{(P_0)}$. If S_M denotes the Satake transform for the group M – we assume that P contains P_0 – we have, with obvious notations:

$$S_M(f^{(P)}) = Sf.$$

We will use the customary notations concerning the L-group of G over F ([8(b)]), except that the split reference torus in G is taken to be H; its dual is ${}^LH^0 \subset {}^LG^0$.

There exists a canonical isomorphism between $\mathcal{H}_H$ and $\mathbf{C}[{}^LH^0]$ ([8(b), §7]); the composite isomorphism, $\mathcal{H}_G \to \mathbf{C}[{}^LH^0]^W$ will be denoted by $f \mapsto f^\vee$. A function $f^\vee \in \mathbf{C}[{}^LH^0]$ can be written $f^\vee = \sum a_\lambda z^\lambda$, where λ runs over $X^*({}^LH^0)$.

For $P = MN$ a parabolic subgroup, let $A = A_M$ be the split component of M. Let $\Delta(G, A)$ denote the roots of G with respect to A. If α is such a root, there is a multiple $m\alpha$ of α which extends to a (unique) rational character χ_α of M. For all α, we choose such a character. Then χ_α restricts to a character of H; by duality, we obtain a cocharacter of ${}^LH^0$. In particular, $\langle\chi_\alpha, \lambda\rangle$ is well-defined if $\lambda \in X^*({}^LH^0)$.

DEFINITION 4.1: *The space of* regular Hecke functions *on G is the space $\mathcal{K} \subset \mathcal{H}$ defined by:*

$$f \in \mathcal{K} \Longleftrightarrow f^\vee = \sum a_\lambda z^\lambda,$$

with $a_\lambda = 0$ if there exists $M \neq G$, $\alpha \in \Delta(G, A_M)$ such that $\langle\chi_\alpha, \lambda\rangle = 0$.

Thus the condition is that only "regular" exponents, for all the parabolic roots, should occur in the expansion of $f^\vee$. The basic property of the space $\mathcal{K}$ will be expressed by the following lemma. For $P = MN$ a parabolic subgroup, let π_N denote the Jacquet module of a representation π of G with respect to N. Let $W(A_M) = W(G, A_M)$, the Weyl group of G with respect to A_M. Let G_{ell} be the set of elliptic regular elements of G. If π is an admissible representation of finite length of G, we will write:

$$\langle \operatorname{trace} \pi, f\rangle_{\text{ell}} = \int_{G_{\text{ell}}} f(g) \operatorname{trace} \pi(g) dg.$$

LEMMA 4.2: *Let π be an admissible irreducible representation of G without K-fixed vector. Assume that π is not a constituent of an unramified principal series representation of G. Then, if $f \in \mathcal{K}$,*

$$\langle \operatorname{trace} \pi, f\rangle_{\text{ell}} = 0.$$

Proof. Since f is in the Hecke algebra and π has no K-fixed vector, $\operatorname{trace}\pi(f) = \int_G f(g)\operatorname{trace}\pi(g)dg = 0$. Using Weyl's integration formula, we may rewrite the elliptic trace of f:

$$-\langle \operatorname{trace}\pi, f\rangle_{\text{ell}} = \sum_{\substack{M \subseteq G \\ M \neq G}} \sum_{\substack{T \text{ elliptic} \subset M \\ T \bmod G}} |W(G,T)|^{-1} \int_T \Delta_G(t)^2 \operatorname{trace}\pi(t)\Phi_f^G(t)dt.$$

The sum, for each M, ranges over elliptic Cartan subgroups in M, modulo G-conjugation. The measure dt is used to define Φ_f^G.

Assume now that T, T' are two elliptic Cartan subgroups of M which are G-conjugate. Since $A = A_M$ is the common split component of T and T', one has $gAg^{-1} = A$, whence $g \in N_G(A) = W(A) \cdot M$. Moreover, $W(G,T)$ preserves $A_T = A$; this gives rise to an inclusion $W(G,T)/W(M,T) \hookrightarrow W(A)$, and $W(G,T)/W(M,T)$ is the stabilizer of T in this action of $W(A)$ on Cartan subgroups of M. Using this fact, and the invariance of $\Delta_G(t)^2 \operatorname{trace}\pi(t)\Phi_f^G(t)$ by G-conjugation, one can easily rewrite the expression above as

$$-\langle \operatorname{trace}\pi, f\rangle_{\text{ell}} =$$
$$\sum_{\substack{M \subseteq G \\ M \neq G}} |W(A_M)|^{-1} \sum_{\substack{T \text{ ell} \subset M \\ T \bmod M}} |W(M,T)|^{-1} \int_T \Delta_G(t)^2 \operatorname{trace}\pi(t)\Phi_f^G(t)dt.$$

We now use the following facts:

(i) A standard integration formula implies that, with obvious notations, $\Phi_f^G(t) = \Delta_{G/M}(t)^{-1}\Phi_{f^{(P)}}^M(t)$ for $t \in T \subset M$ (cf. [29(a), §5]).

(ii) Recall Casselman's Theorem, which to $t \in T$ associates a parabolic subgroup $P_t = M_t N_t \subset G$. Then, for T elliptic in M, the following holds: if $\Phi_f^G(t) \neq 0$, with $f \in \mathcal{K}$, then $M_t = M$.

This is seen as follows. First, if $t \in T \subset M$ (T compact modulo A_M) is such that $|\chi_\alpha(t)| \neq 1$ for all $\alpha \in \Delta(G,A)$, it is easy to check that $M_t = M$. So we have only to show that this condition on the χ_α is satisfied provided that $\Phi_f^G(t) \neq 0$. By (i), we then have $\Phi_{f^{(P)}}^M(t) \neq 0$; so we must have $f^{(P)}(m) \neq 0$ for some $m \in M$, conjugate in M to t. Since χ_α is a character of M, we then have $\chi_\alpha(m) = \chi_\alpha(t)$.

However, the Satake map $\mathcal{H}_M \to \mathcal{H}_H$, which sends g to $g^{(Q_0)}$, Q_0 the minimal parabolic subgroup of M, is injective, and is given by orbital integrals. In particular, if $g^{(Q_0)}(h) = 0$ for all $h \in H$ such that $|\chi_\alpha(h)| = 1$, this implies that $g(m) = 0$ on $\{m \in M : |\chi_\alpha(m)| = 1\}$. Taking $g = f^{(P)}$, we notice that this hypothesis is satisfied for $g^{(Q_0)} = f^{(P_0)}$: it was the definition of regular functions. Thus $f^{(P)}$ vanishes on the kernel of $|\chi_\alpha|$ in M. So, if $\Phi_f^G(t) \neq 0$, we must have $|\chi_\alpha(t)| \neq 1$. This ends the proof of assertion (ii).

(iii) In the expression for $-\langle \operatorname{trace} \pi, f\rangle_{\text{ell}}$, we may now divide each term indexed by M into a sum over all possible unipotent radicals N of parabolic subgroups $P = MN$. The term associated to N is

$$|W(A_M)|^{-1} \sum_{T \text{ ell} \subset M} |W(M,T)|^{-1} \int_{\{t \in T : N_t = N\}} \Delta_G(t)^2 \operatorname{trace} \pi(t) \Phi_f^G(t) dt.$$

We now use the facts listed in (i) and (ii). Moreover, we have $\Delta_G(t) = \Delta_M(t)\Delta_{G/M}(t)$, where

$$\Delta_{G/M}(t)^2 = |\det(\operatorname{Ad}(1-t)|_{\mathfrak{g}/\mathfrak{m}})|_F.$$

If $\mathfrak{g} = \mathfrak{m} \oplus \mathfrak{n} \oplus \mathfrak{n}^-$ is a triangular decomposition, with $\mathfrak{n} = \operatorname{Lie}(N)$, and if $N_t = N$, one may check that all eigenvalues (on some field extension of F, maybe) of t acting on $\mathfrak{n}$ have absolute values smaller than 1. Conversely, the eigenvalues of t acting on $\mathfrak{n}^-$ are larger than 1 in absolute value. This implies, as is easily seen, that

$$\Delta_{G/M}(t) = \left|\det(1-t)|_{\mathfrak{n}^-}\right|_F^{\frac{1}{2}} = \delta_P^{-\frac{1}{2}}(t)$$

where δ_P is the module of P. Thus the term relative to N is the product of $|W(A_M)|^{-1}$ with

$$\sum_{T \text{ ell} \subset M} \int_{\{t \in T : N_t = N\}} |W(M,T)|^{-1} \Delta_M(t)^2 \operatorname{trace}(\delta_P^{-\frac{1}{2}} \pi_N)(t) \Phi_{f^{(P)}}^M(t) dt.$$

Let us write M^+ for $\{m \in M : |\chi_\alpha(m)| < 1 \text{ if } \alpha \text{ is a root of } (N, A_M)\}$, M_{ell}^+ for $M^+ \cap M_{\text{ell}}$. We may then rewrite this term as

$$(4.1) \qquad |W(A_M)|^{-1} \int_{M_{\text{ell}}^+} \operatorname{trace}(\delta_P^{-\frac{1}{2}} \pi_N)(m) f^{(P)}(m) dm.$$

By induction, we may assume Lemma 4.2 to be true for M. (It is clearly true for H.) Thus, if $g \in \mathcal{K}_M$, the analogous space on M, we have

$$\int_{M_{\text{ell}}} \operatorname{trace}(\delta_P^{-\frac{1}{2}} \pi_N)(m) g(m) dm = 0$$

unless $\delta_P^{-\frac{1}{2}} \pi_N$ is a constituent of an unramified principal series, which would imply that π has the same property. This applies to $f^{(P)}$, which lies in $\mathcal{K}_M$ since the definition of $\mathcal{K}$ is clearly transitive. If χ^+ is the characteristic function of M^+, χ^+ is clearly bi-invariant by K_M; moreover, $g \in \mathcal{K}_M \Longrightarrow \chi^+ g \in \mathcal{K}_M$. Applying the induction hypothesis to $g = \chi^+ f^{(P)}$, we see that the term (4.1) vanishes. This proves Lemma 4.2. ∎

We note that we have used only Casselman's Theorem and very simple properties of the integral formulas, orbital integrals, and the Jacquet modules. The theorems necessary to extend this proof to twisted characters are proved in §2 (local integrability: Proposition 2.2; Casselman's Theorem: Proposition 2.3).The descent property for orbital integrals is proved in [29(a), Lemma 8.5]. We just record the result for $G = \mathrm{GL}(n)$. Let E/F be a cyclic extension. We now write $G(E)$ for $\mathrm{GL}(n, E)$. The Weyl integration formula (for σ-conjugation, where σ is a generator of $\mathrm{Gal}(E/F)$) now reads:

$$\int_{G(E)} \phi(g)dg_E = \sum_T |W(G(F), T(F))|^{-1} \int_{T(E)^{1-\sigma}\backslash T(E)} \Delta_G^2(Nt)\Phi_{\phi,\sigma}(t)dt.$$

Here T runs over the conjugacy classes of maximal tori over F in $G(F)$; N is the norm map, $\Phi_{\phi,\sigma}$ is the twisted orbital integral of ϕ. Since, for Nt regular, $T(F)$ can be identified to the σ-centralizer of t, $\Phi_{\phi,\sigma}$ is associated to measures dg_E on $G(E)$ and dt on $T(F)$. The measure on $T(E)^{1-\sigma}\backslash T(E)$ is then defined by dt via the exact sequence

$$1 \to T(E)^{1-\sigma} \to T(E) \underset{N}{\to} T(F).$$

We then define $G(E)_{\mathrm{ell}}$ as the set of regular elements of $G(E)$ with elliptic norms.

Let $B(E) \supset H(E)$ be the standard Borel subgroup and split torus in $G(E)$; $H(E) \cong (E^*)^n$; if X is any character of $H(E)$, let $\Pi(X)$ be the associated (unitarily induced) principal series representation of $G(E)$.

If Π is a σ-invariant representation of $G(E)$ (§2), we extend it to a representation of $G(E) \rtimes \Sigma$ where Σ is the cyclic group generated by σ. If ϕ is a function on $G(E)$, we will then write

$$\langle \mathrm{trace}\,\Pi, \phi \rtimes \sigma\rangle_{\mathrm{ell}} = \int_{G(E)_{\mathrm{ell}}} \phi(g)\,\mathrm{trace}\,\Pi(g,\sigma)dg.$$

It makes sense by the local integrability theorem (§2). Let $K_E = G(\mathcal{O}_E)$.

LEMMA 4.3: *Let* $\Pi \cong \Pi \circ \sigma$ *an irreducible* σ*-stable representation of* $G(E)$*; extend it to* $G(E) \rtimes \Sigma$*. Assume that* Π *has no* K_E*-fixed vector, and is not a* *constituent of a principal series representation* $\Pi(X)$, *with* X *an unramified,* σ*-invariant character of* $(E^*)^n$. *Then, if* $\phi \in \mathcal{K}_E$:

$$\langle \text{trace}\, \Pi, \phi \times \sigma \rangle_{\text{ell}} = 0.$$

Here of course $\mathcal{K}_E$ is the space defined by Definition 4.1 in $\mathcal{H}_E$. The proof is the same as for Lemma 4.2. ∎

4.2. The base change identities

From now on G is GL(n); we assume that E/F is an unramified extension of local fields. Via the Satake isomorphism, we have (cf. Kottwitz [29(a)])

$$\mathcal{H}_E \cong \mathbf{C}[z_1, z_1^{-1}, z_2, z_2^{-1}, \dots z_n, z_n^{-1}]^{\mathfrak{S}_n};$$

the same holds for $\mathcal{H}_F$.

We will denote an element $f^\vee$ of $\mathcal{H}_E$ or $\mathcal{H}_F$ by $f^\vee = \sum a_\lambda z^\lambda$; the sum runs over all multi-indices $\lambda = (\lambda_1, \dots \lambda_n) \in \mathbf{Z}^n$ and $z = (z_1, \dots z_n) \in {}^L H^0 \cong (\mathbf{C}^*)^n$. There is a natural homomorphism $b : \mathcal{H}_E \to \mathcal{H}_F$ which corresponds to the diagonal imbedding of the L-group of G over F into the L-group over F of $\text{Res}_{E/F}\, G$. In terms of the Satake transforms, it associates to $\phi^\vee(z) = \sum a_\lambda z^\lambda$ the function $f^\vee(z) = \phi^\vee(z^\ell) = \sum a_\lambda z^{\ell\lambda}$; here $\ell = \dim_F(E)$.

In terms of the Satake transform, $\mathcal{K} \subset \mathcal{H}$ is identified with

$$\Big\{ f = \sum a_\lambda z^\lambda : a_\lambda = 0 \text{ if } \frac{\lambda_i + \lambda_{i+1} + \cdots + \lambda_{j-1}}{j-i} = \frac{\lambda_j + \cdots + \lambda_{k-1}}{k-j} \text{ for some } i < j < k \Big\}.$$

Note that, with obvious notations, $b(\mathcal{K}_E) \subset \mathcal{K}_F$.

Let $P_0 = HN_0$ be the standard minimal parabolic subgroup of G composed of upper triangular matrices. An unramified character of $H(F)$ is canonically identified with an element z of ${}^L T^0$. In particular, the module $\delta_{P_0}^{\frac{1}{2}}$, regarded as a character of H, is then identified with

$$z = \left(q^{\frac{n-1}{2}}, q^{\frac{n-3}{2}}, \dots, q^{-\frac{(n-1)}{2}} \right), \text{ where } q = |\varpi_F|.$$

Let χ be an unramified character of F^*, which we identify with the complex number $\zeta = \chi(\varpi_F)$. Let $\text{St}(\chi)$ be the Steinberg representation of $G(F)$ such that $\delta_{P_0}^{-1}\, \text{St}(\chi)_{N_0}$ is the character $(\chi, \dots, \chi)$ of $H \cong (F^*)^n$.

Let I_0 denote the interval of integers $[1, 2, \dots n]$. A *partition* of I_0 is a disjoint decomposition $I_0 = I_1 \cup \dots \cup I_k$ with

$$\begin{aligned} I_1 &= [1, \dots n_1], \\ I_2 &= [n_1 + 1, \dots n_2], \\ \dots\ \mathrm{I}_k &= [n_{k-1} + 1, \dots n]. \end{aligned}$$

A *family of nested partitions* (I_{ji}) of I_0 is a family of partitions $I_0 = I_{1,i} \cup \dots \cup I_{k_i,i} (i = 1, \dots N)$ where each new partition is finer than the previous one. It is *complete* if the last partition is given by $I_0 = \{1\} \cup \dots \cup \{n\}$. If $\lambda = (\lambda_1, \dots \lambda_n) \in \mathbf{Z}^n$, we say that λ is positive for (I_{ji}) if the following property is satisfied. If $I = [i, i+1, \dots j]$ is an interval, let $|I|$ be its cardinality and $\lambda(I) = \lambda_i + \lambda_{i+1} + \dots + \lambda_j$. Then, if $I_{j,i} = I_{j_1,i+1} \cup \dots \cup I_{j_k,i+1}$ is given by refining the partition $I_{*,i}$ into $I_{*,i+1}$, λ satisfies:

$$\frac{\lambda(I_{j_1,i+1})}{|I_{j_1,i+1}|} < \frac{\lambda(I_{j_1+1,i+1})}{|I_{j_1+1,i+1}|} < \dots < \frac{\lambda(I_{j_k,i+1})}{|I_{j_k,i+1}|}.$$

LEMMA 4.4: *There exist constants $c_{\mathcal{I}}$, where $\mathcal{I} = (I_{ij})$ runs over the set of complete families of nested partitions of I_0, such that for $f \in \mathcal{K}_F$, $f = \sum a_\lambda z^\lambda$:*

$$\langle \operatorname{trace} \mathrm{St}(\chi), f \rangle_{\mathrm{ell}} = \sum_{\mathcal{I}} c_{\mathcal{I}} \sum_{\substack{\lambda \in \mathbf{Z}^n \\ \lambda > 0 \text{ for } \mathcal{I}}} a_\lambda (\zeta \delta_{P_0}^{\frac{1}{2}})^\lambda.$$

Here

$$\delta_{P_0}^{\frac{1}{2}} \zeta = \left(\zeta q^{\frac{n-1}{2}}, \dots \zeta q^{-\left(\frac{n-1}{2}\right)} \right) \in (\mathbf{C}^*)^n \cong {}^L H^0.$$

The proof of this ugly lemma is simpler than its formulation. The proof of Lemma 4.2 gives an expression of $\langle \operatorname{trace} \mathrm{St}(\chi), f \rangle_{\mathrm{ell}}$ as a sum over M, N (notations of Lemma 4.2) of terms of the form

$$c \int_{M_{\mathrm{ell}} \cap M^+} \operatorname{trace}(\delta_P^{-\frac{1}{2}} \mathrm{St}(\chi)_N)(m) f^{(P)}(m) dm.$$

We may assume, by using the invariance of the character trace $\mathrm{St}(\chi)$, that M is the diagonal block Levi subgroup associated to a partition $n = n_1 + n_2 + \dots + n_r$, and that M^+ is defined by

$$|\det g_1|^{\frac{1}{n_1}} < |\det g_2|^{\frac{1}{n_2}} < \dots < |\det g_r|^{\frac{1}{n_r}},$$

where $g \in M$ is written

$$g = \begin{pmatrix} g_1 & & & \\ & g_2 & & \\ & & \ddots & \\ & & & g_r \end{pmatrix}.$$

By a result of Casselman and Zelevinsky [42, Prop. 3.4],

$$\delta_P^{-\frac{1}{2}} \operatorname{St}(\chi)_N = \delta_P^{\frac{1}{2}} \operatorname{St}(M, \chi)$$

where $\operatorname{St}(M,\chi)$ is the tensor product of the $\operatorname{St}(M_i,\chi)$ ($M = M_1 \times \cdots \times M_r$, $M_i \cong GL(n_i, F)$). Hence, by induction, this term may be written, using Lemma 4.4 on M:

$$c \sum_{\mathcal{I}'} c_{\mathcal{I}'} \sum_{\substack{\lambda>0 \text{ for } \mathcal{I}' \\ \lambda>0 \text{ for } M}} a_\lambda (\zeta \delta_{Q_0}^{\frac{1}{2}} \delta_P^{\frac{1}{2}})^\lambda.$$

The sum now runs over all $\mathcal{I}' = (\mathcal{I}'_1, \dots \mathcal{I}'_r)$ where $\mathcal{I}'_{j+1}$ is a nested sequence in $[n_j + 1, \dots n_{j+1}]$. The condition "$\lambda > 0$ for M" is forced by the fact that we integrate on M^+, and can be written

$$\frac{\lambda(I_1)}{|I_1|} < \frac{\lambda(I_2)}{|I_2|} < \cdots < \frac{\lambda(I_r)}{|I_r|}$$

with $I_{j+1} = [n_1 + \cdots + n_j + 1, \dots, n_1 + \cdots + n_{j+1}]$. These two conditions imply that the $\mathcal{I}'_j$ group together to give a nested sequence $\mathcal{I}$ for $\operatorname{GL}(n)$, and the sum is over the λ positive for $\mathcal{I}$.

We have denoted by Q_0 the minimal parabolic subgroup $P_0 \cap M$ of M; it is easy to check that $\delta_{Q_0}\delta_P = \delta_{P_0}$. This proves Lemma 4.4. ∎

We remark that, by the same proof, the formula in Lemma 4.4 also holds for the twisted character of a representation $\operatorname{St}(X)$ of $\operatorname{GL}(n, E)$, X being an unramified character of E^*. (Note that since E/F is unramified, the character X is then σ-invariant, hence also the representation $\operatorname{St}(X)$.) The constants $c_{\mathcal{I}}$ will be the same. We will use this without further comment.

We are now ready to prove:

THEOREM 4.5: *Let $\phi \in \mathcal{H}_E$, $f = b\phi \in \mathcal{H}_F$. Then, if $\gamma \in G(F)$ is regular semi-simple,*

$$\Phi_f(\gamma) = \begin{cases} 0 & \text{if } \gamma \text{ is not a norm} \\ \Phi_{\phi,\sigma}(\delta) & \text{if } \gamma = N\delta,\ \delta \in G(E). \end{cases}$$

Here the definitions of Φ_f and Φ_ϕ require choices of measures on $G(F)_\gamma$ and $G(E)_{\delta,\sigma}$ – we take them equal after the identification of these two groups, as in §3 – and on $G(E)$ and $G(F)$: on these groups, the measures give mass 1 to $G(\mathcal{O}_E)$ and $G(\mathcal{O}_F)$.

We may now begin the proof. We will often work in fact with $\tilde{G} = \mathrm{PGL}(n)$ rather than G. The homomorphism of Hecke algebras, and the previous results of this paragraph extend in obvious ways to $\tilde{G}$.

Assume first that γ is not elliptic. Thus $\gamma \in M(F)$, where MN is a proper parabolic subgroup of G. Up to σ-conjugation we may assume that $\delta \in M(E)$ if γ is the norm of δ. There is a commutative diagram ([29(a)])

$$\begin{array}{ccc} \mathcal{H}_E & \xrightarrow[b]{} & \mathcal{H}_F \\ \downarrow & & \downarrow \\ \mathcal{H}_E^M & \xrightarrow[b]{} & \mathcal{H}_F^M \end{array}$$

the vertical maps being given by $f \mapsto f^{(P)}$. By induction we may assume the identities of Theorem 4.5 known for M. Using formula (i) in the proof of Lemma 4.2 and the analogous twisted formula ([29(a), Lemma 8.5])

$$\Phi^G_{\phi,\sigma}(\delta) = \Delta_{G(F)/M(F)}(N\delta)^{-1}\Phi^M_{\phi^{(P)},\sigma}(\delta),$$

we see that the theorem for M implies the identities of orbital integrals between δ and γ.

Thus the comparison is reduced to the case of elliptic orbital integrals. If χ is an unramified character such that $\chi^n = 1$, let $\mathrm{St}(\chi)$ be the associated Steinberg representation of $\tilde{G}(F)$. Write $\tilde{\mathcal{K}}_E$, $\tilde{\mathcal{K}}_F$ for the regular Hecke functions on $\tilde{G}(E)$, $\tilde{G}(F)$.

LEMMA 4.6: *Let $f \in \tilde{\mathcal{K}}_F$ be such that*

$$\langle \mathrm{trace}\,\mathrm{St}(\chi), f\rangle_{\mathrm{ell}} = 0$$

for any Steinberg representation of $\tilde{G}(F)$. Then all elliptic orbital integrals of f vanish.

Note that Lemma 4.4 states explicitly what the conditions in Lemma 4.6 mean in terms of $f^\vee$.

Proof. The only discrete series representations of $\mathrm{PGL}(n,F)$ that are constituents of unramified principal series are the Steinberg representations ([42]). Thus, by Lemma 4.2, we have $\langle \mathrm{trace}\,\pi, f\rangle_{\mathrm{ell}}$ for any discrete π. Moreover, let T be a non-elliptic Cartan subgroup of $\tilde{G}(F)$. Then assertion (ii) in the proof of Lemma 4.2 shows that the orbital integrals of f vanish on the maximal compact subgroup of T. These two facts, and Kazhdan's density theorem for orbital integrals – the non-twisted analog of Proposition 2.7, cf. [28, Theorem 1], imply that the elliptic orbital integrals of f vanish. Indeed, let ρ be a faithful rational representation of $\tilde{G}$. Let $P(\rho(g), X) = \sum_{i=1}^{N} a_i(g)X^i$ be the characteristic polynomial of $\rho(g)$. The map $\psi : g \mapsto (a_i(g))$ sends $\tilde{G}$ into the affine N-space. The image of $\tilde{G}(F)_{\text{ell}}$ has a compact closure ω. If Ω is a compact-open neighborhood of ω in F^N, $V = \psi^{-1}(\Omega)$ is open and closed in $\tilde{G}(F)$; for Ω small enough, $V \cap T$ is contained in the maximal compact subgroup of T for any T. Proceeding as in [11(c)], we set $g = \chi_V f$, where χ_V is the characteristic function of V; then $g \in C_c^\infty(\tilde{G}(F))$ and all the non-elliptic orbital integrals of g vanish. It is now clear that $\langle \operatorname{trace} \pi, g\rangle = 0$ for any tempered representation of $\tilde{G}$ (recall that all tempered representations of $\tilde{G}$ are unitarily induced from discrete series). By Kazhdan's theorem, all orbital integrals of g vanish. By construction of g, its elliptic orbital integrals are the same as those of f. Whence the result. ∎

The same argument can be used (using the "twisted" extension of Kazhdan's theorem) with the twisted elliptic orbital integrals.

Recall from §2.3 the notion of σ-discrete representation of $G(E)$. If $\phi \in \tilde{\mathcal{K}}_E$, Lemma 4.3 shows that $\langle \operatorname{trace} \Pi, \phi \rtimes \sigma\rangle_{\text{ell}} = 0$ for any σ-discrete Π, unless Π is a constituent of an unramified principal series. At this point let us revert to $\mathrm{GL}(n)$ rather than $\mathrm{PGL}(n)$ for an instant. A tempered representation is of the form $\Pi = \operatorname{ind}_{MN}(\Pi_M \otimes 1)$ (unitary induction), Π_M a unitary discrete series for M. It will be a constituent of an unramified principal series only if Π_M is, i.e., if Π_M is a Steinberg representation with unramified inducing character. Thus, writing

$$M = \mathrm{GL}(n_1, E) \times \cdots \times \mathrm{GL}(n_r, E),$$

we have $\Pi_M = \mathrm{St}(X_1) \otimes \cdots \otimes \mathrm{St}(X_r)$ where X_i are unramified characters of E^*. But then each $\mathrm{St}(X_i)$ is σ-stable (E/F is unramified!). Thus Π can be σ-discrete only if $r = 1$ and $\Pi = \mathrm{St}(X)$ for unramified X. Applying this, now, to $\mathrm{PGL}(n, E)$, we see that for $\phi \in \tilde{\mathcal{K}}_E$ such that $\langle \operatorname{trace} \mathrm{St}(X), \phi \rtimes \sigma\rangle_{\text{ell}} = 0$ for all unramified X with $X^\ell = 1$, we have $\langle \operatorname{trace} \Pi, \phi \rtimes \sigma\rangle_{\text{ell}} = 0$ for any σ-discrete Π.

The argument of Lemma 4.6 then carries through to give:

LEMMA 4.7: *Let $\phi \in \tilde{\mathcal{K}}_E$ be such that*

$$\langle \operatorname{trace} \mathrm{St}(X), \phi \rtimes \sigma\rangle_{\text{ell}} = 0, \quad X \textit{ unramified}, X^\ell = 1.$$

Then all σ-elliptic orbital integrals of ϕ vanish.

We note now that $\tilde{\mathcal{H}}_F$ may be identified with functions on $G(F)$, bi-invariant by $G(\mathcal{O}_F)$, compactly supported modulo $Z(F)$, where Z is the center of $\mathrm{GL}(m)$, and such that $f(gz) = f(g)$ for $z \in Z(F)$. The same, of course, holds for $\tilde{\mathcal{H}}_E$. Moreover, we could have proved the analog of Lemmas 4.6 and 4.7 for the space $\mathcal{H}_\chi$ of functions satisfying $f(gz) = \chi(z)f(g)$, χ an unramified character of $Z(F)$ (or $Z(E)$.) Lastly, if $f \in \mathcal{H}_F$ (say) and χ is such a character, we get a function transforming under χ by setting $f_\chi(g) = \int_{Z(F)} f(gz)\chi(z)dz$. Using these (trivial) facts, one sees that the identities of Theorem 4.5 for the spaces $\mathcal{H}$ can be deduced from the analogous identities for all the spaces $\mathcal{H}_\chi$; here the characters χ_F of $Z(F)$ and χ_E of $Z(E)$ must be related by $\chi_F \circ N_{E/F} = \chi_E$. Note that for unramified characters this relation is bijective.

For more details, cf. Langlands [30(e), p. 76 ff.].

So let χ_F be unramified, $\chi_E = \chi \circ N_E/F$. We will just write $\mathcal{H}_{F,\chi}$ and $\mathcal{H}_{E,\chi}$ for the corresponding subspaces, and $\mathcal{K}_{F,\chi}$ and $\mathcal{K}_{E,\chi}$ for the regular Hecke functions inside. Combining Lemmas 4.6 and 4.7, we get:

COROLLARY 4.8: *The relations of Theorem 4.5 hold for $\phi \in \mathcal{K}_{E,\chi}$ such that $\langle \mathrm{trace\,St}(\eta_E^{-1}), \phi \rtimes \sigma\rangle_{\mathrm{ell}} = 0$ for all unramified η_E such that $\eta_E^\ell = \chi_E$.*

Indeed, all non-elliptic orbital integrals correspond by the descent argument, and the other ones vanish. ∎

We will now need a result of Kottwitz:

PROPOSITION 4.9: ([29(c), Lemma 8.8]). *The identity of Theorem 4.5 is true when ϕ, f are the units of the Hecke algebras.*

Obviously Proposition 4.9 again extends, in an obvious manner, to the spaces $\mathcal{H}_\chi$.

We will now get Theorem 4.5 for the other functions by an approximation argument using the trace formula. So let k be now a number field, v_0 a place of k such that $k_{v_0} \cong F$. We assume chosen a cyclic extension k' of k of degree ℓ such that $k' \otimes_k k_{v_0} \cong E$; we also assume k totally imaginary.

We now choose a character χ_k of $N_{k'/k}Z(\mathbf{A}_{k'})$ mod $N_{k'/k}Z(k')$ such that the restriction of χ_k to $N_{E/F}Z(E)$ is χ_F; let $\chi_{k'} = \chi_k \circ N_{E/F}$, a character of $Z(\mathbf{A}_{k'})$. Write $Z_1 = N_{k'/k}Z(\mathbf{A}_{k'})$ and $Z_1(k) = Z_1 \cap Z(k) = N_{k'/k}Z(k')$. We will write $\mathbf{A}$ for $\mathbf{A}_k$. Let $C_c^\infty(G(\mathbf{A}), \chi_k^{-1})$ be the space of smooth, compactly supported mod Z_1, functions on $G(\mathbf{A})$, transforming under Z_1 by χ_k^{-1}; it acts on the space $L^2_{\mathrm{cusp}}(Z_1G(k)\backslash G(\mathbf{A}), \chi_k)$. In the same manner, $C_c^\infty(G(\mathbf{A}_{k'}), \chi_{k'}^{-1})$ acts on $L^2_{\mathrm{cusp}}(Z(\mathbf{A}_{k'})G(k')\backslash G(\mathbf{A}_{k'}), \chi_{k'})$; we have denoted by L^2_{cusp} the parabolic spectrum in the L^2-functions transforming according to a given character. Let us write r for the first representation, R for the second; in the case of k', we also have an obvious operator $\psi(g) \mapsto \psi(g^\sigma)$, denoted by I_σ.

We may choose two finite places v_1, v_2, of k, different from v_0, which split in k'.

Let f_1 be a coefficient of a supercuspidal representation of $G(k_{v_1})$ such that $f_1(1) = 1$. On $G(k'_{v_1}) \cong G(k_{v_1})^\ell$, we set $\phi_{v_1} = (f_1, \dots, f_1)$. We set $f_{v_1} = f_1 * \cdots * f_1$. These two functions are associated in the sense of ([30(e), §8]) – see §5.

In $G(k'_{v_2}) \cong G(k_{v_2})^\ell$, we choose an element $\delta = (\delta_1, \dots \delta_\ell)$ such that $N\delta = \delta_1 \dots \delta_\ell$ is regular elliptic; we may even assume that the image of $N\delta$ in $\mathrm{PGL}(n, k_{v_2})$ is strongly regular, i.e., its centralizer is a torus. This implies that in $G(k_{v_2})$ the relation $xN\delta x^{-1} = z \cdot N\delta$, $z \in Z(k_{v_2})$ implies $xN\delta x^{-1} = N\delta$. Then, if $f_i \in C_c^\infty(G(k_{v_2}), \chi_{v_2})$ has support close enough to $\delta_i, \phi_{v_2} = (f_1, \dots f_\ell)$ has support on the σ-regular elements and $f_{v_2} = f_1 * \cdots * f_\ell$ has support on the strongly regular elliptic elements; ϕ_{v_2} and f_{v_2} are associated. We assume, of course, that these functions have the right invariance by the center.

We now construct functions ϕ and f on $G(\mathbf{A}_{k'})$ and $G(\mathbf{A}_k)$ as follows. Let S be a finite (non-empty) set of finite places of k, disjoint from $\{v_0, v_1, v_2\}$ and containing all places where k' ramifies. We choose, at all these places, smooth associated functions ϕ_v and f_v (e.g., with regular support). At all infinite places (which split), we take associated functions of the form $\phi_v = (f_1, \dots, f_\ell)$, $f_v = f_1 * \cdots * f_\ell$ with f_i smooth of compact support. At all other places, including v_0, we take ϕ_v either to be a function in $\mathcal{K}_{k'_v, \chi_v}$ which satisfies the condition of Corollary 4.8 or a unit in $\mathcal{H}_{k'_v, \chi_v}$ and we take $f_v = b\phi_v$. Thus ϕ_v and f_v are associated. To avoid using in a non-obvious way the base change identities in the "intermediary case" (§5), we take ϕ_v, f_v to be units at all places that are neither split nor inert. We assume, of course, that ϕ_v is the unit in $\mathcal{H}_v$ for almost all v. Then we set $\phi = \otimes \phi_v$, $f = \otimes f_v$.

With these assumptions, the Deligne–Kazhdan form of the trace formula applies and yields:

$$\text{(4.4)} \qquad \operatorname{trace}(R_{\text{cusp}}(\phi) I_\sigma) = \sum_{\{\delta\}} \operatorname{meas}(G_{\delta,\sigma}(k) Z(\mathbf{A}_k) \backslash G_{\delta,\sigma}(\mathbf{A})) \Phi_{\phi,\sigma}(\delta).$$

The sum ranges over σ-conjugacy classes of elements $\delta \in G(k')$ such that $N\delta$ is elliptic regular, and in fact has strongly regular image in $\mathrm{PGL}(n)$; $\Phi_{\phi,\sigma}$

is the product of local twisted orbital integrals, defined with the Tamagawa measures.

In the same way, and with analogous notation, we get:

$$\text{trace}(r_{\text{cusp}}(f)) = \sum_{\{\gamma\}} \text{meas}(G_\gamma(k)Z_1\backslash G_\gamma(\mathbf{A}))\Phi_f(\gamma). \tag{4.5}$$

As seen in §1, the map N gives a bijection between the two sets indexing the sums. The assertion of Theorem 4.5 applies to our local functions; moreover, an element of $G(k)$ is a global norm if and only if it is a local norm everywhere (Lemma 1.2). Since $k^*\backslash Z_1 k^*$ has index ℓ in $k^*\backslash Z(\mathbf{A})$, we have

$$\begin{aligned}\text{meas}(Z_1 G_\gamma(k)\backslash G_\gamma(\mathbf{A})) &= \ell\, \text{meas}(Z(\mathbf{A})G_\gamma(k)\backslash G_\gamma(\mathbf{A})) \\ &= \ell\, \text{meas}(Z(\mathbf{A})G_{\delta,\sigma}(k)\backslash G_{\delta,\sigma}(\mathbf{A}))\end{aligned}$$

since $G_{\delta,\sigma}$ and G_γ are isomorphic. Thus, for such functions

$$\ell \cdot \text{trace}(R_{\text{cusp}}(\phi)I_\sigma) = \text{trace}\, r_{\text{cusp}}(f). \tag{4.6}$$

We may regard this as an identity of linear forms on the functions $\phi_v = (f_1, \ldots, f_\ell)$ and $f_v = f_1 * \cdots * f_\ell$, where v is an infinite place. A representation of $G(k'_v) = G(k_v)^\ell$ contributes only if it is of the form $\Pi_v = \pi_v \otimes \cdots \otimes \pi_v$. Then $\text{trace}\, \Pi_v(\phi_v)I_\sigma = \zeta\, \text{trace}\, \pi_v(f_v)$, with ζ an ℓ-th root of unity. Grouping the terms in (4.6) on one side, and putting together the representations equal to π_v or $\pi_v \otimes \cdots \otimes \pi_v$ at the place v, we obtain on identity of the form:

$$\sum a_{\pi_v}\, \text{trace}\, \pi_v(f_v) = 0$$

ranging over unitary representations of $G(k_v)$. By a lemma of Jacquet–Langlands (cf. [25, Theorem 5.2]) all a_{π_v} must be 0. Applying this to all infinite places, we then see that (4.6) can be rewritten as

$$\begin{aligned}&\ell \sum_{\Pi} \text{trace}(\Pi_{S'}(\phi_{S'})I_{\sigma,S'}) \prod_{v\notin S'} \phi_v^\vee(t_{\Pi,v}) \\ &= \sum_{\pi} \text{trace}(\pi_{S'}(f_{S'})) \prod_{v\notin S'} f_v^\vee(t_{\pi,v}).\end{aligned} \tag{4.7}$$

Both sums range over representations in the cuspidal spectra such that their infinite components are equal to $\bigotimes^\ell \pi_\infty$ and π_∞ respectively, where π_∞ is a fixed representation of $G(k_\infty)$. The set S' is the union of S, v_1, v_2 and the infinite places, and $I_{\sigma,S'}$, tensored with the operator equal to 1 on spherical vectors outside S', is the restriction of I_σ on Π. By the fundamental theorem on the finite-dimensionality of spaces of automorphic forms (cf. [20(b), Theorem 1]), the number of cuspidal representations of $G(\mathbf{A})$ which have a vector fixed by a given compact-open subgroup of $G(\mathbf{A}_f)$, the group of points with values in the finite adèles, and have a given infinitesimal character, is finite. In particular, if $f_{S'}$ and $\phi_{S'}$ are fixed, only a finite number of representations appear on both sides of (4.7), independently of the choice of ϕ_v for $v \notin S'$. At this point, we apply the following remark:

LEMMA 4.10: *Let $\tilde{\mathcal{K}} = \tilde{\mathcal{K}}_F$. Let $t_i (i = 1, \dots, N)$ be a finite number of elements of ${}^L\tilde{H}^0 = \{z = (z_1, \dots, z_n) : \Pi z_i = 1\}$, different modulo $W = \mathfrak{S}_n$. Assume that the complex numbers $c_i (i = 1, \dots, N)$ satisfy*

$$\sum_{i=1}^{N} c_i f^\vee(t_i) = 0$$

for all $f \in \tilde{\mathcal{K}}$ such that the conditions of Lemma 4.6 are satisfied. Then $c_i = 0 (i = 1, \dots, N)$.

Proof. By elementary linear algebra, the condition in Lemma 4.10 means that there exist constants c_χ (for χ an unramified character such that $\chi^n = 1$) such that

$$\sum c_i f^\vee(t_i) + \sum_\chi c_\chi \operatorname{trace}\langle \mathrm{St}(\chi), f\rangle_{\mathrm{ell}} = 0 \tag{4.8}$$

for all $f \in \tilde{\mathcal{K}}$. If $\lambda \in \mathbf{Z}^n$, let $f_\lambda^\vee(z) = \sum_{w \in W} z^{w\lambda}$ be the symmetric monomial associated to λ. Then $f_\lambda^\vee \in \tilde{\mathcal{K}}$ if λ does not belong to a finite union of hyperplanes. We will assume that λ is so chosen that $\lambda_1 < \lambda_2 < \cdots < \lambda_n$. Using Lemma 4.4, we may rewrite (4.8) as

$$0 = \sum c_i \sum_{w \in W} t_i^{w\lambda} + \sum_\chi c_\chi \sum_{\mathcal{I}} c_{\mathcal{I}} \sum_{\substack{\nu \in W\lambda \\ \nu > 0 \text{ for } \mathcal{I}}} (\zeta_\chi \delta_{P_0}^{\frac{1}{2}})^\nu; \tag{4.9}$$

ζ_χ is the n-th root of unity associated to χ.

Given $\mathcal{I}$ and $s \in W$, the set of λ such that the term indexed by $\mathcal{I}$ and $\nu = s\lambda$ appears is determined by the positivity of certain linear forms. Consequently, for λ in a certain hyperplane cone $C \subset \mathbf{Z}^n$, that we can take contained in the set $\lambda_1 < \lambda_2 < \cdots < \lambda_n$, this equality may be written

$$0 = \sum_i c_i \sum t_i^{w\lambda} + \sum_\chi c_\chi \sum_{\mathcal{I}} c_{\mathcal{I}} \sum_s (\zeta_\chi \delta_{P_0}^{\frac{1}{2}})^{s\lambda} \tag{4.10}$$

where the set of s depends only on $\mathcal{I}$ and C, and not on $\lambda \in C$. The identity (4.10) is then a linear dependence relation between characters of

$X^*({}^L\tilde{H}^0) \cong \mathbf{Z}^n/\mathbf{Z}$, satisfied for $\lambda \in C$. Therefore it is true for all $\lambda \in X^*({}^L\tilde{H}^0)$.

Let us fix $w \in W$. Assume $c_i \neq 0$. Since the character $\lambda \mapsto t_i^{w\lambda}$ does not appear elsewhere in the sum indexed by i, the t_i being distinct modulo W, it must then appear in the second sum, by the linear independence of characters. Thus:

$$t_i^w = (\zeta_X \delta_{P_0}^{\frac{1}{2}})^s$$

for some $(\mathcal{I}, s)$ such that $s\lambda$ is positive for $\mathcal{I}$ if $\lambda \in C$. If C is contained in $(\lambda_1 < \lambda_2 < \cdots < \lambda_n)$, this obviously implies $s \neq w_0$, where w_0 is the order-reversing permutation. So we have shown that if $c_i \neq 0$, then, for all $w \in W : t_i^w = (\zeta_X \delta_{P_0}^{\frac{1}{2}})^s$ with $s \neq w_0$. This is clearly impossible: if it were true, then

$$t_i^{w \cdot s^{-1} w_0} = (\zeta_X \delta_{P_0}^{\frac{1}{2}})^{w_0}.$$

But $(\zeta_X \delta_{P_0}^{\frac{1}{2}})^{w_0}$ cannot be equal to $(\zeta_X \delta_{P_0}^{\frac{1}{2}})^s$ with $s \neq w_0$, since $\zeta_X \delta_{P_0}^{\frac{1}{2}}$ is regular. This contradiction proves the lemma. ∎

Applying Lemma 4.10, we now see that the identity of traces (4.7) is true when ϕ_{v_0} is *any* function in $\mathcal{H}_E$. We may now take the sum of all such identities over the representations at the infinite places, to get an identity of trace formulas:

$$\sum_{\{\delta\}} \ell \cdot v(\delta) \Phi_{\phi,\sigma}(\delta) = \sum_{\{\gamma\}} v(\gamma) \Phi_f(\gamma). \tag{4.11}$$

The volumes $v(\delta)$ and $v(\gamma)$ are the ones which figure in (4.4) and (4.5). But now, if δ is an element of $G(k')$ whose norm γ in $G(k)$ is regular elliptic, we may, by choosing the functions ϕ and f at the places in S, insure that only $\Phi_{\phi,\sigma}(\delta)$ and $\Phi_f(\gamma)$ appear in (4.9). We may also assume that the orbital integrals at $v \neq v_0$ are non-zero at γ, and that all Hecke functions (except at v_0) are units. If δ_{v_0} and γ_{v_0} are the elements δ and γ considered as elements of $G(E)$ and $G(F)$ respectively, formula (4.11) implies, for $\phi \in \mathcal{H}_E$ and $f = b\phi$:

$$\Phi_{\phi,\sigma}(\delta_{v_0}) = \Phi_f(\gamma_{v_0}).$$

By density of the elements of $G(k')$, this implies the last assertion of Theorem 4.5. The first one is implied by

LEMMA 4.11: *Assume $f \in \mathcal{H}_F$ is in the image of the base change map. Then*

$$\Phi_f(\gamma) = 0 \textit{ if } \gamma \in G(F)_{\text{reg}} \textit{ is elliptic and not a norm.}$$

Proof. Since γ is elliptic regular, γ is a norm if and only if

$$\det \gamma \in N_{E/F}(E^*).$$

Let π be a discrete series representation of $G(F)$. Weyl's formula gives:

$$0 = \operatorname{trace} \pi(f) = \sum_T |W(G,T)|^{-1} \int_{T(F)} \Delta(t)^2 \operatorname{trace} \pi(t) \Phi_f(t) dt.$$

The sum runs over F-maximal tori of G up to conjugacy. If T is non-elliptic, Theorem 4.5, applied inductively, implies that the integral runs only over $N(T(E)) \subset T(F)$.

Let now χ be a character of $F^*/N_{E/F}(E^*)$. We may replace π by $\pi \otimes \chi(\det)$. The sum over the non-elliptic tori remains unchanged. Taking the difference yields:

$$0 = \sum_{T_{\text{ell}}} |W(G,T)|^{-1} \int_{T(F)} \Delta(t)^2 \operatorname{trace} \pi(t) \Phi_f(t)(1 - \chi(\det t)) dt.$$

Since the characters of discrete series form an orthogonal basis on the elliptic set (cf. [33(b)], [15]), this implies that $(1 - \chi \circ \det)\Phi_f$ is zero on the elliptic set. Thus $\Phi_f(t) = 0$ unless $\det(t) \in N_{E/F}E^*$, i.e., t is a norm (Lemma 1.4). This proves Lemma 4.11 and completes the proof of Theorem 4.5. ∎

5. Orbital integrals: non-inert primes

In this section we extend the results of §3–4 about orbital integrals to the case where a place of the small global field is not inert in the field extension. The case of split places was treated by Langlands [30(e), Ch. 8]. We quickly hint at the more general results needed when the degree is not a prime.

We assume that E is an F-algebra of the form $k' \otimes F$, where k is global, F a completion of k – Archimedean or not – and k'/k cyclic of order ℓ. Then $E \cong E_1 \times \cdots \times E_1$ (m factors), with E_1 a cyclic extension of order k, and $\ell = km$. If σ is a generator of $\mathrm{Gal}(k'/k)$, then $\tau = \sigma^m$ generates $\mathrm{Gal}(E_1/F)$; we may write the action of σ on E as

$$\sigma : (x_1, x_2, \ldots x_m) \mapsto (x_2, x_3, \ldots x_{m-1}, x_1^\tau).$$

The action of σ on $G(E) \cong G(E_1) \times \cdots \times G(E_1)$ is described by the same formula; the fixed points of σ compose the group $G(F)$, diagonally embedded into $G(E)$.

The norm $N_{E/F} : G(E) \to G(E)$ is the composite of

$$N_{E/E_1} : (x_1, \ldots x_m) \mapsto (x_1 x_2 \cdots x_m, x_2 x_3 \cdots x_m x_1^\tau, \ldots, x_m x_1^\tau \cdots x_{m-1}^\tau)$$

and $N_{E_1/F}$ which operates componentwise by $x_i \mapsto N_{E_1/F} x_i$.

Note that if $x = (x_1, \ldots x_m)$, then all components of $N_{E/E_1} x$ are τ-conjugate to $x_1 \ldots x_m \in G(E_1)$. We will consider $x_1 \ldots x_m$ as the norm (from E to E_1) of x. We write $x_1 x_2 \cdots x_m = N_1 x$.

Let $\delta = (\delta_1, \ldots \delta_m) \in G(E)$. An easy computation shows that the σ-centralizer $G_{\delta,\sigma}(F) = \{x : x^{-1}\delta x^\sigma = \delta\}$ is given by the equations

$$\begin{aligned} x_1^{-1}\delta_1 x_2 &= \delta_1 \\ x_2^{-1}\delta_2 x_3 &= \delta_2 \\ &\vdots \\ x_m^{-1}\delta_m x_1^\tau &= \delta_m. \end{aligned} \tag{5.1}$$

In particular, setting $\xi = N_1\delta$, we obtain $x_1^{-1}\xi x_1^\tau = \xi$, so $G_{\delta,\sigma}(F)$ is described by $x_1 \in G_{\xi,\tau}(F)$ and the $(m-1)$ first equations giving the $x_i (i > 1)$ in terms of x_1.

Let $\phi = \phi_1 \otimes \phi_2 \otimes \cdots \otimes \phi_m$ be a function in $C_c^\infty(G(E))$ which is a tensor product. Then

$$\int_{G_{\delta,\sigma}(F)\backslash G(E)} \phi(x^{-1}\delta x^\sigma)dx = \int \phi_1(x_1^{-1}\delta_1 x_2)\phi_2(x_2^{-1}\delta_2 x_3) \cdots \phi_m(x_m^{-1}\delta_m x_1^\tau)d\dot{x}$$

where $d\dot{x}$ is the quotient measure. Let us define the new variables $y_1, \dots y_m$ by

$$\begin{aligned} y_1 &= x_1 \\ y_2 &= x_2^{-1}\delta_2\delta_3\cdots\delta_m x_1^\tau \\ &\vdots \\ y_{m-1} &= x_{m-1}^{-1}\delta_{m-1}\delta_m x_1^\tau \\ y_m &= x_m^{-1}\delta_m x_1^\tau. \end{aligned} \tag{5.2}$$

Using (5.1), the integral may be rewritten as

$$\int\limits_{G_{\delta,\sigma}(F)\backslash G(E)} \phi_1(y_1^{-1}\xi y_1^\tau y_2^{-1})\phi_2(y_2 y_3^{-1})\cdots\phi_{m-1}(y_{m-1}y_m^{-1})\phi_m(y_m)d\dot{y}.$$

In the y-variables, $G_{\delta,\sigma}(F)$ is defined by the equation $y_1 \in G_{\xi,\tau}(F)$; the values of $y_2, \dots y_m$ are then fixed. Thus the integral can actually be written, up to a change of variable, as

$$\int\limits_{G_{\xi,\tau}(F)\backslash G(E)} \phi_1(y_1^{-1}\xi y_1^\tau y_m^{-1}\cdots y_2^{-1})\phi_2(y_2)\cdots\phi_m(y_m)d\dot{y}_1 dy_2\cdots dy_m$$

where $G_{\xi,\tau}$ is embedded into $G(E)$ through the first component, and $dy_i (i \geq 2)$ is the Haar measure on $G(E_1)$. This is in turn equal to

$$\int\limits_{G_{\xi,\tau}(F)\backslash G(E_1)} \psi(y_1^{-1}\xi y_1^\tau)d\dot{y}_1$$

where $\psi = \phi_1 * \phi_2 * \cdots * \phi_m$, the convolution product on $G(E_1)$.

Thus the σ-twisted orbital integrals of $\phi_1 \otimes \cdots \otimes \phi_m$ on $G(E)$ coincide with the τ-twisted orbital integrals of ψ on $G(E_1)$; if $f \in C_c^\infty(G(F))$ is associated to ψ in the sense of Proposition 3.1, we see that the σ-twisted orbital integrals of ϕ coincide with the orbital integrals of f. The quotient measures must be normalized in obvious ways.

Assume now that the extension E/F of non-Archimedean fields comes as indicated above from an extension k'/k of global fields, and is unramified. Let $\mathcal{H}_E, \mathcal{H}_{E_1}, \mathcal{H}_F$ be the Hecke algebras of $G(E), G(E_1), G(F)$ with respect to the standard compact groups. The homomorphism of L-groups given by base change then yields, as is easily checked, the homomorphism

$$b : \mathcal{H}_E \to \mathcal{H}_F$$

given on decomposed functions in $H_E \cong \mathcal{H}_{E_1} \otimes \cdots \otimes \mathcal{H}_{E_1}$ by

$$\phi_1 \otimes \cdots \otimes \phi_m \mapsto b_{E_1/F}(\phi_1 * \phi_2 * \cdots * \phi_m),$$

$b_{E_1/F}$ being the base change homomorphism $\mathcal{H}_{E_1} \to \mathcal{H}_F$ described on the Satake transform by $f(z) \mapsto f(z^k)$ (§4). Consequently, the previous computation and the results of §4 show that ϕ and $b\phi$ have associated (twisted) orbital integals.

6. Base change lifting of local representations

6.1. In this section we will obtain the lifting, by base change, of admissible representations of local linear groups by a cyclic extension. Thus E/F is an extension of local non-Archimedean fields of degree ℓ; $\Sigma = \mathrm{Gal}(E/F)$ is generated by σ.

DEFINITION 6.1: (Shintani). *Let π, Π be irreducible, admissible representations of $G(F), G(E)$ respectively. Assume that $\Pi \cong \Pi \circ \sigma$. We say that Π is a (base change) lift of π if, for $g \in G(E)$ such that $\mathcal{N}g$ is regular:*

$$\mathrm{trace}(\Pi(g)I_\sigma) = \mathrm{trace}\,\pi(\mathcal{N}g).$$

Here I_σ is the canonical intertwining operator of §2. The values of the characters are well defined (cf. Proposition 2.2).

The basic results concerning local base change are contained in the following theorem. We first consider *tempered* representations only; for the general case, see §6.4. We denote by Z the center of G. Note that the central character $\omega_\pi : Z(F) \to \mathbf{C}^*$ of a tempered representation is unitary.

THEOREM 6.2: *Let π, Π denote irreducible tempered representations of $G(F), G(E)$ respectively.*

(a) Any tempered irreducible representation π of $G(F)$ has a unique lift Π to $G(E)$. The representation Π is tempered.

(b) Conversely, assume $\Pi \cong \Pi \circ \sigma$ is an irreducible tempered representation of $G(E)$. Then there is at least one representation π of $G(F)$ such that Π lifts π; π is tempered.

(c) The notion of local lifting does not depend on the choice of σ.

(d) Let ω_π, ω_Π be the central characters of π and Π. Then if Π lifts π, they satisfy

$$\omega_\Pi(z) = \omega_\pi(N_{E/F}z) \qquad z \in Z(E) \cong E^*.$$

(e) If E, F are Galois extensions of a subfield L, and $\tau \in \mathrm{Gal}(E/L)$, then, if Π lifts π, Π^τ lifts π^τ where $\mathrm{Gal}(E/L)$ acts on $G(E), G(F)$ in the obvious manner.

6.2. Reduction to the discrete case

We start the proof of Theorem 6.2 by reducing it to the case of π belonging to the discrete series of $G(F)$ and *σ-discrete* representations Π of $G(E)$ (§2).

We first remark that the uniqueness in part (a) of the theorem is obvious.

LEMMA 6.3: *If two σ-stable representations Π and Π' are irreducible and non-isomorphic, their twisted characters are linearly independent.*

Proof. We may choose a compact-open subgroup $K \subset G(E)$, stable by σ, small enough that Π and Π' have non-zero vectors fixed by K. If $\mathcal{H}_K = C_c^\infty(K\backslash G(E)/K)$, $\mathcal{H}_K$ acts on the space of K-fixed vectors in Π (resp. Π') and this finite-dimensional representation is irreducible and determines Π (resp. Π') ([7(a)]). Since non-isomorphic, finite-dimensional representations of $\mathcal{H}_K$ have linearly independent coefficients, the functions on $\mathcal{H}_K$ defined by $\phi \mapsto \mathrm{trace}(\Pi(\phi)I_\sigma)$ and $\phi \mapsto \mathrm{trace}(\Pi'(\phi)I'_\sigma)$ are independent. ∎

LEMMA 6.4: *Assume $\Pi \cong \Pi \circ \sigma$ is tempered and irreducible. Then there is a parabolic F-subgroup $P = MN$ of G, and a σ-discrete representation Π_M of $M(E)$ such that*

$$\Pi = \mathrm{ind}_{M(E)N(E)}^{G(E)} (\Pi_M \otimes 1).$$

Remark. It follows from the orthogonality relations between σ-discrete characters (Proposition 6.6) that M is then unique up to conjugacy and Π well defined up to $W(M, A_M)$.

Proof. Any tempered Π can be written as induced from a standard parabolic subgroup $P = MN$ (thus P defined over F), with

$$M \cong \mathrm{GL}(n_1) \times \cdots \times \mathrm{GL}(n_r),$$

of a discrete unitary representation $\delta = \delta_1 \otimes \cdots \otimes \delta_r$ of $M(E)$.

If $\Pi \cong \Pi \circ \sigma$, we must have, by standard results:

$$(6.1) \qquad (\delta_1^\sigma, \ldots, \delta_r^\sigma) = s(\delta_1, \ldots, \delta_r)$$

where $s \in W(G, A_M)$ can be seen as a permutation of $(1, \ldots, r)$ which preserves the ranks n_i. Moreover, if s leaves stable a subpartition of $(1, \ldots, r)$, Π may be induced from a tempered, σ-stable representation of a smaller group, and Π is not σ-discrete.

Considering the orbits of the group generated by s in $(1, \ldots, r)$, we may write, in the obvious way Π as induced from an induced, σ-stable representation of a Levi subgroup; the inducing representation is then σ-discrete (cf. Lemma 2.8). ∎

We now assume Theorem 6.2 in the discrete case and deduce it for other representations. For (a), assume that $\pi = \mathrm{ind}_{M(F)N(F)}^{G(F)} \pi_M$, where π_M

belongs to the discrete series of $M(F)$ and is unitary. We may assume that π_M has a base change lift Π_M, a σ-stable representation of $G(E)$. Moreover, we will see (§6.3) that Π_M is σ-discrete and unitary.

By the proof of Lemma 6.4, we may write Π_M as a tensor product $\Pi_1 \otimes \cdots \otimes \Pi_r$ (where $M = \mathrm{GL}(n_1) \times \cdots \times \mathrm{GL}(n_r)$), each Π_i being of the form $\mathrm{ind}(\delta \otimes \delta^\sigma \otimes \cdots \otimes \delta^{\sigma^{s-1}})$ with $\delta^{\sigma^s} \cong \delta$ and $\delta^{\sigma^i} \not\cong \delta$ $(i < s)$. The central character of Π_i is then $\omega_\delta \omega_\delta^\sigma \cdots \omega_\delta^{\sigma^{s-1}}$; since Π_i is unitary, this implies that ω_δ, the central character of δ, is unitary. Thus Π_i is induced from a unitary discrete series representation, i.e., Π_i, and therefore Π_M, is tempered.

By a well-known result of Bernstein [4] this implies that

$$\Pi = \mathrm{ind}_{P(E)}^{G(E)}(\Pi_M \otimes 1)$$

is irreducible and tempered. But then the Atiyah–Bott Theorem [11(b), Theorem 2] shows that, if we define I_σ, the intertwining operator in the space of Π, by inducing the operator $I_{\sigma,M}$ for Π_M, we have

$$\mathrm{trace}\, \pi(\mathcal{N}g) = \mathrm{trace}(\Pi(g) I_\sigma).$$

By Lemma 2.1, I_σ is the normalized operator for Π. This proves (a).

Likewise, assume given $\Pi \cong \Pi \circ \sigma$, irreducible and tempered. If Π is not σ-discrete, write $\Pi = \mathrm{ind}_{P(E)}^{G(E)}(\Pi_M \otimes 1)$ with Π_M σ-discrete. Then Π_M lifts at least one representation π_M of $M(F)$. Moreover, Lemma 2.12 implies that π_M belongs to the discrete series and is unitary. This implies that $\pi = \mathrm{ind}_{P(F)}^{G(F)} \pi_M$ is irreducible and tempered; again, the Atiyah–Bott Theorem shows that π is a lift of Π. This proves (b).

Notice that (d) is an obvious consequence of the character identities since $\mathrm{trace}\, \pi(zg) = \omega_\pi(z)\, \mathrm{trace}\, \pi(g)$, $z \in Z(F)$, the analog holds for $G(E)$, and $\mathcal{N}(zg) = (\mathcal{N}z)\mathcal{N}g$ for $z \in Z(E)$. If (c) is true for the inducing representations, it is true for π and Π.

As for (e), assume that

$$\mathrm{trace}(\Pi(g) I_\sigma) = \mathrm{trace}\, \pi(\mathcal{N}g).$$

Replacing g by $\tau(g)$, we obtain

$$\mathrm{trace}((\Pi \circ \tau(g)) I_\sigma) = \mathrm{trace}\, \pi(\mathcal{N}(\tau g)).$$

Now

$$\mathcal{N}(\tau g) = \tau g \sigma(\tau g) \sigma^2(\tau g) \cdots \sigma^{\ell-1}(\tau g) = \tau(\mathcal{N}_1 g),$$

where

$$\mathcal{N}_1 g = g \sigma_1(g) \cdots \sigma_1^{\ell-1}(g)$$

and $\sigma_1 = \tau^{-1}\sigma\tau$ is a new generator of Σ. On the other hand, I_σ satisfies

$$\Pi(g)I_\sigma = I_\sigma \Pi(\sigma^{-1}g)$$

from which we infer

$$(\Pi \circ \tau(g))I_\sigma = I_\sigma(\Pi \circ \tau)(\tau^{-1}\sigma^{-1}\tau g).$$

This shows that I_σ intertwines $(\Pi \circ \tau)$ and $(\Pi \circ \tau) \circ \sigma_1$. The equation

$$\mathrm{trace}((\Pi \circ \tau)(g)I_\sigma) = \mathrm{trace}(\pi \circ \tau)(\mathcal{N}_1 g)$$

then shows that $\Pi \circ \tau$ lifts $\pi \circ \tau$. This proves (e).

6.3. Discrete case

We still have to treat the case of π discrete, or Π σ-discrete. We will use the following result, to be proved in Chapter 3. We fix a global field k, and a cyclic extension k' of k of degree ℓ. Notations are as in §2. If v is a prime of k, let $t_{\pi,v} \in (\mathbf{C}^*)^n$ be the element associated to a representation π of $G(\mathbf{A}_k)$ unramified at v : thus $t_{\pi,v}$ is defined up to permutation of the coordinates.

THEOREM III.3.1: *Let π, π' be cuspidal automorphic representations of $G(\mathbf{A}_k)$. If v is a finite prime of k, let f_v be the residual degree of k' over v. Let S be a finite set of primes of k, containing the infinite primes and the places where k', π or π' is ramified. Assume that, for $v \notin S$:*

$$(t_{\pi,v})^{f_v} = (t_{\pi',v})^{f_v}. \tag{6.1}$$

Then $\pi' \cong \pi \otimes \chi$, for some character χ of $k^* N(A^*_{k'})\backslash \mathbf{A}^*_k$.

We now prove Theorem 6.2(a) in the discrete case. We assume k'/k so chosen that, for a place v_0, $k_{v_0} \cong F$ and the extension k'_{v_0} is isomorphic to E. Let $v_1 \neq v_2$ be two places where k' splits and $v_3 \neq v_0$ be an inert place of k. Let π_0 be a discrete series representation of $G(F)$; without restricting generality, we will assume that its central character is 1.

LEMMA 6.5: *There is a cuspidal representation π of $G(\mathbf{A})$ such that*

$\pi_{v_0} \cong \pi_0$,

π_{v_1} *is a given supercuspidal representation of* $G(k_{v_1})$,

π_{v_3} *is a Steinberg representation,*

and such that π_v is unramified for any finite place $v \notin \{v_0, v_1, v_2, v_3\}$.

Proof. This follows easily from the Deligne–Kazhdan trace formula (Lemma 2.4): by [26, Theorem K], there are compactly supported functions f_{v_0}, f_{v_3} such that $\operatorname{trace}\pi_{v_i}(f_{v_i}) = 1 (i = 0, 3)$ and that their trace is 0 in any other tempered (or generic) representation. We take for f_{v_1} a coefficient of π_{v_1}. We take f_v unramified for other finite places $v \neq v_2$. The choice of the Archimedean factors is arbitary. Taking f_{v_2} as in Lemma 2.4, we obtain a formula

$$\sum_{\pi \text{ cuspidal}} \operatorname{trace}\pi(f) = \sum_{\{\gamma\}} v(\gamma)\Phi_f(\gamma).$$

By [26, Theorem K], the elliptic orbital integrals of $f_{v_i} (i = 0, 1, 3)$ are equal to the character. In particular, they do not identically vanish, and using f_{v_2} and the f_v for v infinite, we may arrange to have exactly one non-zero term in the right-hand side. Since cuspidal representations have generic components [37], this proves the existence of π as in Lemma 6.5. ∎

We will assume now that all infinite places of k split in k'. Assume that the function f on $G(\mathbf{A})$ satisfies the conditions in the proof of Lemma 6.5, except that f_{v_0} is now arbitrary, and f satisfies the vanishing conditions of Prop. 3.1; and let ϕ on $G(\mathbf{A}_{k'})$ be associated to f. By formula (4.6), we have:

$$\ell \operatorname{trace}(R_{\text{cusp}}(\phi) I_\sigma) = \operatorname{trace} r_{\text{cusp}}(f).$$

(We assume, as we may, that ϕ satisfies the conditions of Lemma 2.5; at the finite places w above $v \notin \{v_0, v_1, v_2, v_3\}$, ϕ_w is unramified.) Separating the representations, as in §4, by using their components at infinity and their Hecke eigenvalues, we obtain the identity

$$\ell \operatorname{trace}(\Pi(\phi) I_\sigma) = \sum_{\pi'} \operatorname{trace}\pi'(f), \tag{6.2}$$

a finite sum for ϕ, f given. The left-hand side of (6.2) is composed of the unique cuspidal representation of $G(\mathbf{A}_{k'})$ determined by the Hecke eigenvalues of π, composed with the norm maps for Hecke algebras. The right-hand side contains all representations of $G(\mathbf{A})$ verifying conditions (6.1) at the finite places $v \notin \{v_0, v_1, v_2, v_3\}$.

By Theorem III.3.1, the representations π' are of the form $\pi \otimes \chi$. Since π_{v_3} is a Steinberg representation, the relation $\pi_{v_3} \otimes \chi_3 \cong \pi_{v_3}$, χ_3 a character of $k_{v_3}^*$, implies $\chi_3 = 1$. (Consider the Jacquet module for the Borel subgroup!) Thus the condition $\pi \cong \pi \otimes \chi$ implies $\chi_{v_3} = 1$, whence $\chi = 1$ since v_3 is inert. So the representations π' are all the distinct representations $\pi \otimes \chi$, χ ranging over the Abelian class field characters of $\mathbf{A}_k^*$ associated to k'. Since then $\operatorname{trace}\pi'(f) = \operatorname{trace}\pi(f)$ for any f in the image of the base change correspondence, we may rewrite (6.2) as

$$\operatorname{trace}(\Pi(\phi)I_\sigma) = \operatorname{trace}\pi(f). \tag{6.3}$$

This proves that Π_{v_0} lifts π_0, except for the value of the normalizing constant. This is provided by the theory of Whittaker models. By [37], [31], there is, up to a scalar, a unique linear form λ on the space of Π such that, for w in the space of Π :

$$\lambda(\Pi(n)w) = \theta(n)\lambda(w), \quad n \in N(\mathbf{A}_{k'}). \tag{6.4}$$

It is given, on the function w by

$$\lambda(w) = \int_{N(k')\backslash N(\mathbf{A}_{k'})} w(n)\overline{\theta(n)}dn.$$

Here θ is a σ-invariant character of $N(\mathbf{A}_{k'})$ defined, as in §2, by a σ-invariant character ψ of $k'\backslash\mathbf{A}_{k'}$. It is clear that $\lambda(I_\sigma w) = \lambda(w)$.

This implies that we can write λ and I_σ as tensor products:

$$\lambda = \bigotimes_v \lambda_v, \quad I_\sigma = \bigotimes_v I_{\sigma,v}$$

over the places of k, in such a fashion that at each v, $I_{\sigma,v}\lambda_v = \lambda_v$. In other terms, I_σ is the tensor product of the normalized intertwining operators. (We let the reader fill the gaps at the non-inert places.) We now remark that the Shintani identities (with the right constant) are obviously true at the split places; obviously also, they hold for unramified representations by §6.2; finally, it is easy to check, using the construction of the Steinberg representation given by Casselman ([10(b)], see also [9]) that they hold for the Steinberg representation. Since the normalization, then, is correct at all places except v_0, it is also correct at v_0. This proves Theorem 6.2(a). Note that the elliptic character of $\Pi_0 \rtimes \sigma$ is non-zero, and then Lemma 2.11 implies that Π *is σ-discrete.*

Assume now that $\Pi_0 \cong \Pi_0 \circ \sigma$ is σ-discrete. (Again, we will assume all central characters trivial.) Let k'/k be as above; let $\phi_{v_0} \in C_c^\infty(G(E), X)$ be a pseudo-coefficient of $\Pi_0 \rtimes \sigma$ (Corollary 2.10). Recall that the σ-elliptic twisted orbital integrals of ϕ_{v_0}, then, are not identically 0 (Lemma 2.12). As in the proof of (a), we may construct a function ϕ on $G(\mathbf{A}_{k'})$ such that ϕ is unramified for any finite place w above a place $v \neq \{v_0, v_1, v_2, v_3\}$, and such that

$$\operatorname{trace}(R_{\text{cusp}}(\phi)I_\sigma) \neq 0.$$

Therefore, there is a representation Π of $G(\mathbf{A}_{k'})$ in the space of cusp forms, such that $\Pi_{v_0} \cong \Pi_0$, $\Pi_{v_1} \cong (\pi_{v_1})^{\otimes \ell}$ is supercuspidal, $\Pi_{v_2} \cong (\pi_{v_2})^{\otimes \ell}$ (since Π_{v_2} is σ-stable), and Π_{v_3} is a Steinberg representation. (Here we have used the fact that a Steinberg representation stable by σ is σ-discrete and admits a twisted pseudo-coefficient.) Again, the comparison of traces yields an identity of the form:

$$\ell \operatorname{trace}(\Pi(\phi)I_\sigma) = \sum_{\pi'} \operatorname{trace} \pi'(f). \tag{6.5}$$

By Theorem III.3.1, there are at most ℓ representations on the right-hand side; they are all of the form $\chi \otimes \pi$, for a unique cuspidal π. On the image of the norm, their local characters coincide. Consequently, we have an equality at the place v_3 of the form:

$$\operatorname{trace} \pi_{v_3} \circ N = c \operatorname{trace}(\Pi_{v_3} I_\sigma)$$

for some constant c. Thus the character of π_{v_3} is equal to a Steinberg character on the image of N; it is easy to show that π_{v_3} is then a Steinberg representation. But then $\pi_{v_3} \otimes \chi_{v_3} \not\cong \pi_{v_3}$, where χ_{v_3} is the class field character associated to the extension k'_{v_3}/k_{v_3}. This implies that $\pi \otimes \chi \not\cong \pi$ unless $\chi = 1$. So there are ℓ terms in the right-hand side of (6.5). Repeating the arguments used to prove (a), we then obtain part (b) of Theorem 6.2. Also note that all representations π_{v_0} lifting Π_0 are obtained from one of them by twisting by some power of the local class field character.

We now remark that (c) follows from the fact that the local lifting has been constructed globally. Indeed, given the local representation π_0 of $G(F)$ (π_0 discrete), we have constructed a global representation π of $G(k)$; then π lifts to Π, which restricts at v_0 to Π_0 lifting π_0. Since the notion of global lifting is independent of the choice of σ, the local lifting is also. Given (c), (e) has been proved in §6.2; (d) is clear. Theorem 6.2 is complete.

6.4. Properties of local base change

We now list some properties of local base change deduced from Theorem 6.2 and its proof. The proofs are easy and are only sketched. The first result concerns the orthogonality relations for σ-discrete representations. Fix a unitary character X of $Z(E)$, with $X \cong X \circ \sigma$. If Θ_1, Θ_2 are two functions on $G(E)$, invariant by σ-conjugation, and such that $\Theta_i(zg) = X(z)\Theta_i(g)$, set

(6.6)

$$\langle\Theta_1,\Theta_2\rangle_{\sigma-\mathrm{ell}} = \sum_{T\mathrm{ell}} |W(G(F),T(F))|^{-1} \int_{Z(E)T(E)^{1-\sigma}\backslash T(E)} \Delta_G^2(Nt)\Theta_1(t)\cdot\Theta_2(t)dt$$

(cf. the Weyl integration formula in §4.1).

Here dt denotes the Haar measure on $Z(E)T(E)^{1-\sigma}\backslash T(E) \cong NZ(E)\backslash T(F)$ of total mass 1.

Let us denote by ξ the character of F^* associated to the extension E/F : thus $\xi^\ell = 1$. Let $\Xi \cong \mathbf{Z}/\ell\mathbf{Z}$ be the group generated by ξ : it acts on representations by $\pi \mapsto \pi \otimes (\eta \circ \det)$, $\eta \in \Xi$.

The proof of Proposition 6.6 uses a global result from Chapter 3. This proposition will not be used in the rest of the paper.

PROPOSITION 6.6: *Let π, Π denote discrete series (resp. σ-discrete) representations of $G(F), G(E)$.*

(i) Assume that Π lifts π. Write

$$\Pi = \mathrm{ind}_{M(E)N(E)}^{G(E)}((\Pi_M \otimes \Pi_M^\sigma \otimes \cdots \otimes \Pi_M^{\sigma^{g-1}}) \otimes 1)$$

with g minimal such that $\Pi_M^{\sigma^g} \cong \Pi_M$, and Π_M a discrete series representation of $\mathrm{GL}(n/g, E)$.

Then g is equal to the order of the stabilizer of π in Ξ. In particular, Π belongs to the discrete series if and only if $\pi \otimes \xi \not\cong \pi$. We write $g = g(\Pi)$.

(ii) If Π_1, Π_2 are σ-discrete, we have, writing $\Theta_{i,\sigma}$ for their twisted characters $(i = 1, 2)$:

$$\langle\Theta_{1,\sigma},\Theta_{2,\sigma}\rangle_{\sigma-\mathrm{ell}} = \begin{cases} 0 & \text{if } \Pi_1 \not\cong \Pi_2 \\ g(\Pi_1)^{-1} & \text{if } \Pi_1 \cong \Pi_2. \end{cases}$$

Moreover, the σ-discrete characters form an orthogonal basis for the invariant functions on $G(E)_{\sigma-\mathrm{ell}}$ with the scalar product (6.6).

Proof. Assume that Π lifts π and $\pi \cong \pi \otimes \eta$ with $\eta = \xi^i$ and i is minimal. We may imbed π in a cusp form $\pi_{\mathbf{A}}$ such that $\pi_{\mathbf{A}} \cong \pi_{\mathbf{A}} \otimes \eta_{\mathbf{A}}$, where $\eta_{\mathbf{A}}$ is analogously defined; $\eta_{\mathbf{A}}$ is then minimal. Under these assumptions, we will see in Lemma III.6.6 that $\pi_{\mathbf{A}}$ lifts to a representation induced from cuspidal: $\Pi_{\mathbf{A}} = \mathrm{ind}(\Pi_{\mathbf{A},M} \otimes \cdots \otimes \Pi_{\mathbf{A},M}^{\sigma^{g-1}})$, $g = \frac{n}{i}$, $\Pi_{\mathbf{A},M}$ cuspidal. The local component of $\Pi_{\mathbf{A}}$ at v_0 is then as stated in (i); $\Pi_{v_0,M}$ has to be discrete, since otherwise one easily checks that the induced representation is not σ-discrete. This proves (i). As for (ii), we may use the lifting identities to rewrite the scalar product in terms of π_1 and π_2, assumed to lift Π_1

and Π_2. Using Lemma 1.4, the computation is then easily reduced to the orthogonality relations on the group $G(F)_1 = \{x \in G(F) : \xi \circ \det(x) = 1\}$ (cf. e.g. [19, Lemma 1.9]). ∎

Using computations of Jacquet modules, we will show (Lemma 6.10) that in Proposition 6.6, π is supercuspidal if and only if Π_M is. Thus base change preserves representations unitarily induced from supercuspidal.

PROPOSITION 6.7: (π, Π tempered). *Assume* Π *lifts* π. *Write*

$$\pi = \mathrm{ind}(\pi_1 \otimes \cdots \otimes \pi_r),$$

π_r *belonging to the discrete series of* $\mathrm{GL}(n_r, F)$. *Then the other representations lifting to* Π *are those of the form*

$$\pi' = \mathrm{ind}((\pi_1 \otimes \eta_1) \otimes \cdots \otimes (\pi_r \otimes \eta_r) \quad \textit{with} \quad \eta_i \in \Xi.$$

Proof. It is obvious that these representations are lifted by Π; conversely, assume π' is lifted by Π : π' then has the same character as π on the image of the norm map. Using the formulas for induced characters [11(b)] it is then easy to show that π is induced from a discrete series representation $\sigma_1 \otimes \cdots \otimes \sigma_r$ of the same parabolic subgroup, and that $\sigma_1 \otimes \cdots \otimes \sigma_r$ and a Weyl conjugate of $\pi_1 \otimes \cdots \otimes \pi_r$ have the same character on the norms. The orthogonality relations then imply the result.

We now record the obvious property of lifting:

PROPOSITION 6.8: (π, Π tempered). *Let* $\tilde{\pi}, \tilde{\Pi}$ *denote the contragredient representations.*

(i) *If* π *lifts to* Π, $\tilde{\pi}$ *lifts to* $\tilde{\Pi}$.

(ii) *If* π *lifts to* Π, $\pi \otimes \omega$ *lifts to* $\Pi \otimes (\omega \circ N_{E/F})$, ω *being a character of* F^*.

Lastly, we will relate the L-functions of representations associated by base change.

We first remark that at this point we have obtained the base change correspondence between *all* representations, tempered or not, of $G(E)$ and $G(F)$. Indeed, by the Langlands classification, any representation π of $G(F)$ can be realized as the unique quotient of a representation

$$\mathrm{ind}_{M(F)N(F)}^{G(F)}(\pi_M \otimes 1)$$

with π_M essentially tempered and dominant ([24(c), §3.3]). Then π_M has a unique lift Π_M to $M(E)$ – the previous results obviously extend to essentially tempered representations; Π_M is again dominant, thus $\mathrm{ind}(\Pi_M \otimes 1)$ has a unique quotient Π; Π is the base change lift of π (it is clearly σ-stable). Conversely, given $\Pi \cong \Pi \circ \sigma$, we may realize it as the Langlands quotient of $\mathrm{ind}(\Pi_M \otimes 1)$. The uniqueness of the Langlands datum then shows that Π_M is σ-stable; if π_M is any representation of $M(F)$ lifted by Π_M, its Langlands quotient π has as base change lift Π. With these definitions, Proposition 6.8 and the obvious adaptation of Proposition 6.7 still hold. In general, of course, the base change correspondence is not given by character identities.

Let us now recall the notion of L-function of pairs of representations. Let G_n denote $\mathrm{GL}(n)$. If π, τ are irreducible representations of $G_n(F)$ and $G_m(F)$ respectively, Jacquet, Piatetskii–Shapiro and Shalika [26(b)] define a local L-function, denoted by $L(s, \pi \times \tau)$. To conform to our general notations, we will denote it by $L(s, \pi \otimes \tau)$. It can be expressed as $P(q^{-s})^{-1}$, where P is a polynomial with constant coefficient 1, and q the cardinality of the residue field.

There is an associated ε-factor ([26(b), §2.7]) $\varepsilon(s, \pi \otimes \tau, \psi)$, where ψ is an additive character of F; ε is a monomial in q^{-s}. We will also need the λ-constants of Langlands. Let E/F be an extension of local fields, ψ_F an additive character of F, $\psi_E = \psi_F \circ \mathrm{tr}_{E/F}$. In [14, p. 549] are defined numbers $\lambda(E/F, \psi_F, dx_E, dx_F)$ where dx_E and dx_F are Haar measures on E and F. If we take dx_E and dx_F to be the self-dual measures associated to ψ_E and ψ_F, we obtain Langlands' factor $\lambda(E/F, \psi_F)$.

It has the following property, which we could take in our case for definition. Assume now that E/F is cyclic of order ℓ. Let χ_F be a character of F^*, $\chi_E = \chi_F \circ N_{E/F}$. Let Ξ be, as in the beginning of 6.4, the group of characters of F^* vanishing on NE^*. Then, with ψ_E and ψ_F related as above:

$$\prod_{\eta \in \Xi} \varepsilon(\chi_F \eta, \psi_F) = \lambda(E/F, \psi_F) \varepsilon(\chi_E, \psi_E). \tag{6.7}$$

This is an immediate consequence of the behavior of λ-factors under induction ([14, 5.6.2]), and the fact that, if we identify characters of a local field K^* and 1-dimensional representations of the Weil group W_F, we have

$$\mathrm{ind}_{W_E}^{W_F}(\chi_E) = \bigoplus_{\eta \in \Xi} \chi_F \eta.$$

PROPOSITION 6.9: *Assume E/F is cyclic of order ℓ. Let π, τ be irreducible representations of $G_n(F), G_m(F)$ and Π, T their base change lifts to $G_n(E), G_m(E)$. Then*

(i) $L(s, \Pi \otimes \mathrm{T}) = \prod_{\eta \in \Xi} L(s, \pi \otimes \tau \otimes \eta)$.

(ii) $\varepsilon(s, \Pi \otimes \mathrm{T}, \psi_E) = \lambda(E/F, \psi_F)^{-mn} \prod_{\eta \in \Xi} \varepsilon(s, \pi \otimes \tau \otimes \eta, \psi_F)$.

In the right-hand sides of (i) and (ii), $L(s, \pi \otimes \tau \otimes \eta)$ may be interpreted as $L(s, \pi \otimes (\tau \otimes \eta))$ where $\tau \otimes \eta$ denotes τ twisted by the character η of the determinant, or as $L(s, (\pi \otimes \eta) \otimes \tau)$. These two L-functions coincide, as can be extracted from [26(b)]. The same applies to the ε-factors.

Proof. First notice that it is enough to consider *generic* representations π, τ. If π, for example, is any irreducible representation, it may be realized as the Langlands quotient of some representation $\bar{\pi}$, and $\bar{\pi}$ is induced from an essentially square integrable representation of a Levi subgroup. Write $\bar{\pi} = \pi_1 \times \cdots \times \pi_s$, where π_i is an essentially square integrable representation of $G_{n_i}(F)$, if $\bar{\pi}$ is induced from the representation $\pi_1 \otimes \cdots \otimes \pi_s$ of the corresponding Levi subgroup. We may write, analogously, $\bar{\tau} = \tau_1 \times \cdots \times \tau_s$. We then have (26(b), §9])

$$L(s, \pi \otimes \tau) = \prod_{i,j} L(s, \pi_i \otimes \tau_j).$$

Analogous considerations apply to Π, τ. Since, as we observed after Proposition 6.8, this construction is compatible in an obvious way with base change, we may deduce the identity (i) for π, τ from the analogous identity for the factors $L(s, \pi_i \otimes \tau_j)$ and their lifts. (One also has to observe that, π_i being essentially square-integrable, its lift Π_i is essentially tempered and, therefore, generic.)

The same argument applies to the ε-factors. Therefore, we may assume that π, τ are essentially square-integrable. Twisting by a character of the determinant, we may even assume that π, τ are square-integrable (i.e., in addition, unitary). We first make the following simple remarks:

(6.8) The identity of L-functions (6.9(i)) is true for π, τ unramified and the extension E/F unramified. This is clear by the expression of the L-functions in that case, cf. [27(a), §2].

(6.9) It is enough to consider the case of E/F cyclic of *prime* order ℓ. Indeed, (1) and (2) in Proposition 6.9 can be obtained by repeated lifting.

According to Bernstein and Zelevinsky [7b, 42], the square-integrable representations of $G_n(F)$ are obtained as follows. Let $n = ar$, $a, r \in \mathbf{N}$. If ω is a unitary supercuspidal representation of $G_r(F)$, let $St(\omega, a)$ denote the unique submodule of

$$\mathrm{ind}_{MN}^{G_n}(\omega|\ |^{\frac{a-1}{2}}, \omega|\ |^{\frac{a-3}{2}}, \ldots, \omega|\ |^{\frac{1-a}{2}}),$$

where MN is the parabolic subgroup of type $(r, \ldots, r)$. Then all square-integrable representations are of this type, and $St(\omega, a)$ is isomorphic to $St(\omega', b)$ if and only if $a = b$, $\omega \cong \omega'$.

We will now need the following lemma, which is of interest in itself. Let ω be a supercuspidal representation of $G_r(F)$. Its lift Ω to $G_r(E)$ (Thm. 6.2) is σ-discrete, and therefore Ω is square-integrable or equal to the induced representation $\Gamma \times \Gamma^\sigma \times \cdots \times \Gamma^{\ell-1}$, where Γ is a square-integrable representation of $G_t(E)$ $(t = \frac{r}{\ell})$ and $\Gamma \not\cong \Gamma^\sigma$ (Lemma 2.8).

LEMMA 6.10: *Let ω be a supercuspidal representation of $G_r(F)$.*

(i) Assume the lift Ω of ω is discrete. Then it is supercuspidal.

(ii) Assume ω lifts to $\Omega = \Gamma^\sigma \times \cdots \times \Gamma^{\sigma^{\ell-1}}$, Γ discrete, $\Gamma \not\cong \Gamma^\sigma$. Then Γ is supercuspidal.

Proof.

(i) Assume Ω is not supercuspidal. Then $\Omega = St(\Delta, c)$, for $c|r$, $c \neq 1$, and Δ a supercuspidal representation of $G_{r/c}(E)$. By the uniqueness of the Bernstein-Zelevinsky classification, $\Delta^\sigma \cong \Delta$. Let N be the unipotent radical of the standard parabolic subgroup of G_r of type $(r/c, \ldots, r/c)$. Then, as is well-known [42], the (unnormalized) Jacquet module of Ω for N is

$$\Omega_N = \Delta|\ |^{c-1} \otimes \Delta|\ |^{c-2} \otimes \cdots \otimes \Delta|\ |^{1-c}.$$

As the twisted character of Δ is not identically zero on the σ-elliptic set, the twisted Casselman theorem (Prop. 2.3) implies that the twisted character of Ω does not vanish on points $g \in G_r(E)$ such that $\mathcal{N}(g) = h \in G_r(F)$ and $N_h = N$. This contradicts the fact that the lift ω of Ω is cuspidal.

(ii) Assume Γ is not supercuspidal, and write $\Gamma = St(\Delta, c)$ for $c|(r/\ell)$, $c \neq 1$, and Δ supercuspidal. Since $\Gamma \not\cong \Gamma^\sigma$, $\Delta \not\cong \Delta^\sigma$. Set $r = c\ell d$: thus Δ is a representation of $G_d(E)$. We have:

$$\Omega = \mathrm{ind}_{M_1 N_1}^{G_r}(St(\Delta, c) \otimes \cdots \otimes St(\Delta^{\sigma^{\ell-1}}, c))$$

where M_1 is the standard Levi subgroup of type $(dc, \ldots, dc)$. We consider the Jacquet module Ω_{N_2}, where $P_2 = M_2 N_2$ is of type $(d\ell, \ldots d\ell)$. Note that M_2 and M_1 both contain the Levi subgroup M_3 of type $(d, \ldots, d)$, and the representation $\Pi = \Gamma \otimes \cdots \otimes \Gamma^{\sigma^{\ell-1}}$ of M_1 is a submodule of a representation induced to M_1 from a supercuspidal representation of M_3.

An easy extension of a theorem of Bernstein-Zelevinsky ([7b]; see also [42]) and Casselman [10 a] then gives the following description of Ω_{N_2}. Let $W = W(G, A_3) \cong \mathfrak{S}_{c\ell}$ be the Weyl group of (G, A_3), where A_3 is the split component of M_3; then Π_{N_2}, a representation of M_2, is isomorphic (in the Grothendieck group of representations of M_2) to a sum indexed by the subset $W_{1,2}$ of W determined by the following conditions:

$$\begin{array}{llll} \text{(a)} & wi < wj \quad \text{for all} & i,j \in I_k = & [kc+1,\ldots,(k+1)c], \\ & k = 0,\ldots,\ell-1 & & \\ \text{(b)} & w^{-1}i < w^{-1}j \quad \text{for all} & i,j \in J_r = & [r\ell+1,\ldots,(r+1)\ell], \\ & r = 0,\ldots,c-1. & & \end{array}$$

These conditions are equivalent to

$$\begin{array}{ll} \text{(A)} & w\,.\,N_1 \subset N_3 \\ \text{(B)} & w^{-1}\,.\,N_2 \subset N_3 \end{array}$$

where $w\,.$ denotes the action of w by conjugation.

For $w \in W_{1,2}$, the corresponding constituent V_w of Ω_{N_2} is equal to

$$\operatorname{ind}_{w.P_1\cap M_2}^{M_2}(w\,.\,\Pi_{(w^{-1}.N_2\cap M_1)})$$

(here $w^{-1}\,.\,N_2 \cap M_1 \subset N_3 \cap M_1$ is the unipotent radical of a parabolic subgroup of M_1, and we take the corresponding Jacquet module†, a representation of $w^{-1}\,.\,M_2 \cap M_1$; composing this with $Ad(w)$ yields a representation of $M_2 \cap w\,.\,M_1$, a Levi subgroup of $M_2 \cap w\,.\,P_1$; finally, we induce this representation to M_2 by unitary induction). Set $N_w = w^{-1}N_2 \cap M_1$.

Recall that $\Pi = St(\Delta, c)\otimes\cdots\otimes St(\Delta^{\sigma^{\ell-1}}, c)$; Δ is a representation of G_d, and the unipotent radical N_w has blocks of length divisible by d. Using the known formulas for the Jacquet modules of Steinberg representation [42], we see that Π_{N_w} is a tensor product of representation of the blocks G_{da_i} of type $St(\Delta^{\sigma^x}, y)$ for some integers x, y—at least up to an unramified twist by some half integral power of $|\det|$. Therefore V_w is induced to M_2 of a representation of this type.

We are going to consider the twisted trace of Ω_{N_2} and, therefore, we are only interested in the V_w that have σ-stable subquotients. Consider a block $G_{d\ell}$ of M_2. Let $\ell = \ell_1 + \cdots + \ell_s$ be a partition of ℓ, and assume that

$$\operatorname{ind}(St(\Delta^{\sigma^{x_1}}, y_1) \otimes \cdots \otimes St(\Delta^{\sigma^{x_s}}, y_s))$$

†Here the reader must beware of the following fact. In Casselman's theorem (cf. Prop. 2.3), *unnormalized* Jacquet modules are used; the Jacquet modules used in [42, 7(b)] are normalized Jacquet modules, deduced form the unnormalized ones by a twist by $\delta_P^{-\frac{1}{2}}$ where P is the parabolic subgroup in question. The Jacquet module occurring here is normalized. In the arguments that follow this distinction will be unimportant.

where $St(\Delta^{\sigma^{x_i}}, y_i)$ is a representation of $G_{d\ell_i}(E)$, has a σ-stable subquotient. The fundamental disjointness theorem for representations induced from supercuspidal ([7b, Thm. 2.0]; [10a]) easily implies that all conjugates of Δ by the Galois group must be involved: therefore $\ell_i = 1$, and the representation must be equal to $\Sigma = \mathrm{ind}(\Delta \otimes \Delta^\sigma \otimes \cdots \otimes \Delta^{\sigma^{\ell-1}})$ (in fact, the factors $St(\Delta^{\sigma^{x_i}}, y_i)$ may be twisted by half-integral powers of $|\det|$: this does not change the argument).

Finally, considering $M_2 = G_{d\ell} \times \cdots \times G_{d\ell}$, we see that any σ-stable subquotient of Ω_{N_2} must occur in $\Sigma|\ |^{x_1} \otimes \cdots \otimes \Sigma|\ |^{x_c}$ where the x_i are half-integers (note that Σ is irreducible since, up to a twist, it is induced from a unitary representation).

We now notice that this can occur only for one element $w \in W_{1,2}$. Indeed, by the preceding arguments, we see that w must send distinct elements of I_k into distinct intervals J_r; by (a), we must have $w(kc+i) \in J_i$; by (b) we have therefore $J_i = \{wi, w(c+i), \ldots, w((\ell-1)c+i)\}$. But these conditions completely determine w.

We have shown that the twisted trace of Ω_{N_2} coincides with the twisted trace—for the action of σ canonically defined on the Jacquet module—of a unique representation $\Sigma|\ |^{x_1} \otimes \cdots \otimes \Sigma|\ |^{x_c}$, where $\Sigma = \mathrm{ind}(\Delta \otimes \cdots \otimes \Delta^{\sigma^{\ell-1}})$. Since Σ is σ-discrete ($\Delta \not\cong \Delta^\sigma$), we know that its twisted character does not vanish identically on the set of elements with elliptic norms. We can now argue as in the proof of case (a) to show that the twisted character of Ω does not vanish identically on elements whose norm does not belong to the compact part of $G_r(F)$, which contradicts the cuspidality of ω. Lemma 6.10 is proved. ∎

We will now give the proof of Proposition 6.9 (i) using the following facts about global L-functions. Let π, τ be cuspidal representation of $G_n(\mathbb{A}_k)$, $G_m(\mathbb{A}_k)$ where k is a number field. For v an infinite place of k, define $L(s, \pi_v \bigotimes \tau_v)$ as the L-function of the tensor product representation of the Weil group: it is a product of Γ-factors. Set

$$L(s, \pi \otimes \tau) = \prod_v L(s, \pi_v \otimes \tau_v).$$

Let $\tilde{\pi}$, $\tilde{\tau}$ be the contragredient representations. Then the L-functions extend meromorphically to the whole plane and satisfy a functional equation:

$$(6.10) \qquad L(s, \pi \otimes \tau) = \epsilon(s, \pi \otimes \tau)\, L(1-s, \tilde{\pi} \otimes \tilde{\tau}).$$

Here $\epsilon(s, \pi \otimes \tau) = \prod_v \epsilon(s, \pi_v \otimes \tau_v)$, the ϵ-factors being defined above for finite places and via the Langlands classification for infinite places; the ϵ-factor is 1 for almost all v.

This function equation is announced, but not completely proven, in [26(a),(b)]. Let us assume it for the moment.

Now let k'/k be an extension of global fields, chosen as in §4. Specifically, we assume that, at some place v_0 of k, k'_{v_0}/k_{v_0} is isomorphic to E/F, that some finite places v_1, v_2 split in k' while another finite place v_3 remains inert.

By the arguments for Lemma 6.5, we may choose cuspidal representations π_k, τ_k of $G_n(\mathbf{A}_k)$, $G_m(\mathbf{A}_k)$ such that π_{k,v_0} (resp. τ_{k,v_0}) is isomorphic to π (resp. τ). We will first consider the case where π, τ are *supercuspidal.* We assume that π_k (resp. τ_k) is unramified at any place $v \notin \{v_0, v_1, v_2, v_3\} \cup S'_{p_0}$, where S'_{p_0} is the set of places v of F dividing p_0, the prime divisor of v_0, and different from v_0; we assume that v_1, v_2, v_3 do not divide p_0. We may further assume that all places in S'_{p_0} split in E. Finally, for $v \in S'_{p_0}$, we assume that π_0 is supercuspidal. By the arguments in §6.3, π_k and τ_k then lift to two cuspidal representations $\Pi_{k'}$ and $\mathrm{T}_{k'}$ of the adèlic groups over k'. Consider the two L-functions:

$$L_1(s) = L(s, \Pi_{k'} \otimes \mathrm{T}_{k'})$$

and

$$L_2(s) = \prod_{\eta \in \Xi} L(s, \pi_k \otimes \tau_k \otimes \eta),$$

which we consider as Euler products over the rational primes. By (6.8), their Euler factors coincide at almost all primes. They both satisfy functional equations of the usual type. We now apply the following well-known principle, a precise version of which is given in Vignéras [41(b)]:

LEMMA 6.11: *Assume given four L-functions L_1, $\tilde{L}_1$, L_2, $\tilde{L}_2$ given by Euler products over the rational primes, including the real prime. Assume they admit a meromorphic continuation to* $\mathbf{C}$ *and verify functional equations $L_i(s) = \epsilon_i(s)\tilde{L}_i(1-s)$ $(i = 1, 2)$ with $\epsilon_i(s) = c_i e^{b_i s}$, $c_i, b_i \in \mathbf{C}$.*

Assume that (a) *the Euler factor of L_1 (resp. $\tilde{L}_1$) is equal to that of L_2 (resp. $\tilde{L}_2$) at almost all primes.*

(b) *At a prime p we have*

$$L_{1,p}(s) = \prod_i \frac{1}{1 - a_i^1 p^{-s}}, \quad \tilde{L}_{1,p}(s) = \prod_j \frac{1}{1 - b_j^1 p^{-s}}$$

with $|a_i^1| \neq p|b_j^1|$ for all i, j; the same applies to $L_{2,p}$ and $\tilde{L}_{2,p}$.

Then the Euler factors of L_1 and L_2 coincide at p.

We apply this to $L_1(s)$ and $L_2(s)$, the functions $\tilde{L}_1(s)$ and $\tilde{L}_2(s)$ being the ones figuring in equation (6.11) for L_1 and L_2. We need only check condition (b). By Lemma 6.10, and our assertions on π_v for $v|p_0$, the L-functions occurring for all places of F or E over p_0 are associated to pairs of representations that are (unitary) supercuspidal, or unitarily induced from such. By the results of Jacquet, Piatetski–Shapiro, Shalika [26(b)] (see formula (6.11) below), the associated Euler factors are products of terms $(1-\chi(\bar{\omega})q^{-s})^{-1}$, where q is a power of p_0 and $|\chi(\bar{\omega})|=1$. In particular, the reciprocal roots of the p_0-factor have absolute value one, whence (b). By Lemma 6.11, we see that the p_0-Euler factor of L_1 and L_2 coincide. Since $L_{1,v}(s)=L_{2,v}(s)$, trivially, for $v\in S'_{p_0}$, we deduce that $L_{1,v_0}=L_{2,v_0}$, proving Proposition 6.9(i) in the supercuspidal case.

To treat general discrete series representations, we will need to know how generalized Steinberg representations behave under base change:

LEMMA 6.12: *Let $\pi=St(\omega,a)$ be a generalized Steinberg representation of $G_n(F)$, $n=ar$.*

(i) Assume ω lifts to a supercuspidal representation Ω of $G_r(E)$. Then π lifts to $\Pi=St(\Omega,a)$.

(ii) Assume ω lifts to $\Gamma^\sigma\times\cdots\times\Gamma^{\sigma^{\ell-1}}$, with $\Gamma\not\cong\Gamma^\sigma$ supercuspidal. Then π lifts to

$$St(\Gamma,a)\times St(\Gamma^\sigma,a)\times\cdots\times St(\Gamma^{\sigma^{\ell-1}},a).$$

Of course, Lemma 6.10 implies that the assumptions (i), (ii) are the only two possibilities.

Proof. Consider first the case (i). We know that π lifts to a σ-discrete representation Π. We first show that Π is discrete (note that we cannot use Proposition 6.6, which relies on the global results of Chapter III, which will require Proposition 6.9!). Assume that Π is not discrete, whence $\Pi=St(\Delta,c)\times\cdots\times St(\Delta^{\sigma^{\ell-1}},c)$ for $n=c\ell d$, $\Delta\not\cong\Delta^\sigma$ a supercuspidal representation of $G_d(E)$. Consider the parabolic subgroup $P_2=M_2N_2$ of type $d\ell,\ldots,d\ell)$. The Jacquet module Π_{N_2} has been described in the proof of Lemma 6.10(ii): its σ-stable part is equal to $\Sigma|\ |^{x_1}\times\cdots\times\Sigma|\ |^{x_c}$, where $\Sigma=\Delta\times\Delta^\sigma\times\cdots\times\Delta^{\sigma^{\ell-1}}$. In particular, its twisted trace is non-zero on elliptic elements: the character identities then imply that $\pi_{N_2}\neq 0$, whence $r|d\ell$. Moreover, π_{N_2} is then, up to unramified twists, a tensor product of representations of the blocks $G_{d\ell}(F)$ of type $St(\omega,\frac{d\ell}{r})$ ([42]) and therefore $St(\omega,\frac{d\ell}{r})$ would (again up to twists) lift to Σ. However, Σ has clearly no

σ-stable Jacquet modules. Therefore we must have $d\ell = r$, and Σ lifts ω. But this contradicts our assumption on ω: therefore Π is discrete.

Write, then, $\Pi = St(\Lambda, b)$ for $n = bt$, Λ supercuspidal. We must have $\Lambda \cong \Lambda^\sigma$. If N is the unipotent radical of the parabolic subgroup P of type $(t, \ldots, t)$, we have

$$\Pi_N \cong \Lambda|\ |^{b-1} \otimes \Lambda|\ |^{b-2} \otimes \cdots \otimes \Lambda$$

(uninormalized Jacquet module).Clearly this is σ-stable, so by Proposition 2.3, Λ being σ-discrete, the twisted character of Π does not vanish identically on elements $g \in G_n(E)$ such that $\mathcal{N}g \in G_n(F)$ satisfies $P_{\mathcal{N}g} = P$.

By the identities of characters and Casselman's theorem for π, we see that $\pi_N \neq 0$. This implies that $b|a$. If $b < a$, N is strictly contained in the unipotent radical N_1 of type $(r, \ldots, r)$. Since Λ is supercuspidal, $\Pi_{N_1} = 0$ while $\pi_{N_1} \neq 0$; this contradicts again the identity of traces. Therefore $a = b$. But now the identity of traces, and the twisted and non-twisted Casselman theorems, are easily seen to imply that $\Theta_{\pi_N} \circ \mathcal{N} = c\Theta_{\Pi_N, \sigma}$, at least on elliptic elements, c being a (non-zero) constant coming from the action on the space of Π_N of the normalized intertwining operator for Π. Therefore Λ lifts ω, at least up to a constant and if we consider characters on elliptic elements. If λ is a (discrete) representation lifted by Λ, the orthogonality relations (cf. proof of Prop. 6.6) imply that $\lambda \cong \omega \otimes \eta$ for some $\eta \in \Xi$, whence the result.

We now consider case (ii). We first show that Π, the lift of π, is not discrete. Assume it were. Write $\Pi = St(\Lambda, b)$ with Λ a supercuspidal representation of $G_t(E)$, $n = bt$, $\Lambda \cong \Lambda^\sigma$. Then, if N is the unipotent radical of type $(t, \ldots, t)$, we have $\Pi_N = \Lambda|\ |^{b-1} \otimes \Lambda|\ |^{b-2} \otimes \cdots \otimes \Lambda$. The identity of characters then implies that $\pi_N \neq 0$, whence $b|a$. Moreover (up to a twist) Λ lifts the Jacquet module $St(\omega, \frac{a}{b})$. Since the Jacquet modules of Λ are null, this implies that $a = b$, and Λ lifts ω. But this contradicts our assumption on ω.

Therefore π lifts to $\Pi_1 \times \Pi_1^\sigma \times \cdots \times \Pi_1^{\sigma^{\ell-1}}$ with Π_1 discrete, $\Pi_1 \not\cong \Pi_1^\sigma$. Set $\Pi_1 = St(\Delta, b)$ with Δ supercuspidal, $\Delta \not\cong \Delta^\sigma$. Let $P = MN$ be the parabolic subgroup of type $(t, \ldots, t)$ where $n = bt$. In the proof of Lemma 6.10(ii) we showed that the only σ-stable subquotient of Π_N is the module

$$\Sigma|\ |^{b-1} \otimes \Sigma|\ |^{b-2} \otimes \cdots \otimes \Sigma,$$

where $\Sigma = \Delta \times \Delta^\sigma \times \cdots \times \Delta^{\sigma^{\ell-1}}$ (in fact, we did not compute the precise unramified twists of Σ that occur: but this is easily deduced from Frobenius reciprocity). Now the identity of characters implies, in the usual manner, that $\pi_N \neq 0$, whence $b|a$.

Now we show that $b = a$. Indeed, $\Delta \times \Delta^\sigma \times \cdots \times \Delta^{\sigma^{\ell-1}}$ has no σ-stable (non-trivial) Jacquet modules. Assume $b < a$, whence $r < t$, and consider the unipotent radical N_1 of type $(r, \ldots, r)$ in G_t. Then Π_{N_1} has a vanishing twisted trace, whereas the corresponding Jacquet module for π is non-zero, and in fact square-integrable. This contradicts again the equality of traces.

Therefore, $b = a$, $\Pi_1 = St(\Delta, a)$. We must show that $\Delta \approxeq \Gamma$. Considering the σ-stable part $\Sigma|\ |^{b-1} \otimes \cdots \otimes \Sigma$ of Π_N, we easily deduce that the twisted character of $\Sigma = \Delta \times \cdots \times \Delta^{\sigma^{\ell-1}}$ is equal, on the elliptic norms, to the character of ω composed with $\mathcal{N}$, up to a non-zero constant. We finish the proof as in (i). ∎

We now finish the proof of Proposition 6.9(i). We assume that π, τ are square-integrable. Write $\pi = St(\omega, a)$ for ω supercuspidal, $n = ar$. Similarly, let $\tau = St(\delta, b)$, $m = bt$.

Under these assumptions, the local L-function $L(s, \pi \otimes \tau)$ has been computed by Jacquet, Piatetski–Shapiro and Shalika. Their result is as follows ([26(b), Prop. 8.1 and Thm. 8.2]):

(i) Consider the supercuspidal representations ω, δ of $G_r(F)$, $G_t(F)$. Then

$$L(s, \omega \otimes \delta) = \prod_{\chi} (1 - \chi(\varpi) q^{-s})^{-1} \tag{6.11}$$

where the product ranges over all unramified characters χ of $F^\times$ such that $\pi \otimes \chi \cong \tilde{\delta}$, $\tilde{\delta}$ being the contragredient of δ. In particular, if $L(s, \omega \otimes \delta) \neq 1$, we have $r = t$.

(ii) Let $\pi = St(\omega, a)$ and $\tau = St(\delta, b)$. Then

$$L(s, \pi \otimes \tau) = 1 \quad \text{unless} \quad r = t. \tag{6.12}$$

(iii) If $r = t$, assume $m \leq n$. Then

$$L(s, \pi \otimes \tau) = \prod_{i=1}^{b} L(s + \frac{a+b}{2} - 1 - i, \omega \otimes \delta). \tag{6.13}$$

In proving Proposition 6.9(i), we now distinguish between cases (i) and (ii) occurring in Lema 6.10. The identity to be proved is

$$L(s, \Pi \times T) = \prod_{\eta \in \Xi} L(s, \pi \otimes \tau \otimes \eta). \tag{6.14}$$

Using Lemmas 6.10 and 6.12, Proposition 6.9(i) for π and τ can now be deduced from the supercuspidal case. For instance, assume that ω, δ lift to supercuspidal Ω, Δ (case (i) of Lemma 6.10). Then π, τ lift to $St(\Omega, a)$ and $St(\Delta, b)$ and (6.12) and (6.13) reduce the identity (6.14) to the supercuspidal case. The two other cases are analogous.

To avoid using the unpublished results announced in [26(a),(b)], we rely on Shahidi's work. Let S be a finite set of places of k containing all the ramified places for k'/k, π_k, τ_k and the Archimedian places. If ψ_v denotes a non-trivial character of k_v ($v \in S$), Shahidi defines in [36(b)] local coefficients $C(s, \psi_v, \pi_{k,v} \times \tau_{k,v})$ (... denoted there by $C_{\chi_v}(s, \pi_{k,v} \times \tau_{k,v})$ where χ_v is the non-degenerate character of the upper nilpotent group defined by ψ_v). Now assume ψ is a character of $\mathbf{A}_k$; choose S so that ψ_v is unramified for $v \notin S$. Then, writing L^S for the Euler product *outside* S:

$$(6.15)\qquad L^S(s, \pi_k \otimes \tau_k) = \left(\prod_{v \in S} C(s, \psi_v, \pi_{k,v} \times \tau_{k,v}) \right) L^S(1-s, \tilde{\pi}_k \otimes \tilde{\tau}_k)$$

([36(b), Thm. 4.1]). Moreover, it is shown in [36(d),(e)] that these local coefficients at the p-adic and real places of S are equal to the corresponding γ-factors. Therefore, equation (6.15) implies the functional equation 6.10. The proof of Proposition 6.9(1) is complete.

To prove the identity of ϵ-factors, we choose a global extension k'/k of number fields as in §4. Specifically, we assume that, at some place v_0 of k, k'/k is isomorphic to E/F, and that some places v_1, v_2 split in k'. We assume moreover that k'/k splits at all infinite places. As in Lemma 6.5, we may find a cuspidal representation π_k of $G_n(\mathbb{A}_k)$ such that π_{k,v_1} is a given supercuspidal representation of $G_n(k_{v_1})$, $\pi_{k,v_0} \cong \pi$, and $\pi_{k,v}$ is unramified for finite $v \notin \{v_0, v_1, v_2\}$. The identity of traces then yields a cuspidal representation $\pi_{k'}$ of $G_n(\mathbf{A}_{k'})$ verifying the identity (6.2):

$$\ell \ \mathrm{trace}(\pi_{k'}(\varphi) I_\sigma) = \sum_{\pi'_k} \mathrm{trace}\, \pi'_k(f)$$

for associated functions φ, f. The sum on the right runs over all π' twisted from π by a power of the class field character associated to k'/k.

This identity implies that each local component of π_k lifts to the corresponding component of $\pi_{k'}$—*a priori* up to a scalar, but this scalar must be equal to 1 since the local lifting has already been proven for generic representations, and twisted characters are linearly independent (note that this implies that there are ℓ terms on the right: this will also follow from the global results of Chapter III).

The same construction can be applied to $\tau = \tau_{k,v_0}$. Let $\psi = \bigotimes_v \psi_v$ be a non-degenerate character of $\mathbf{A}_k$, and $\psi' = \psi \circ \mathrm{trace}_{k'/k}$ the associated character of $\mathbf{A}_{k'}$. We may assume that ψ_{v_0} is the non-degenerate character ψ_F we consider. We now have equations

$$(6.15) \qquad \prod_\eta L(s, \pi \otimes \tau \otimes \eta) = \prod_\eta [\epsilon(s, \pi \otimes \tau \otimes \eta) L(1-s, \tilde{\pi} \otimes \tau \otimes \eta)]$$

where for simplicity we write π for π_k, τ for τ_k ... (the representation of local groups we consider are then $\pi_0 = \pi_{v_0}$ and $\tau_0 = \tau_{v_0}$). Analogously,

$$(6.16) \qquad L(s, \Pi \otimes \mathrm{T}) = \epsilon(s, \Pi \otimes \mathrm{T}) L(1-s, \tilde{\Pi} \otimes \tilde{\mathrm{T}}).$$

By Proposition 6.9(i), we know that the L-functions figuring on the two sides of (6.15) and (6.16) are equal at all places (at the Archimedean primes, it is clear since the extension splits). Therefore

$$(6.17) \qquad \prod_\eta \epsilon(s, \pi \otimes \tau \otimes \eta) = \epsilon(s, \Pi \otimes \mathrm{T}).$$

We now write the ϵ-factors as products; for example,

$$\epsilon(s, \pi \otimes \tau) = \prod_v \epsilon(s, \pi_v \otimes \tau_v, \psi_v).$$

We now remark:

LEMMA 6.13: *Assume $v = v_0$.*

(i) If v is inert,

$$\epsilon(s, \Pi_v \otimes \mathrm{T}_v, \psi'_v) = \lambda(k'_v/k_v, \psi_v)^{-mn} \prod_\eta \epsilon(s, \pi_v \otimes \tau_v \otimes \eta_v, \psi_v)$$

(ii) If v is split,

$$\epsilon(s, \Pi_v \otimes \mathrm{T}_v, \psi'_v) = \epsilon(s, \pi_v \otimes \tau_v, \psi_v)^\ell.$$

Proof. Part (ii) is trivial. For (i), note that v is non-archimedian and all representations are unramified. Therefore the ϵ-factors are just products of ϵ-factors associated to characters; the identity in that case is just (6.7). ∎

We now use the obvious product formula for the λ-factors:

$$\prod_{v \text{ inert}} \lambda(k'_v/k_v, \psi_v) = 1$$

which again follows directly from their definition (6.7). Now dividing the right-hand side by $\prod_{v \text{ inert}} \lambda(k'_v/k_v, \psi_v)^{mn}$ and using the equations of Lemma 6.13 at the places $v \neq v_0$, we are left with the identity of Proposition 6.9(ii). ∎

7. Archimedean case

In this paragraph, we will rapidly treat the case of Archimedean fields: thus the only interesting case is the extension $\mathbf{C}/\mathbf{R}$, further extensions being treated by the methods of §5. The local base change results have been proved in that case by Shintani and Repka [30]; we will only refer to them, and also to [11(a)] when necessary. We want here to prove the results about the Paley–Wiener Theorem and orbital integrals which will be needed for the trace formula. Let σ be the generator of $\mathrm{Gal}(\mathbf{C}/\mathbf{R})$.

Recall the parametrization of the generalized principal series of $\mathrm{GL}(n,\mathbf{R})$ ([30(b)], [10(d)]). Set $n = 2n_2 + n_1$; let $\chi_i (i = 1, \dots n_2)$ be ramified characters of $\mathbf{C}^*$ (so that $\chi_i(z) \neq \chi_i(\bar{z})$) and let $\xi_j (j = 1, \dots n_1)$ be characters of $\mathbf{R}^*$. By the Langlands classification, χ_i defines a discrete series representation $\pi(\chi_i)$ of $\mathrm{GL}(2,\mathbf{R})$; we have $\pi(\chi_i) = \pi(\chi_i^\sigma)$ where $\chi_i^\sigma(z) = \chi_i(\bar{z})$. Let $P = MN$ be the standard parabolic subgroup with n_2 2-blocks and n_1 1-blocks. We write $\pi(\chi_1, \chi_2, \dots \xi_1, \dots \xi_{n_1})$ for the representation induced from $\pi(\chi_1) \otimes \cdots \otimes \pi(\chi_{n_2}) \otimes \xi_1 \otimes \cdots \otimes \xi_{n_1}$: it is a generalized principal series representation of $\mathrm{GL}(n,\mathbf{R})$. All generalized principal series are of this form, and the isomorphisms between them are the obvious ones.

It will be convenient to parametrize them by "discrete" and "continuous" parameters: we may write

$$\chi_i(z) = z^p(\bar{z})^q,\ p - q \in \mathbf{Z} - \{0\},\ p + q = s_i \in \mathbf{C}$$
$$\xi_j(x) = (\operatorname{sgn} x)^{\varepsilon_j} \cdot |x|^{s_j},\ s_j \in \mathbf{C}.$$

Then $(s_1, \dots s_{n_2+n_1}) \in \mathbf{C}^{n_2+n_1} \cong \mathfrak{a}_M^*$. The Weyl group of M, W_M, acts on $\mathfrak{a}_M$ as $\mathfrak{S}_{n_2} \times \mathfrak{S}_{n_1}$.

The σ-stable lifts of these representations are the representations induced from the Borel subgroup $B(\mathbf{C})$:

$$\Pi(\chi_1, \chi_2, \dots \xi_1, \dots \xi_{n_1}) = \mathrm{ind}_{B(\mathbf{C})}^{G(\mathbf{C})}(\chi_1, \chi_1^\sigma, \chi_2, \chi_2^\sigma, \dots, \xi_1 \circ N, \dots \xi_{n_1} \circ N)$$

where $N = N_{\mathbf{C}/\mathbf{R}}$. (They are obviously σ-stable.) The lifting thus obtained coincides, via the Langlands classification, with restriction on the Weil group side [11(a)]. We remark that the base change identities extend to all values (tempered or not) of the parameters provided we consider the full induced representations and not their Langlands quotients.

We wish to remark on the normalization of intertwining operators. In [11(a)], operators A_σ are defined, using Vogan's theory; they introduce a sign $\varepsilon(M)$, equal with our data to $(-1)^{n_2}$, in the Shintani formulas. On the other hand, we may define I_σ as in §2; using the trace formula as in §6, one can see that the Shintani formulas hold without a sign for the I_σ. In what follows we use the operators I_σ.

We will call *σ-stable data* the data parametrizing the σ-stable generalized principal series. Since $\xi_i \circ N$ does not depend on the sign of ξ_i, we will write σ-stable data in the form $(\chi_1, \dots \chi_{n_2}, \xi_1, \dots \xi_{n_1})$ where χ_i is a ramified character and ξ_i is now just an element of $\mathbf{C}$. Thus $\xi_i(z) = |Nz|^{s_i}$, $s_i \in \mathbf{C}$.

We will first prove, for twisted representations, the analogue of the Paley–Wiener Theorem of [12(a)]. We assume familiarity with the results of this paper. The proof in our case is analogous and will only be sketched. Let $K_{\mathbf{C}}$ be a maximal compact subgroup of $G(\mathbf{C})$, stable by σ – e.g. $K_{\mathbf{C}} = U(n)$; let $C_c^\infty(G(\mathbf{C}), K_{\mathbf{C}})$ denote the smooth functions, $K_{\mathbf{C}}$-finite on both sides.

Again, the σ-stable representations are parametrized by "discrete" and "continuous" parameters, the latter being parametrized by the spaces $\mathfrak{a}_M$ for cuspidal parabolic subgroups.

PROPOSITION 7.1: *Assume given scalar-valued functions on the set of σ-stable data:*

$$(\chi, \xi) = (\chi_1, \chi_2, \dots \chi_{n_2}, \xi_1, \xi_2, \dots \xi_{n_1}) \mapsto F(\chi, \xi) \in \mathbf{C}.$$

Then there is a function $\phi \in C_c^\infty(G(\mathbf{C}), K_{\mathbf{C}})$ such that

$$F(\chi, \xi) = \mathrm{trace}(\Pi(\chi, \xi) I_\sigma)$$

for all χ, ξ if and only if

(i) $F(\chi, \xi)$ has finite support in the discrete data.

(ii) The function on $\mathfrak{a}_M$:

$$(s_1, \dots s_{n_2+n_1}) = F(\chi_1 \mid |_{\mathbf{C}}^{s_1}, \dots \xi_{n_1} | \, |_{\mathbf{C}}^{s_{n_2}+n_1})$$

is of Paley–Wiener type on $\mathfrak{a}_M$.

(iii) For $w \in W_M = \mathfrak{S}_{n_2} \times \mathfrak{S}_{n_1}$:

$$F(w(\chi, \xi)) = F(\chi, \xi).$$

Proof. The fact that the traces of $\phi \in C_c^\infty(G(\mathbf{C}), K_{\mathbf{C}})$ satisfy (i)–(iii) is straightforward ([12(a), §2]). We prove the converse. The same argument as in [12(a), §2] reduces the proof to an assertion concerning only one "series" of representations at a time, the analogue of Proposition 1 in [12(a)]. The "discrete" part of χ_i is expressed by $p_i - q_i = r_i$. We may assume that $r_i > 0$. Let χ^0 be the character of $(\mathbf{C}^*)^n$:

$$(z_1, \dots z_n) \mapsto \left((z_1/\overline{z_1})^{\frac{r_1}{2}}, (z_1/\overline{z_1})^{-\frac{r_1}{2}}, \dots, (z_{n_2}/\overline{z_{n_2}})^{-\frac{r_{n_2}}{2}}, 1, \dots 1\right).$$

For $s \in \mathfrak{a}_M^*$, we may then consider the character $\chi^0 \otimes s$ defined in the obvious manner. It yields, by induction, a σ-stable representation.

LEMMA 7.2: *Let μ be the minimal $K_{\mathbf{C}}$-type in the induced representation $\Pi(\chi^0 \otimes s) = \mathrm{ind}_{B(\mathbf{C})}^{G(\mathbf{C})}(\chi^0 \otimes s)$. Let $F(s)$ be a function of Paley–Wiener type on $\mathfrak{a}_M^*$. Assume $F(s)$ is invariant by $(W_M)_{\chi^0}$, the stabilizer of χ^0 in W_M. Then there exists $\phi \in C_c^\infty(G(\mathbf{C}), K_{\mathbf{C}})$, transforming under μ on the right and left, such that $F(s) = \mathrm{trace}(\Pi(\chi^0 \otimes s)(\phi) I_\sigma)$.*

Here $W_M \cong \mathfrak{S}_{n_1} \times \mathfrak{S}_{n_2}$ acts in the obvious way on the discrete parameters.

Proof. Note that I_σ preserves the space of μ; since μ has multiplicity 1, a function ϕ of type (μ, μ) acts there as a scalar, and therefore the twisted trace coincides with the ordinary trace.

Now let $\mathfrak{a}_0$ be the vector space associated to $B(\mathbf{C})$: thus $\mathfrak{a}_0 \cong \mathbf{C}^n$. The stabilizer of the character χ^0 of the compact part of $(\mathbf{C}^*)^n$ in $W = W(\mathfrak{a}_0) = \mathfrak{S}_n$ is then isomorphic to

$$(\mathfrak{S}_{m_1})^2 \times (\mathfrak{S}_{m_2})^2 \times \cdots \times (\mathfrak{S}_{m_k})^2 \times \mathfrak{S}_{n_1},$$

with $m_1 + m_2 + \cdots + m_k = n_2$, and m_i is the multiplicity of a given ramified character $(z_i/\overline{z_i})^{r_i/2}$ in the ramified part of χ^0. (Note that the ramified characters occur by pairs.) By Proposition 1 of [12(a)], we know that any Paley–Wiener function on $\mathfrak{a}_0^*$, invariant by W_{χ^0}, is the value on the minimal $K_{\mathbf{C}}$-type of a function ϕ. Thus, to prove Lemma 7.2, it suffices (using Lemmas 7 and 8 of [12(a)]) to check that the restriction map:

$$S(\mathfrak{a}_0^*)^{W_{\chi^0}} \to S(\mathfrak{a}_M^*)^{(W_M)_{\chi^0}}$$

is onto. In our case this is easily checked (see also [12(b), Theorem 2.2]). ∎

With this the proof of Proposition 7.1 is complete. ∎

We can now use the Paley–Wiener Theorem to compare orbital integrals. Recall from §3 the definition of associated functions: they satisfy the conditions of Proposition 3.1(i). Let $K_{\mathbf{R}}$ be maximal compact in $G(\mathbf{R})$.

LEMMA 7.3:

(i) Assume $\phi \in C_c^\infty(G(\mathbf{C}), K_{\mathbf{C}})$. Then there is $f \in C_c^\infty(G(\mathbf{R}), K_{\mathbf{R}})$ associated to ϕ.

(ii) Conversely, if $f \in C_c^\infty(G(\mathbf{R}), K_{\mathbf{R}})$ has vanishing orbital integrals on the regular elements not in the image of $\mathcal{N}$, there is ϕ associated to f.

Proof. We use the methods of [12(a), Appendix]. Assume ϕ given. Then its twisted traces satisfy the conditions of Proposition 7.1. Therefore there is (by the ordinary Paley–Wiener Theorem for $\mathrm{GL}(n, \mathbf{R})$) $f \in C_c^\infty(G(\mathbf{R}))$, $K_{\mathbf{R}}$-finite, such that

$$\operatorname{trace}(\Pi(\phi) I_\sigma) = \operatorname{trace} \pi(f)$$

when Π lifts π and both are generalized principal series.

By a theorem of Shelstad ([38(b), Corollary 4.5.2]) we know that there is a function f^* in the Schwartz space of $G(\mathbf{R})$ with orbital integrals matching those of ϕ. But then, by the Weyl integration formula and the identities of characters,

$$\operatorname{trace}(\Pi(\phi) I_\sigma) = \operatorname{trace} \pi(f^*)$$

for tempered π. Thus f and f^* have the same traces in tempered representations, and therefore the same orbital integrals. This proves (i).

Conversely, assume f given. The vanishing condition on the orbital integrals implies that $\operatorname{trace} \pi(f) = \operatorname{trace} \pi'(f)$ if π, π' lift to the same Π, since then $\operatorname{trace} \pi$ and $\operatorname{trace} \pi'$ differ only on elements not in $\mathcal{N}G(\mathbf{C})$. Then the assignment $\Pi \mapsto \pi(\Pi) \mapsto \langle \operatorname{trace} \pi(\Pi), f \rangle$ defines a family of functions on σ-stable Π as in Proposition 7.1; thus there is a ϕ such that

$$\operatorname{trace}(\Pi(\phi) I_\sigma) = \operatorname{trace} \pi(f)$$

for π associated to Π. By part (i) of the lemma, there is a function $f^* \in C_c^\infty(G(\mathbf{R}), K_{\mathbf{R}})$ associated to ϕ; then we have

$$\operatorname{trace} \pi(f) = \operatorname{trace} \pi(f^*)$$

for any π, so f and f^* have the same orbital integrals. This proves (ii). ∎

CHAPTER 2

The Global Comparison

1. Preliminary remarks

The goal of Chapter 2 is a full comparison of trace formulas. The immediate purpose of this is to extract global information about automorphic representations. Along the way, we shall also gain some insight into the rather mysterious local objects which appear in the general trace formula.

We shall treat the problems of base change and inner twisting simultaneously. For this reason it will be convenient to revert to the notation of the introduction, which is more streamlined for dealing with the general trace formula. For example, we will be letting G stand for a connected component of an algebraic group, while G' will denote the endoscopic group $\mathrm{GL}(n)$. The norm map and the local correspondence of functions will be written $\gamma \to \gamma'$ and $f \to f'$ respectively, instead of $\delta \to \gamma$ and $\phi \to f$ as in Chapter 1. It will also be useful to make the distinction between a well-defined function, such as f, and a function such as f' which is determined only by its characters or orbital integrals. In this paragraph we outline our assumptions and notation for G in some detail. We shall also recapitulate the local results, established for base change in Chapter 1 and for inner twistings in [15].

Let G be a connected component of a reductive algebraic group. We assume that G is defined over a number field F, and that $G(F)$ is not empty. We shall write G^+ for the reductive group generated by G, and G^0 for the identity component of G^+. Let M_0 be a fixed minimal Levi subgroup of G^0, defined over F, and let $\mathcal{L}$ denote the finite collection of Levi subsets M of G such that M^0 contains M_0. We shall routinely adopt the notation of Sections 1 and 2 of [1(e)]. In particular, for any $M \in \mathcal{L}$, we have the lattice $X(M)_F$ of rational characters of M^+, and the real vector space

$$\mathfrak{a}_M = \mathrm{Hom}(X(M)_F, \mathbf{R}).$$

We also have various other objects, such as $\mathcal{L}^M, \mathcal{L}(M), \mathcal{P}(M)$ and A_M, which were defined in §1 of [1(e)].

As always, GL(n) stands for the general linear group of rank n over F. Fix a positive integer ℓ. As a simple example, consider the component

$$G^* = \underbrace{(\mathrm{GL}(n) \times \cdots \times \mathrm{GL}(n))}_{\ell} \rtimes \theta^*,$$

where θ^* is the permutation

$$(1, \ldots, \ell) \to (2, \ldots, \ell, 1).$$

Then $(G^*)^+$ is the semi-direct product of ℓ copies of GL(n) with the cyclic group of order ℓ generated by θ^*. Our fundamental assumption on G is that it is an inner twist of G^*. In other words, there is a morphism

$$\eta : G \to G^*,$$

which extends to an isomorphism from G^+ onto $(G^*)^+$, such that for every $\sigma \in \mathrm{Gal}(\overline{F}/F)$, $\eta^{-1}\eta^\sigma$ equals a conjugation by an element in G^+. We shall let E denote the smallest extension of F over which the image of this cocycle in G^+/G^0 splits. Then E is a cyclic extension of F whose degree is a divisor of ℓ. We can choose η so that $\eta(M_0)$ contains the standard minimal Levi subgroup of $(G^*)^0$, and so that the restriction of η to A_{M_0} is defined over F. Fix such an η, and set $\theta = \eta^{-1}(\theta^*)$. Then

$$G = G^0 \rtimes \theta.$$

Set

$$G' = \mathrm{GL}(n)$$

and embed G' diagonally in $(G^*)^0$. We shall write $\mathcal{L}'$ for the set of Levi subgroups of G' which contain the group of diagonal matrices. The map

$$M \to M' = \{m' \in \eta(M^0) : (\theta^*)^{-1} m' \theta^* = m'\}, \qquad M \in \mathcal{L},$$

is then an injection of $\mathcal{L}$ into $\mathcal{L}'$. If γ is any element in G, the centralizer of γ in G^0 is connected. As in §2 of [1(e)], we shall denote it by G_γ. Observe that in this notation,

$$\eta : G_\theta \to G' = G^*_{\theta^*}$$

is an inner twist.

The norm map may be described as follows. If $\{\gamma\}$ is a $G^0(F)$-orbit in $G(F)$, the intersection of $\{\eta(\gamma)^\ell\}$ with $G'(F)$ is a $G'(F)$-conjugacy class. We shall write γ' to denote this $G'(F)$-conjugacy class, or by abuse of notation, for an element in the class. Suppose that σ is a semisimple element in G.

The centralizer G_σ of σ in G^0 is the group of units in a product of central simple algebras over F. Given η and σ, there is a canonical inner twist

$$\eta_\sigma : G_\sigma \to G'_{\sigma'},$$

which is uniquely determined up to $G'(F)$-conjugacy. We shall let $\mu \to \mu_{\sigma'}$ denote the associated map from conjugacy classes in $G_\sigma(F)$ to conjugacy classes in $G'_{\sigma'}(F)$. One checks easily that

$$(\sigma\mu)' = \sigma' \mu^\ell_{\sigma'}, \qquad \mu \in G_\sigma(F). \tag{1.1}$$

Suppose that S is a finite set of valuations of F. Then similar remarks apply if γ and σ are points in

$$G(F_S) = \prod_{v \in S} G(F_v).$$

The local results of Chapter 1 could probably be established for the group $G^+(F_v)$. However, they are more limited as they stand, and we must impose an additional condition on G. We shall assume that the image of the cocycle $\eta^{-1}\eta^\sigma$ is contained in either $(G^*)^0$ or the group generated by θ^*. In the first instance

$$G(F) \cong \underbrace{(A^*(F) \times \cdots \times A^*(F))}_{\ell} \rtimes \theta^*,$$

where A is a central simple algebra of degree n over F. This is essentially the case of inner twisting of $\mathrm{GL}(n)$, studied in [15]. In the second case,

$$G(F) \cong \underbrace{(\mathrm{GL}(n, E) \times \cdots \times \mathrm{GL}(n, E))}_{\ell_1} \rtimes \theta^*_E,$$

where if σ is a generator of $\mathrm{Gal}(E/F)$, θ^*_E is the cyclic permutation of order

$$\ell_1 = \ell \deg(E/F)^{-1}$$

given by

$$(g_1, \ldots, g_{\ell_1}) \to (g_2, \ldots, g_{\ell_1}, \sigma g_1).$$

This is the base change situation considered in Chapter 1. (The cyclic permutations θ^*_E and θ^* here are of no consequence. The reader, if so inclined, could eliminate them by making the further assumption that $\deg(E/F) = \ell$.) In what follows, we will generally not refer explicitly to the additional condition on G. Indeed, most of the techniques of Chapter 2 apply to the more general setting.

Suppose that S is a finite set of valuations of F. In §1.2 we used the theory of Whittaker models to extend any irreducible, ad(θ)-invariant representation π^0 of $G^0(F_S)$ in a canonical way to an irreducible representation π of $G^+(F_S)$. Let $\Pi^+(G(F_S))$ be the set of (equivalence classes of) irreducible representations of $G^+(F_S)$ obtained in this way. Let $\Pi^+_{\text{temp}}(G(F_S))$ and $\Pi^+_{\text{unit}}(G(F_S))$ be the subsets of $\Pi^+(G(F_S))$ which are respectively tempered and unitary. The local correspondence of representations can be described as an injection $\pi \to \Pi'(\pi)$ from $\Pi^+(G(F_S))$ onto a collection of finite disjoint subsets of $\Pi(G'(F_S))$ which is dual to the map $\gamma \to \gamma'$. (We suppress the superscript $+$ in denoting sets of representations of $G'(F_S)$.) To describe the associated character identity, set

$$e_S = \prod_{v \in S} e_v,$$

where $e_v = e_v(G_\theta)$ is the sign associated to the group G_θ by Kottwitz [29(b)]. (Recall that if v is nonArchimedean,

$$e_v(G_\theta) = (-1)^{r_v(G_\theta) - r_v(G')} = (-1)^{r_v(G_\theta) - n},$$

where $r_v(G_\theta)$ is the F_v-split rank of G_θ; if v is Archimedean,

$$e_v(G_\theta) = (-1)^{\frac{1}{2}(q_v(G_\theta) - q_v(G'))},$$

where $q_v(G_\theta)$ is the dimension of the symmetric space associated to $G_\theta(F_v)$. It is clear that e_S remains unchanged if it is defined with respect to an element $M \in \mathcal{L}$ instead of G.) The character Θ_π of any representation $\pi \in \Pi^+_{\text{temp}}(G(F_S))$ then satisfies

$$\Theta_\pi(\gamma) = e_S \Theta_{\pi'}(\gamma') \tag{1.2}$$

for any $\pi' \in \Pi'(\pi)$ and any $\gamma \in G_{\text{reg}}(F_S)$, the set of regular elements in $G(F_S)$. For if $G = G^0$, the sets $\Pi'(\pi)$ each contain one element. The correspondence is just the injection from the representations of a central simple algebra to those of GL(n). (See [15].) In the base change situation, the sets $\Pi'(\pi)$ consist of the representations of $G'(F_S)$ which lift to a given representation π. (See §1.6.) In this case e_S equals 1, and $\Pi(G'(F_S))$ is the disjoint union of the sets $\Pi'(\pi)$.

We can also introduce the set $\Sigma^+(G(F_S))$ of standard representations. Recall first that if $M \in \mathcal{L}$ and $\pi \in \Pi^+(M(F_S))$, we can form the induced representation

$$\mathcal{I}_P(\pi) = \mathcal{I}_P^G(\pi) = \mathcal{I}_{P^+}^{G^+}(\pi), \qquad P \in \mathcal{P}(M).$$

It is often denoted simply by π^G. More generally, for each valuation v, let $\mathcal{L}_v \supset \mathcal{L}$ be the finite collection of Levi subsets defined over F_v which contain a chosen minimal one, and consider representations of the form

$$\bigotimes_{v\in S}\pi_v^G, \qquad \pi_v \in \Pi^+(M_v(F_v)),\ M_v \in \mathcal{L}_v.$$

Then $\Sigma^+(G(F_S))$ is the set of all such representations for which each π_v is tempered modulo $A_{M_v}(F_v)$. By analytic continuation from the tempered case, we obtain an injection $\rho \to \Sigma'(\rho)$ from $\Sigma^+(G(F_S))$ to a collection of finite disjoint subsets of $\Sigma(G'(F_S))$ for which the character identity above holds. The reader is reminded, however, that the character identity does not hold for arbitary representations $\pi \in \Pi^+(G(F_S))$. We will look at this difficulty more closely in §8, where we will introduce a substitute for the character identity (Proposition 8.2) that applies in general.

As always, $\mathbf{A} = \mathbf{A}_F$ denotes the adèle ring of F. Let $K = \prod_v K_v$ and $K' = \prod_v K'_v$ be maximal compact subgroups of $G^0(\mathbf{A})$ and $G'(\mathbf{A})$, endowed with the usual properties. In particular, it is understood that K' is the standard maximal compact subgroup of $G'(\mathbf{A}) = \mathrm{GL}(n, \mathbf{A})$, that K is θ-stable, and that K'_v is the fixed point set of θ^* in $\eta(K_v)$ for any unramified place v. Having chosen K, we can form the Hecke space $\mathcal{H}(G(F_S))$ of smooth, compactly supported functions on $G(F_S)$ which are finite under $K_S = \prod_{v\in S} K_v$.

For any $f \in \mathcal{H}(G(F_S))$ and $M \in \mathcal{L}$, we set

$$f_M(\pi) = \mathrm{tr}(\pi^G(f)) = \mathrm{tr}(\mathcal{I}_P(\pi, f)), \quad \pi \in \Pi^+_{\mathrm{temp}}(M(F_S)),\ P \in \mathcal{P}(M).$$

Consider the case that $M = G$. The trace Paley–Wiener Theorem (Proposition I.7.1, [12(a)], [33(c)], [6]) holds in all cases under consideration, and this allows us to characterize the image space

$$\mathcal{I}(G(F_S)) = \{f_G : f \in \mathcal{H}(G(F_S))\}$$

of functions on $\Pi^+_{\mathrm{temp}}(G(F_S))$. (See §1 of [1(g)].) Now, suppose that θ is a continuous linear map from $\mathcal{H}(G(F_S))$ to another topological vector space $\mathcal{V}$. Recall that θ is *supported on characters* if it vanishes on any function f with $f_G = 0$. For example, the map $f \to f_M$ from $\mathcal{H}(G(F_S))$ to $\mathcal{I}(M(F_S))$ has this property. It factors through a map $\phi \to \phi_M$ from $\mathcal{I}(G(F_S))$ to $\mathcal{I}(M(F_S))$. In general, if θ is supported on characters, there is a unique continuous map

$$\hat{\theta} : \mathcal{I}(G(F_S)) \to \mathcal{V}$$

such that

$$\hat{\theta}(f_G) = \theta(f), \qquad f \in \mathcal{H}(G(F_S)).$$

In the papers [1(g)] and [1(h)], it was established that the various invariant distributions and maps obtained from the trace formula for G were all supported on characters. We shall use this fact repeatedly throughout Chapter 2.

The basic invariant distributions are of course the (invariant) orbital integrals. In this chapter it will be convenient to follow the conventions of [1(e)], and to normalize them with the discriminant

$$D(\gamma) = D^G(\gamma) = \det(1 - \mathrm{Ad}(\gamma))_{\mathfrak{g}/\mathfrak{g}_\sigma}, \qquad \gamma \in G.$$

Here, σ is the semisimple component of γ while $\mathfrak{g}$ and $\mathfrak{g}_\sigma$ are the Lie algebras of G^0 and G_σ.

LEMMA 1.1: *Suppose that $\gamma \in G$ is semisimple. Then*

$$D^G(\gamma) = \ell^{\dim G_\gamma} D^{G'}(\gamma').$$

Proof. The function $D^G(\gamma)$ depends only on the conjugacy class of $\mathrm{Ad}(\gamma)$ in the general linear group of $\mathfrak{g}$. We may therefore assume that

$$G = G^* = (\mathrm{GL}(n) \times \cdots \times \mathrm{GL}(n)) \rtimes \theta^*,$$

and that the isomorphism η is the identity. Replacing γ by a G^0-conjugate if necessary, we may also assume that

$$\gamma = (\delta, 1, \dots, 1) \rtimes \theta^*, \qquad \gamma \in \mathrm{GL}(n),$$

and

$$\gamma' = (\delta, \dots, \delta).$$

Then $\mathfrak{g}_\gamma$ equals $\mathfrak{g}'_{\gamma'}$. If $\mathfrak{h}_\delta$ denotes the Lie algebra of $\mathrm{GL}(n)_\delta$, each of these will equal the diagonal subalgbra of

$$\mathfrak{g}_{\gamma'} = \underbrace{\mathfrak{h}_\delta \oplus \cdots \oplus \mathfrak{h}_\delta}_{\ell}.$$

In particular, we can write

$$D^G(\gamma) = \det(1 - \mathrm{Ad}(\gamma))_{\mathfrak{g}/\mathfrak{g}_{\gamma'}} \cdot \det(1 - \mathrm{Ad}(\theta^*))_{\mathfrak{g}_{\gamma'}/\mathfrak{g}_\gamma}.$$

It is a simple exercise in linear algebra, which we leave to the reader, to show that

$$\det(1 - \mathrm{Ad}(\gamma))_{\mathfrak{g}/\mathfrak{g}_{\gamma'}} = \det(1 - \mathrm{Ad}(\gamma'))_{\mathfrak{g}'/\mathfrak{g}_{\gamma'}} = D^{G'}(\gamma')$$

and

$$\det(1 - \mathrm{Ad}(\theta^*))_{\mathfrak{g}_{\gamma'}/\mathfrak{g}_\gamma} = \ell^{\dim G_\gamma}.$$

The lemma follows. ▮

Suppose that $\gamma = \prod_{v \in S} \gamma_v$ is an arbitrary point in $G(F_S)$, with Jordan decomposition

$$\gamma = \sigma u = \prod_{v \in S} \sigma_v u_v.$$

Since G_{σ_v} is the multiplicative group of a product of central simple algebras, the unipotent element $u_v \in G_{\sigma_v}$ is contained in the Richardson orbit of a parabolic subgroup

$$P_{\sigma_v} = M_{\sigma_v} N_{\sigma_v}$$

of G_{σ_v}. We shall write A_{σ_v} for the F_v-split component of M_{σ_v}. It is a simple consequence of the definition of P_{σ_v} that

$$\dim(G_{\gamma_v}) = \dim(M_{\sigma_v}).$$

Define

$$\Lambda^G(\gamma) = \prod_{v \in S} \Lambda^G(\gamma_v) = \prod_{v \in S} \left(|\ell|_v^{\frac{1}{2}\dim G_{\gamma_v}} e^G(\gamma_v) \right),$$

where, as in Chapter 1,

$$e^G(\gamma_v) = e(G_{\sigma_v}) = e(M_{\sigma_v})$$

is the sign introduced by Kottwitz [29(b)]. In the special case that γ is semisimple, we set

$$I_G(\gamma, f) = |D^G(\gamma)|^{\frac{1}{2}} \int_{G_\gamma(F_S)\backslash G^0(F_S)} f(x^{-1}\gamma x)dx, \quad f \in \mathcal{H}(G(F_S)),$$

where $|D(\gamma)| = \prod_{v \in S} |D(\gamma_v)|_v$ and $G_\gamma(F_S) = \prod_{v \in S} G_{\gamma_v}(F_v)$. Once defined for semisimple γ, the distribution is determined in the general case by a limit

$$I_G(\gamma, f) = \lim_{a \to 1} I_G(\sigma a, f), \qquad a \in \prod_{v \in S} A_{\sigma_v}(F_v). \tag{1.3}$$

Observe that if a is a small regular point in $\prod_v A_{\sigma_v}(F_v)$, then

$$G_{\sigma a}(F_S) = \prod_{v \in S} M_{\sigma_v}(F_v).$$

The distribution $I_G(\gamma)$ depends implicitly on a choice of Haar measure on this group as well as one on $G^0(F_S)$. We use the inner twist η_σ to transfer the former to a Haar measure on

$$G'_{(\sigma a)'}(F_S) = \prod_{v\in S} M_{\sigma'_v}(F_v).$$

Combined with a fixed Haar measure on $G'(F_S)$, it allows us to define the distribution $I_{G'}(\gamma')$ on $G'(F_S)$.

We shall write

$$f \to f_{G'} = f'$$

for the map from $\mathcal{H}(G(F_S))$ to $\mathcal{I}(G'(F_S))$ constructed in [15] and Chapter 1 by transferring orbital integrals. We claim that

$$I_G(\gamma, f) = \Lambda^G(\gamma)\hat{I}_{G'}(\gamma', f'), \qquad \gamma \in G(F_S). \tag{1.4}$$

If γ is semisimple, this follows from Lemma 1.1 and Lemma I.3.6. If γ is arbitrary, the formula follows from the semisimple case, the formula (1.3) and the definition of $\Lambda^G(\gamma)$. (See Corollary I.3.13.) Consider the special case that γ is G-regular. Then the sign $e^G(\gamma)$ equals 1, and (1.4) becomes

$$I_G(\gamma, f) = \left(\prod_{v\in S} |\ell|_v^{n/2}\right)\hat{I}_{G'}(\gamma', f').$$

We combine this formula with the character identity (1.2) and the Weyl integration formula. In stating the Weyl integation formula in I.4, we used a Haar measure on the torus

$$G'_{\gamma'_v}(F_v), \qquad v \in S,$$

distinct from that obtained from $G_{\gamma_v}(F_v)$ under η_{γ_v}. The discrepancy between the two measures is just the factor $|\ell|_v^{n/2}$. It follows that

$$\operatorname{tr}\pi(f) = e_S f'(\pi'), \qquad \pi \in \Pi^+_{\text{temp}}(G(F_S)), \pi' \in \Pi'(\pi), \tag{1.5}$$

for any $f \in \mathcal{H}(G(F_S))$.

Bear in mind that (1.4) and (1.5) both come with supplementary vanishing properties. If ζ is an element in $G'(F_S)$ which is not of the form γ', $\gamma \in G(F_S)$, then $\hat{I}_{G'}(\zeta, f') = 0$. If τ is a representation in $\Pi_{\text{temp}}(G'(F_S))$ which does not belong to one of the image sets $\Pi'(\pi)$, then $f'(\pi') = 0$. In particular, the map $f \to f'$ is supported on characters. It follows that if

$$\theta' : \mathcal{H}(G'(F_S)) \to \mathcal{V}'$$

is any map which is supported on characters, then the map $f \to \hat{\theta}'(f')$ from $\mathcal{H}(G(F_S))$ to $\mathcal{V}'$ is also supported on characters. We will apply this later, without further comment, to the invariant distributions in the trace formula for G'.

There is a related point on measures that we should address. Fix a Euclidean norm $\|\cdot\|$ on the space $\mathfrak{a}_{M_0}$ which is invariant under $W_0 = W_0^G$, the Weyl group of (G^0, A_{M_0}). We then take the associated Euclidean measures on each of the spaces $\mathfrak{a}_M$, $M \in \mathcal{L}$. The measure on $\mathfrak{a}_M$ together with a given invariant measure on $M(\mathbf{A})$ then provides an invariant measure on the space $M(\mathbf{A})^1$. (Recall that $M(\mathbf{A})^1$ is the kernel of the usual map

$$H_M : M(\mathbf{A}) \to \mathfrak{a}_M.)$$

Observe that the map which sends any $\chi \in X(M)_F$ to the rational character

$$m' \to \chi(\eta^{-1}(m')), \qquad m' \in M',$$

gives an injection of $X(M)_F$ into $X(M')_F$. The dual map

$$\mathfrak{a}_{M'} = \mathrm{Hom}(X(M')_F, \mathbf{R}) \to \mathrm{Hom}(X(M)_F, \mathbf{R}) = \mathfrak{a}_M$$

is an isomorphism, and we use it to identify the real vector spaces $\mathfrak{a}_M$ and $\mathfrak{a}_{M'}$. With this identification, we transfer the Euclidean measure from $\mathfrak{a}_M$ to one on $\mathfrak{a}_{M'}$. Since $M'(\mathbf{A})^1$ is the kernel of the map

$$H_{M'} : M'(\mathbf{A}) \to \mathfrak{a}_{M'},$$

we can then associate a Haar measure on $M'(\mathbf{A})^1$ to one on $M'(\mathbf{A})$.

For each $M \in \mathcal{L}$, we have just identified the spaces $\mathfrak{a}_{M'}$ and $\mathfrak{a}_M$ in a certain way. The norm provides a second natural isomorphism between the two spaces. Consider the map from $\mathfrak{a}_M$ to $\mathfrak{a}_{M'} \xrightarrow{\sim} \mathfrak{a}_M$ defined by

$$H \to H' = \ell H, \qquad H \in \mathfrak{a}_M.$$

Then one can check that

$$H_M(m)' = H_{M'}(m'), \qquad m \in M(F_S). \tag{1.6}$$

Let

$$\lambda \to \lambda' = \ell^{-1}\lambda, \qquad \lambda \in \mathfrak{a}^*_{M,\mathbf{C}},$$

be the adjoint map. If $\pi \in \Pi^+(M(F_S))$ and $\lambda \in \mathfrak{a}^*_{M,\mathbf{C}}$, the representation

$$\pi_\lambda(m) = \pi(m)e^{\lambda(H_M(m))}, \qquad m \in M^+(F_S),$$

also belongs to $\Pi^+(M(F_S))$. Suppose that π' is a representation in $\Pi'(\pi)$. It is then a direct consequence of (1.6) that the representation $\pi'_{\lambda'}$ belongs to $\Pi'(\pi_\lambda)$.

We shall say that a finite set S of valuations of F has the closure property if for each $M \in \mathcal{L}$,

$$\mathfrak{a}_{M,S} = \{H_M(m) : m \in M^+(F_S)\}$$

is a closed subgroup of $\mathfrak{a}_M$. If S contains any Archimedean place, it automatically has the closure property. If not, S has the closure property if and only if it contains only valuations which divide a fixed rational prime. Assume that S does have the closure property. We define

$$i\mathfrak{a}^*_{M,S} = i\mathfrak{a}^*_M / i\,\mathrm{Hom}(\mathfrak{a}_{M,S}, \mathbf{Z}).$$

This abelian group has a natural measure $d\lambda$, which is obtained from the Euclidean measure on $i\mathfrak{a}^*_M$ dual to our measure on $\mathfrak{a}_M$. It is convenient to identify any $\phi \in \mathcal{I}(M(F_S))$ with the function

$$\phi(\pi, X) = \int_{i\mathfrak{a}^*_{M,S}} \phi(\pi_\lambda) e^{-\lambda(X)} d\lambda, \quad \pi \in \Pi^+_{\text{temp}}(M(F_S)), X \in \mathfrak{a}_{M,S}.$$

In a similar way, we identify $\mathcal{I}(M'(F_S))$ with a space of functions on $\Pi^+_{\text{temp}}(M'(F_S)) \times \mathfrak{a}_{M,S}$. If ϕ belongs to $\mathcal{I}(M(F_S))$, define

$$\begin{aligned} &\phi'(\pi', X') = \ell^{-(\dim A_M)} e_S \phi(\pi, X), \\ &\pi \in \Pi^+_{\text{temp}}(M(F_S)), X \in \mathfrak{a}_{M,S}, \pi' \in \Pi'(\pi). \end{aligned} \tag{1.7}$$

By defining ϕ' to be zero on the remaining points in $\Pi_{\text{temp}}(M'(F_S)) \times \mathfrak{a}_{M,S}$, we obtain a function in $\mathcal{I}(M'(F_S))$. This is compatible with our earlier definition. For suppose that $\phi = h_M$ for some function $h \in \mathcal{H}(M(F_S))$. Then

$$\begin{aligned} \phi'(\pi', X') &= \ell^{-(\dim A_M)} e_S \phi(\pi, X) \\ &= \ell^{-(\dim A_M)} e_S \int_{i\mathfrak{a}^*_{M,S}} \mathrm{tr}(\pi_\lambda(h)) e^{-\lambda(X)} d\lambda \\ &= \ell^{-(\dim A_M)} e_S^2 \int_{i\mathfrak{a}^*_{M,S}} h'((\pi_\lambda)') e^{-\lambda(X)} d\lambda \\ &= \ell^{-(\dim A_M)} \int_{i\mathfrak{a}^*_{M,S}} h'(\pi'_{\lambda'}) e^{-\lambda'(X')} d\lambda. \end{aligned}$$

Since $\ell^{-(\dim A_M)} d\lambda$ equals $d\lambda'$, we obtain

$$\phi'(\pi', X') = \int_{i\mathfrak{a}^*_{M,S}} h'(\pi'_{\lambda'}) e^{-\lambda'(X')} d\lambda'.$$

In other words, ϕ' is the function on $\Pi_{\text{temp}}(M'(F_S)) \times \mathfrak{a}_{M,S}$ which is identified with h'.

It is clear that we can define the spaces $\mathcal{H}(G(\mathbf{A}))$, $\mathcal{I}(G(\mathbf{A}))$ and the sets $\Pi^+(G(\mathbf{A}))$, $\Pi^+(G(\mathbf{A})^1)$, etc., as above. If π belongs to $\Pi^+_{\text{temp}}(G(\mathbf{A}))$, the obvious analogue of (1.5) holds. However, if π belongs to $\Pi^+_{\text{temp}}(G(\mathbf{A})^1)$, the right-hand side of (1.5) must be multiplied by ℓ. This is because of our choice of Haar measure on $G'(\mathbf{A})^1$. On the other hand, our Haar measures are compatible with various earlier formulas of descent. (See for example Remark 1 following Theorem 8.2 of [1(d)].) As an exercise in such things, the reader could try comparing the Poisson summation formula for the F-idèles of norm 1 with its twisted analogue for E.

2. Normalization factors and the trace formula

Our tool for the global comparison is the full trace formula, for which the main references are [1(g)] and [1(h)]. The trace formula is in invariant form, and its terms depend on a normalization of the intertwining operators between induced representations. Since we are going to compare the trace formulas for G and G', we shall want to choose the normalization for these two groups in a compatible way.

Let S be a finite set of valuations with the closure property. Fix an element $M \in \mathcal{L}$ and a representation $\pi = \bigotimes_{v\in S} \pi_v$ in $\Pi^+(M(F_S))$. Associated to parabolic subsets P and Q in $\mathcal{P}(M)$ there are intertwining operators

$$J_{Q|P}(\pi_\lambda), \qquad \lambda \in \mathfrak{a}^*_{M,\mathbf{C}},$$

between the induced representations $\mathcal{I}_P(\pi_\lambda)$ and $\mathcal{I}_Q(\pi_\lambda)$. Each of these is defined by an integral over $N_Q(F_S) \cap N_{\overline{P}}(F_S)$, and so depends upon a choice of Haar measure on this group. In order to put the trace formula into invariant form, it is necessary to define meromorphic scalar valued functions

$$r_{Q|P}(\pi_\lambda) = \prod_{\alpha \in \Sigma_Q \cap \Sigma_{\overline{P}}} r_\alpha(\pi, \lambda(\alpha^\vee)), \qquad \lambda \in \mathfrak{a}^*_{M,\mathbf{C}}, \tag{2.1}$$

so that the normalized operators

$$R_{Q|P}(\pi_\lambda) = r_{Q|P}(\pi_\lambda)^{-1} J_{Q|P}(\pi_\lambda)$$

satisfy the conditions of Theorem 2.1 of [1(f)]. Here Σ_P denotes the set of roots α of (P, A_M), and for each α,

$$r_\alpha(\pi, s) = \prod_{v \in S} r_\alpha(\pi_v, s), \qquad s \in \mathbf{C},$$

is a meromorphic function of one complex variable. In [1(f)] we saw that such normalizing factors could be chosen for any group. However, to show that this can be done in a compatible way for G and G', we must use the more precise results of Shahidi.

The intertwining operators and the normalizing factors are given by products over $v \in S$, so we can work with a given valuation. For the moment, then, we shall suppose that S consists of a single valuation v. Let ψ_v be a fixed nontrivial additive character on F_v. We shall consider first the special case that $G = G' = \mathrm{GL}(n)$. To define the normalizing factors, it is enough to define the function $r_\alpha(\pi, s)$ for any root α of (G, A_M) and any

$\pi \in \Pi(M(F_v))$. There is an isomorphism

$$M \xrightarrow{\sim} \prod_{i=1}^{r} \mathrm{GL}(n_i),$$

where $(n_1, \ldots, n_r)$ is a partition of n. The root α is associated to an ordered pair (p, q) of distinct integers between 1 and r, and π corresponds to a representation

$$\pi_1 \times \cdots \times \pi_r, \qquad \pi_i \in \Pi(\mathrm{GL}(n_i, F_v)).$$

If π is tempered, define

$$(2.2) \qquad r_\alpha(\pi, s) = L(s, \pi_p \otimes \tilde{\pi}_q)\varepsilon(s, \pi_p \otimes \tilde{\pi}_q, \psi_v)^{-1} L(s+1, \pi_p \otimes \tilde{\pi}_q)^{-1}.$$

Then $r_\alpha(\pi, s)$ is a meromorphic function of s with the the property that

$$r_\alpha(\pi_\lambda, s) = r_\alpha(\pi, \lambda(\alpha^\vee) + s), \qquad \lambda \in i\mathfrak{a}_M^*.$$

By meromorphic continuation in λ the definition can then be extended to standard representations. Finally, if π is an arbitrary representation in $\Pi^+(M(F_v))$, let ρ be the standard representation of which π is the Langlands quotient, and set

$$r_\alpha(\pi, s) = r_\alpha(\rho, s).$$

It follows from the results of Shahidi [36(d)] that there are Haar measures on the groups $N_Q(F_v) \cap N_{\overline{P}}(F_v)$, depending on ψ_v, such that the normalizing factors (2.1) have all the right properties. (See also §4 of [1(f)].)

We return to the general case, with G as in §1. For $G' = \mathrm{GL}(n)$, we fix the normalizing factors as above. We shall show that the normalizing factors for G can be defined in terms of those for G'. Suppose that α is a root of (G, A_M). Then α' is a root of $(G', A_{M'})$. There is an isomorphism

$$M' \xrightarrow{\sim} \prod_{i=1}^{r} \mathrm{GL}(n_i)$$

and as above, α' corresponds to a pair (p, q). Set

$$\lambda_{\alpha,v} = \lambda(E_v/F_v, \psi_v)^{n_p n_q}.$$

We shall also write

$$\zeta_\alpha(m') = \eta_{E/F}(\det m_p)\eta_{E/F}(\det m_q)^{-1},$$

if

$$m' \to (m_1, \ldots, m_r), \qquad m_i \in \mathrm{GL}(n_i, F_v),$$

is any point in $M'(F_v)$ and $\eta_{E/F}$ is a primitive Grössencharacter associated to E/F by class field theory. Then ζ_α is a character on $M'(F_v)$ which depends on our choice of $\eta_{E/F}$. Given the representation $\pi \in \Pi^+(M(F_v))$, we set

$$r_\alpha(\pi, s) = \lambda_{\alpha,v} \prod_{j=1}^{\ell} r_{\alpha'}(\zeta_\alpha^j \pi', s), \qquad s \in \mathbf{C}, \tag{2.3}$$

where π' is any representation in $\Pi'(\pi)$ and

$$(\zeta_\alpha^j \pi')(m') = \zeta_\alpha(m')^j \pi'(m'), \qquad m' \in M'(F_v).$$

Then $r_\alpha(\pi, s)$ is a meromorphic function of s which is independent of our choice of π' and $\eta_{E/F}$.

LEMMA 2.1: *The normalizing factors $r_{Q|P}(\pi_\lambda)$ defined by (2.1) and (2.3) satisfy all the properties of Theorem 2.1 of [1(f)].*

Proof. Many of the required properties follow from standard properties of the operators $J_{Q|P}(\pi)$ and the general form (2.1) of the normalizing factors. There is, in fact, only one condition to verify. We must show that

$$r_{P|\overline{P}}(\pi_\lambda) r_{\overline{P}|P}(\pi_\lambda) = \mu_M(\pi_\lambda)^{-1}, \qquad P \in \mathcal{P}(M), \lambda \in \mathfrak{a}_{M,\mathbf{C}}^*, \tag{2.4}$$

for any $\pi \in \Pi_{\text{temp}}^+(M(F_v))$. Here $\mu_M(\pi_\lambda)$ denotes Harish–Chandra's μ-function. As explained in [1(f)], all of the required properties of the normalized operators will follow from (2.4).

There are two cases to consider. Suppose first that $E = F$. Then G is obtained from a central simple algebra. We may as well assume that $\ell = 1$ and $G = G^0$. The definition (2.3) then simplifies to

$$r_\alpha(\pi, s) = r_{\alpha'}(\pi', s),$$

where π' is the unique representation in $\Pi'(\pi)$. Therefore

$$r_{Q|P}(\pi_\lambda) = r_{Q'|P'}(\pi'_{\lambda'})$$

by (2.1). Since (2.4) is true for G', we have only to show that

$$\mu_M(\pi) = \mu_{M'}(\pi') \tag{2.5}$$

for any $\pi \in \Pi_{\text{temp}}(M(F_v))$. Let $\Pi_{\text{disc}}(M(F_v))$ be the subset of representations in $\Pi_{\text{temp}}(M(F_v))$ which are square integrable modulo $A_M(F_v)$. Any $\pi \in \Pi_{\text{temp}}(M(F_v))$ is obtained by induction from a representation

$\pi_1 \in \Pi_{\text{disc}}(M_1(F_v))$, where M_1 is a Levi component of parabolic subgroup of M over F_v. By definition

$$\mu_M(\pi) = \mu_{M_1}(\pi_1).$$

Therefore, if we are willing to assume that M is defined only over F_v, we need only prove (2.5) for $\pi \in \Pi_{\text{disc}}(M(F_v))$. We shall do so by comparing the Plancherel formulas for G and G'.

Let f be a function in $\mathcal{H}(G(F_v))$ whose character vanishes on any irreducible tempered representation which is not equivalent to some

$$\mathcal{I}_P(\pi), \qquad P \in \mathcal{P}(M), \pi \in \Pi_{\text{disc}}(M(F_v)).$$

The Plancherel formula for G ([20(e)], [20(f)]) provides a constant γ_M such that

$$f(1) = \gamma_M \int_{\Pi_{\text{disc}}(M(F_v))} d^M(\pi)\mu_M(\pi)f_M(\pi)d\pi.$$

Here $d^M(\pi)$ is the formal degree of π, and is not to be confused with $d\pi$. The latter stands for the measure on $\Pi_{\text{disc}}(M(F_v))$ which is obtained from our Haar measure on $i\mathfrak{a}^*_{M,v}$ and the free action

$$\pi \longrightarrow \pi_\lambda, \qquad \lambda \in i\mathfrak{a}^*_{M,v}.$$

By (1.4),

$$f(1) = I_G(1, f) = e_v I_{G'}(1, f').$$

Moreover, f' vanishes for any tempered representation of $G'(F_v)$ which is not equivalent to one of the form

$$\Pi' = \mathcal{I}_{P'}(\pi'), \qquad P \in \mathcal{P}(M), \pi \in \Pi_{\text{disc}}(M(F_v)).$$

But by (1.5),

$$f'(\Pi') = f'_{M'}(\pi') = e_v^{-1} f_M(\pi).$$

Combined with the Plancherel formula for G', these observations tell us that

$$f(1) = \gamma_{M'} \int_{\Pi_{\text{disc}}(M(F_v))} d^{M'}(\pi')\mu_{M'}(\pi')f_M(\pi)d\pi.$$

We choose Haar measures on the groups $N_Q(F_v) \cap N_{\overline{P}}(F_v)$ to match those on $N_{Q'}(F_v) \cap N_{\overline{P}'}(F_v)$ under the isomorphism η. The constants γ_M and $\gamma_{M'}$ are defined in terms of certain integrals on these groups, and are therefore equal. By varying f, and taking note of the trace Paley–Wiener theorems ([6], [12(a)]) we see that

$$d^{M'}(\pi')\mu_{M'}(\pi') = d^M(\pi)\mu_M(\pi), \qquad \pi \in \Pi_{\text{disc}}(M(F_v)).$$

In the special case that M equals G, the functions μ_G and $\mu_{G'}$ both equal 1. It follows that $d^{G'}(\pi')$ equals $d^G(\pi)$. The same formula of course holds for arbitrary M. We therefore obtain the formula (2.5) for any representation π in $\Pi_{\text{disc}}(M(F_v))$. This establishes the lemma when $G = G^0$.

The second case is that $E \neq F$. Then we are in the base change setting. The intertwining operators for G depend only on the connected component G^0. But

$$G^0(F_v) \cong \text{GL}(n, E_v)^{\ell/\ell_v},$$

where

$$\ell_v = \deg(E_v/F_v).$$

We could therefore define the functions

$$r_\alpha(\pi, s), \qquad \pi \in \Pi^+(M(F_v)),$$

in terms of the formula (2.2), with ψ_v replaced by

$$\psi_{E_v} = \psi_v \circ \text{tr}_{E_v/F_v} .$$

It is an immediate consequence of Proposition I.6.9 that the same functions also satisfy (2.3). Since we are dealing with the general linear group, we know that the resulting normalizing factors have all the right properties. This completes the proof of the lemma. ∎

Remarks 1. The proof of the lemma in case $E \neq F$ is somewhat unsatisfactory. It would be preferable to have a proof based on local harmonic analysis.

2. If the valuation v is Archimedean, Lemma 2.1 is essentially a special case of the general results of [1(f),§3].

We have thus defined the local normalizing factors for G. They depend on the additive characters ψ_v. We assume that each ψ_v is the local component of a fixed additive character ψ of $\mathbf{A}/F$. This allows us to build global normalizing factors from infinite products of local ones. We shall return to the study of normalizing factors, both local and global, in §11.

Having chosen normalizations for the intertwining operations, we can then write down the full trace formula for G ([1(h), §3,4,7]). It may be regarded as an identity

$$\begin{aligned} I(f) &= \sum_{M \in \mathcal{L}} |W_0^M||W_0^G|^{-1} \sum_{\gamma \in (M(F))_{M,S}} a^M(S,\gamma) I_M(\gamma, f) \\ &= \sum_t \sum_{M \in \mathcal{L}} |W_0^M||W_0^G|^{-1} \int_{\Pi(M,t)} a^M(\pi) I_M(\pi, f) d\pi, \end{aligned}$$

in which a certain linear functional I on $\mathcal{H}(G(\mathbf{A}))$ is expressed in two different ways. Both sides break up into constituents which are of either a local or a global nature. We shall discuss these separately in Paragraphs 3, 5, 8 and 9.

3. The distributions $I_M(\gamma)$ and $I_M^{\mathcal{E}}(\gamma)$

For the next few sections we shall study the geometric side of the trace formula. It is a sum of terms which are indexed by orbits, and which can be separated naturally into local and global constituents. We shall look first at the local constituents.

As will always be the case in what follows, M denotes an element in $\mathcal{L}$ and S is a finite set of valuations of F with the closure property. The local terms on the geometric side of the trace formula of G are invariant distributions

$$I_M(\gamma, f), \qquad \gamma \in M(F_S), f \in \mathcal{H}(G(F_S)),$$

which depend only on the $M^0(F_S)$-orbit of γ. If $M_\gamma = G_\gamma$, they can be defined fairly directly in terms of weighted orbital integrals. In general, however, they must be defined by a formula

$$(3.1) \qquad I_M(\gamma, f) = \lim_{a \to 1} \sum_{L \in \mathcal{L}(M)} r_M^L(\gamma, a) I_L(a\gamma, f),$$

in which a takes small regular values in $A_M(F_S)$. (This is formula (2.2) of [1(g)]. The functions $r_M^L(\gamma, a)$ are obtained from a certain (G, M) family, which is defined in §5 of [1(e)].) We shall recall some properties of these distributions.

The first property relates the distributions to orbital integrals on M. Suppose that σ is a semisimple element in $M(F_S)$. Consider two functions ϕ_1 and ϕ_2 which are defined on an open subset Σ of $\sigma M_\sigma(F_S)$ whose closure contains an $M_\sigma(F_S)$-invariant neighborhood of σ. We write

$$\phi_1(\gamma) \overset{(M,\sigma)}{\sim} \phi_2(\gamma), \qquad \gamma \in \Sigma,$$

if the difference is an orbital integral on $M(F_S)$ for γ near σ. That is, if there is a function $h \in C_c^\infty(M(F_S))$ and a neighborhood U of σ in $M(F_S)$ such that

$$\phi_1(\gamma) - \phi_2(\gamma) = I_M^M(\gamma, h), \qquad \gamma \in \Sigma \cap U.$$

Now, suppose that G_σ equals M_σ. Then according to ([1(g)], (2.3)), we have

$$(3.2) \qquad I_M(\gamma, f) \overset{(M,\sigma)}{\sim} 0, \qquad \gamma \in M_\sigma(F_S),$$

for any $f \in \mathcal{H}(G(F_S))$. The next property is one of descent. Suppose that M_1 is a Levi subset in $\mathcal{L}$, with $M_1 \subset M$, and that γ is an element in $M_1(F_S)$

such that $M_{1,\gamma} = M_\gamma$. Then

$$I_M(\gamma, f) = \sum_{L\in\mathcal{L}(M_1)} d^G_{M_1}(M, L)\hat{I}^L_{M_1}(\gamma, f_L), \tag{3.3}$$

where $d^G_{M_1}(M, L)$ is a constant which vanishes unless the map

$$\mathfrak{a}^M_{M_1} \oplus \mathfrak{a}^L_{M_1} \to \mathfrak{a}^G_{M_1}$$

is an isomorphism ([1(g)], Corollary 8.3). There is also a splitting property. Suppose that S is a disjoint union of S_1 and S_2, and that $f = f_1 f_2$ and $\gamma = \gamma_1\gamma_2$ are corresponding decompositions. Then by ([1(g)], Proposition 9.1),

$$I_M(\gamma, f) = \sum_{L_1,L_2\in\mathcal{L}(M)} d^G_M(L_1, L_2)\hat{I}^{L_1}_M(\gamma_1, f_{1,L_1})\hat{I}^{L_2}_M(\gamma_2, f_{2,L_2}). \tag{3.4}$$

We shall also make use of properties that apply to particular fields. To state them assume that $S = \{v\}$, so that $F_S = F_v$ is a local field. First take the case that F_v is Archimedean. Then for every element $z \in \mathcal{Z}(G(F_v))$, the center of the universal enveloping algebra of the complexified Lie algebra of $G(F_v)$, we have

$$I_M(\gamma, zf) = \sum_{L\in\mathcal{L}(M)} \partial^L_M(\gamma, z_L)I_L(\gamma, f), \quad \gamma \in M(F_v) \cap G_{\mathrm{reg}}. \tag{3.5}$$

Here z_L is the image of z under the natural map from $\mathcal{Z}(G(F_v))$ to $\mathcal{Z}(L(F_v))$, and $\partial^L_M(\gamma, z_L)$ is a linear differential operator on

$$M(F_v) \cap L_{\mathrm{reg}}$$

which is invariant under conjugation by $M^0(F_v)$. (See formula (2.6) of [1(g)].) Next, suppose that F_v is non-Archimedean. Then for every semisimple element $\sigma \in M(F_v)$ there is a germ expansion

$$I_M(\gamma, f) \overset{(M,\sigma)}{\sim} \sum_{L\in\mathcal{L}(M)} \sum_{\delta\in\sigma(\mathcal{U}_{L_\sigma}(F_v))} g^L_M(\gamma, \delta)I_L(\delta, f), \tag{3.6}$$

for $\gamma \in \sigma M_\sigma(F_v) \cap G_{\mathrm{reg}}$. Here $(\mathcal{U}_{L_\sigma}(F_v))$ is the set of conjugacy classes of unipotent elements in $L_\sigma(F_v)$, and $g^L_M(\gamma, \delta)$ is a certain (M, σ)-equivalence class of functions defined on the L-regular elements $\gamma \in \sigma M_\sigma(F_v)$. (See formula (2.5) of [1(g)].)

Now, consider the group G'. There are of course similar distributions on $\mathcal{H}(G'(F_S))$. They possess a key vanishing property that we should recall. We have the injection $M \to M'$ of $\mathcal{L}$ into $\mathcal{L}'$. For each $v \in S$, let

$M'(F_v)^M$ denote the subgroup of elements $m \in M'(F_v)$ such that for every $\chi \in X(M)_{F_v}$, $\chi(m)$ belongs to $\mathrm{Norm}_{E_v/F_v}(E_v^*)$. Then

$$M'(F_S)^M = \prod_{v \in S} M'(F_v)^M$$

is a subgroup of finite index in $M'(F_S)$. Suppose that f is a function in $\mathcal{H}(G(F_S))$. The vanishing property is then

$$(3.7) \qquad I_{M'}(\delta, f') = 0, \qquad \delta \in M'(F_S)^M \backslash \{\gamma' : \gamma \in M(F_S)\}.$$

That is, $I_{M'}(\delta, f')$ vanishes for any element $\delta \in M'(F_S)^M$ which does not come from $M(F_S)$. (See [1(g)], Proposition 10.3.)

Our overall strategy will be to pull objects on G' back to G, where they can be compared with the corresponding objects on G. We shall systematically denote objects on G which have been obtained from G' by a superscript $\mathcal{E}$. ($\mathcal{E}$ stands for "endoscopic.") In particular, if γ is an element in $M(F_S)$ such that $M_\gamma = G_\gamma$, we define

$$(3.8) \qquad I_M^{\mathcal{E}}(\gamma, f) = \Lambda^M(\gamma) \hat{I}_{M'}(\gamma', f'), \qquad f \in \mathcal{H}(G(F_S)).$$

Any γ is of course of this form in the special case that $M = G$, and formula (1.4) becomes

$$(3.9) \qquad I_G^{\mathcal{E}}(\gamma, f) = I_G(\gamma, f).$$

More generally, suppose that γ is an arbitary element in $M(F_S)$. For any small regular point $a \in A_M(F_S)$ we have $M_{a\gamma} = G_{a\gamma}$. Consequently, for any $L \in \mathcal{L}(M)$, the distribution $I_L^{\mathcal{E}}(a\gamma, f)$ is defined.

LEMMA 3.1: *The expression*

$$\sum_{L \in \mathcal{L}(M)} r_M^L(\gamma, a) I_L^{\mathcal{E}}(a\gamma, f)$$

extends to a continuous function of $a \in A_M(F_S)$ is a neighborhood of the identity.

Proof. Let $\gamma = \sigma u$ be the Jordan decomposition of γ. Then by (1.1) we have

$$(a\gamma)' = \sigma' u_{\sigma'}^{\ell} a_{\sigma'}^{\ell} = \gamma' a_{\sigma'}^{\ell}.$$

(Remember, that $\mu \to \mu_{\sigma'}$ denotes the map from the conjugacy classes of M_σ to the conjugacy classes of its quasi-split form $M'_{\sigma'}$.) By the definition (3.8) we have

$$I_L^{\mathcal{E}}(a\gamma, f) = \Lambda^L(a\gamma) \hat{I}_{L'}((a\gamma)', f') = \Lambda^M(\gamma) \hat{I}_{L'}(\gamma' a_{\sigma'}^{\ell}, f').$$

We shall relate $r_M^L(\gamma, a)$ with the functions $r_{L_1'}^{L'}(\gamma', a_{\sigma'}^\ell)$.

Suppose that $\sigma = \prod_{v\in S} \sigma_v$, $u = \prod_{v\in S} u_v$, and $a = \prod_{v\in S} a_v$. Then $M_\sigma = \prod_{v\in S} M_{\sigma_v}$. The function $r_M^L(\gamma, a)$ is obtained from the (G, M) family

$$r_P(\nu, \gamma, a) = \prod_{v\in S}\prod_\beta r_\beta\left(\frac{1}{2}\nu, u_v, a_v\right), \qquad P \in \mathcal{P}(M), \nu \in i\mathfrak{a}_M^*,$$

where the inside product is taken over the roots of $(P_{\sigma_v}, A_{M_{\sigma_v}})$ and for each such β,

$$r_\beta(\nu, u_v, a_v) = |a_v^\beta - a_v^{-\beta}|_v^{\rho(\beta,u_v)\nu(\beta^\vee)}.$$

(See (3.4) and (5.1) of [1(e)]. For the basic properties of (G, M) families, we refer the reader to §6 of [1(b)].) The real number $\rho(\beta, u_v)$ was defined in §3 of [1(e)]. It depends only on the geometric conjugacy class of u_v in M_{σ_v}. In particular,

$$\rho(\beta, u_v) = \rho(\beta', (u_v)_{\sigma_v'}) = \rho(\beta', (u_v)_{\sigma_v'}^\ell),$$

where β' is the root of G' associated to β. It follows without difficulty that

$$r_{L_1'}^{L'}(\gamma', a_{\sigma'}^\ell) = r_{L_1}^L(\gamma, a^\ell), \qquad L \supset L_1 \supset M.$$

On the other hand, it is not hard to relate $r_M^L(\gamma, a)$ to the function $r_{L_1}^L(\gamma, a^\ell)$. For

$$r_P(\nu, \gamma, a) = c_P(\nu, \gamma, \ell, a) r_P(\nu, \gamma, a^\ell),$$

where

$$c_P(\nu, \gamma, \ell, a) = \prod_v\prod_\beta |(a_v^{\ell-1})^\beta + (a_v^{\ell-3})^\beta + \cdots + (a_v^{-(\ell-1)})^\beta|_v^{-\frac{1}{2}\rho(\beta,u_v)\nu(\beta^\vee)}$$

is another (G, M)-family. By Lemma 6.5 of [1(b)],

$$r_M^L(\gamma, a) = \sum_{L_1\in\mathcal{L}^L(M)} c_M^{L_1}(\gamma, \ell, a) r_{L_1}^L(\gamma, a^\ell).$$

Observe that $c_P(\nu, \gamma, \ell, a)$ is continuous at $a = 1$, and that

$$c_P(\nu, \gamma, \ell, 1) = \prod_v\prod_\beta |\ell|_v^{-\frac{1}{2}\rho(\beta,u_v)\nu(\beta^\vee)}.$$

In particular, $c_M^{L_1}(\gamma, \ell, a)$ is continuous at $a = 1$.

Combining these observations, we have

$$\sum_{L\in\mathcal{L}(M)} r_M^L(\gamma,a)I_L^{\mathcal{E}}(a\gamma,f)$$

$$=\Lambda^M(\gamma)\sum_{L\in\mathcal{L}(M)}\sum_{L_1\in\mathcal{L}^L(M)} c_M^{L_1}(\gamma,\ell,a)r_{L_1}^L(\gamma,a^\ell)\hat{I}_{L'}(\gamma' a_{\sigma'}^\ell,f')$$

$$=\Lambda^M(\gamma)\sum_{L_1\in\mathcal{L}(M)} c_M^{L_1}(\gamma,\ell,a)\Big(\sum_{L\in\mathcal{L}(L_1)} r_{L_1'}^{L'}(\gamma',a_{\sigma'}^\ell)\hat{I}_{L'}(\gamma' a_{\sigma'}^\ell,f')\Big).$$

By formula (2.2*) of [1(g)], the function

$$\sum_{L\in\mathcal{L}(L_1)} r_{L_1'}^{L'}(\gamma',a')\hat{I}_{L'}(\gamma' a',f'),$$

defined for regular elements a' in $A_{M'}(F_S)$, extends to a continuous function around $a'=1$. Its value at $a'=1$ is just $\hat{I}_{L_1'}(\gamma_1',f')$, where $\gamma_1=\gamma^{L_1}$ is the induced orbit. (For the definition of γ^{L_1}, see §6 of [1(e)].) Therefore, the original expression extends to a continuous function around $a=1$. Observe that its value at $a=1$ is just

$$\Lambda^M(\gamma)\sum_{L_1\in\mathcal{L}(M)} c_M^{L_1}(\gamma,\ell,1)\hat{I}_{L_1'}(\gamma_1',f').$$

The lemma is proved. ∎

If γ is any element in $M(F_S)$, we define

$$(3.1)^{\mathcal{E}}\qquad I_M^{\mathcal{E}}(\gamma,f)=\lim_{a\to 1}\sum_{L\in\mathcal{L}(M)} r_M^L(\gamma,a)I_L^{\mathcal{E}}(a\gamma,f).$$

Then $I_M^{\mathcal{E}}(\gamma)$ is an invariant distribution on $\mathcal{H}(G(F_S))$. Much of our effort will go towards comparing $I_M^{\mathcal{E}}(\gamma)$ with $I_M(\gamma)$.

In the lemma we used the (G,M) family

$$c_P(\nu,\gamma,\ell)=c_P(\nu,\gamma,\ell,1)=\prod_{v\in S}\prod_{\beta}|\ell|_v^{-\frac{1}{2}\rho(\beta,u_v)\nu(\beta^\vee)},\qquad P\in\mathcal{P}(M).$$

In the proof of the lemma we established

Corollary 3.2: *For any $\gamma\in M(F_S)$, we have*

$$I_M^{\mathcal{E}}(\gamma,f)=\Lambda^M(\gamma)\sum_{L\in\mathcal{L}(M)} c_M^L(\gamma,\ell)\hat{I}_{L'}((\gamma^L)',f'),\qquad \gamma\in M(F_S).\ ∎$$

Consider the special case that $\ell=1$. Then $G=G^0$. Each function $c_P(\nu,\gamma,\ell)$ equals 1. It follows from the basic properties of (G,M) families that

$$c_M^L(\gamma,\ell) = \begin{cases} 1, & L = M \\ 0, & \text{otherwise.} \end{cases}$$

Since $\Lambda^M(\gamma) = e^M(\gamma)$ in this case, the last corollary reduces to

COROLLARY 3.3: *If $\ell = 1$ we have*

$$I_M^{\mathcal{E}}(\gamma, f) = e^M(\gamma)\hat{I}_{M'}(\gamma', f'), \qquad \gamma \in M(F_S). \quad \blacksquare$$

COROLLARY 3.4: *Suppose that S contains all the Archimedean and ramified places, and that $\gamma \in M(F_S)$. Then*

$$I_M^{\mathcal{E}}(\gamma, f) = \hat{I}_{M'}(\gamma', f').$$

Proof. By the nature of S and γ, $e^M(\gamma) = 1$. Moreover, the rationality of $\gamma = \sigma u$ implies that $M_{\gamma_v} = M_\gamma$ for each v in S. Since

$$|\ell|_S = \prod_{v \in S} |\ell|_v = 1,$$

we have

$$\Lambda^M(\gamma) = e^M(\gamma)\prod_v |\ell|_v^{\frac{1}{2}\dim M_{\gamma_v}} = |\ell|_S^{\frac{1}{2}\dim M_\gamma} = 1.$$

Observe also that the numbers

$$\rho(\beta, u_v) = \rho(\beta, u)$$

are independent of v. Therefore

$$c_P(\nu, \gamma, \ell) = \prod_\beta |\ell|_S^{-\frac{1}{2}\rho(\beta,u)\nu(\beta^\vee)} = 1, \qquad P \in \mathcal{P}(M).$$

Therefore,

$$c_M^L(\gamma,\ell) = \begin{cases} 1, & L = M \\ 0, & \text{otherwise.} \end{cases}$$

Corollary 3.4 then follows from Corollary3.2. $\blacksquare$

Before going on, we make note of a property of the numbers $c_M^L(\gamma,\ell)$. Suppose that v is a valuation of F and that σ is a semisimple element in $M(F_v)$. The centralizer M_σ is of course a reductive group defined over F_v. Letting F_v play the role of F, we define the real vector space $\mathfrak{a}_{M_\sigma}$ as in §1.

LEMMA 3.5: *Assume that $\mathfrak{a}_{M_\sigma} = \mathfrak{a}_M$, and let γ be any element in $M(F_v)$ with Jordan decomposition σu. Then $c_M^L(\gamma,\ell)$ equals $c_{M_\sigma}^{L_\sigma}(u,\ell)$ if $\mathfrak{a}_L = \mathfrak{a}_{L_\sigma}$ and is 0 otherwise.*

Proof. The proof is an exercise in (G, M) families. It is identical to the proof of Lemma 8.2 of [1(e)], so we shall not reproduce it. ∎

Our ultimate goal is to prove that $I_M^{\mathcal{E}}(\gamma, f)$ equals $I_M(\gamma, f)$. For a start, we shall list those properties that $I_M^{\mathcal{E}}(\gamma, f)$ evidently shares with $I_M(\gamma, f)$.

Suppose that σ is a semisimple element in $M(F_S)$ such that $G_\sigma = M_\sigma$. Then

$$(3.2)^{\mathcal{E}} \qquad I_M^{\mathcal{E}}(\gamma, f) \overset{(M,\sigma)}{\sim} 0, \qquad \gamma \in M(F_S).$$

This follows from the characterization (Proposition I.3.1) of orbital integrals on $M(F_S)$, the vanishing property (3.7), and the property (3.2) applied to G'. We also have the descent property

$$(3.3)^{\mathcal{E}} \qquad I_M^{\mathcal{E}}(\gamma, f) = \sum_{L \in \mathcal{L}(M_1)} d_{M_1}^G(M, L) \hat{I}_{M_1}^{L,\mathcal{E}}(\gamma, f_L),$$

for elements $\gamma \in M_1(F_S)$ with $M_{1,\gamma} = M_\gamma$, and the splitting property

$$(3.4)^{\mathcal{E}} \quad I_M^{\mathcal{E}}(\gamma, f) = \sum_{L_1, L_2 \in \mathcal{L}(M)} d_M^G(L_1, L_2) \hat{I}_M^{L_1,\mathcal{E}}(\gamma_1, f_{1,L_1}) \hat{I}_M^{L_2,\mathcal{E}}(\gamma_2, f_{2,L_2}),$$

for $\gamma = \gamma_1\gamma_2$ and $f = f_1 f_2$ as in (3.4). If γ is regular, these two properties follow directly from the analogous properties for G' and the fact that

$$d_{M_1}^G(M, L) = d_{M_1'}^{G'}(M', L').$$

For general γ, the argument is slightly more complicated, requiring Corollary 3.2 and the formula [1(g), (7.1)]. (For a similar argument, see the last stage of the proof of Theorem 8.1 of [1(g)].) The analogues of the differential equation (3.5) and the germ expansion (3.6) are more difficult. They will have to be established later.

LEMMA 3.6: *Suppose that f is a function in $\mathcal{H}(G(F_S))$ such that*

$$I_M(\gamma, f) = I_M^{\mathcal{E}}(\gamma, f)$$

for every element $\gamma \in M(F_S)$ which is G-regular (and semisimple). Then the same formula holds for any element $\gamma \in M(F_S)$.

Proof. Suppose that δ is an element in $M(F_S)$, with semisimple component σ, such that $G_\sigma = M_\sigma$. The orbital integral at δ of any function on $M(F_S)$ is completely determined by its orbital integrals at elements $\gamma \in \sigma M_\sigma(F_S)$ which are in general position and near to σ. It follows from (3.2) and $(3.2)^{\mathcal{E}}$ that

$$I_M(\delta, f) = I_M^{\mathcal{E}}(\delta, f).$$

Now, suppose that γ is an arbitrary element in $M(F_S)$. If a is a small point in general position in $A_M(F_S)$, $\delta = \gamma a$ is as above, so that

$$I_M(\gamma a, f) = I_M^{\mathcal{E}}(\gamma a, f).$$

It follows from (3.1) and $(3.1)^{\mathcal{E}}$ that

$$I_M(\gamma, f) - I_M^{\mathcal{E}}(\gamma, f) = \lim_{a \to 1} \sum_{L \in \mathcal{L}(M)} r_M^L(\gamma, a)(I_L(\gamma a, f) - I_L^{\mathcal{E}}(\gamma a, f)) = 0,$$

as required. ∎

4. Convolution and the differential equation

We shall pause to look more closely at a special case. Suppose that S consists of one valuation v and for the moment assume that G splits at v. Then we can identify $G^0(F_S)$ with

$$\underbrace{G'(F_v) \times \cdots \times G'(F_v)}_{\ell},$$

and the automorphism θ acts by the permutation

$$(x_1, x_2, \ldots, x_\ell) \to (x_2, \ldots, x_\ell, x_1), \quad x_i \in G'(F_v).$$

The group $G'(F_v)$ is embedded diagonally in $G^0(F_v)$. Suppose that

$$\gamma = (\gamma_1, \ldots, \gamma_\ell) \rtimes \theta, \qquad \gamma_i \in G'(F_v).$$

Then if

$$\eta = (1, \gamma_2 \cdots \gamma_\ell, \gamma_3 \cdots \gamma_\ell, \ldots, \gamma_\ell),$$

we have

$$\eta^{-1}\gamma\eta = (\gamma_1 \cdots \gamma_\ell, 1, \ldots, 1) \rtimes \theta.$$

It follows that the norm γ' equals the conjugacy class of $\gamma_1\gamma_2\cdots\gamma_\ell$ in $G'(F_v)$. We shall simply write

$$\gamma' = \gamma_1\gamma_2\cdots\gamma_\ell.$$

Suppose that f is a function in $\mathcal{H}(G(F_v))$ of the form

$$(x_1, \ldots, x_\ell) \rtimes \theta \to f_1(x_1)\cdots f_\ell(x_\ell), \quad f_i \in \mathcal{H}(G'(F_v)), x_i \in G'(F_v).$$

Then the function

$$f_1 * f_2 * \cdots * f_\ell$$

also belongs to $\mathcal{H}(G'(F_v))$. We will denote it by f', since its image in $\mathcal{I}(G'(F_v))$ coincides with the function we denoted above by f'. (See §1.5.)

Suppose that $M \in \mathcal{L}$. Then

$$M(F_v) = \underbrace{(M'(F_v) \times \cdots \times M'(F_v))}_{\ell} \rtimes \theta.$$

Let γ' be a semisimple element in $M'(F_v)$ such that $M'_{\gamma'}$ equals $G'_{\gamma'}$, and set

$$\gamma = (\gamma', 1, \ldots, 1) \rtimes \theta.$$

Then G_γ equals the group $M'_{\gamma'}$, embedded diagonally. In particular, G_γ equals M_γ. We shall investigate the distributions $I_M(\gamma, f)$ with f as above.

We must first look at the weighted orbital integral $J_M(\gamma, f)$. By definition ([1(e)], §2),

$$J_M(\gamma, f) = |D^G(\gamma)|_v^{\frac{1}{2}} \int\limits_{G_\gamma(F_v)\backslash G^0(F_v)} f(x^{-1}\gamma x)v_M(x)dx,$$

where $v_M(x)$ is the number obtained from the (G, M) family

$$e^{\nu(H_P(x))}, \qquad P \in \mathcal{P}(M),\ \nu \in i\mathfrak{a}_M^*.$$

Consider the integral

$$\int\limits_{G_\gamma(F_v)\backslash G^0(F_v)} f(x^{-1}\gamma x)v_M(x)dx. \tag{4.1}$$

This is just the integral over $(x_1, \ldots, x_\ell)$ in the space of cosets of the group $G'_{\gamma'}(F_v)$, embedded diagonally in $(G'(F_v))^\ell$, of

$$f_1(x_1^{-1}\gamma' x_2)f_2(x_2^{-1}x_3)\cdots f_\ell(x_\ell^{-1}x_1)v_M(x_1, \ldots, x_\ell).$$

In this integral, introduce new variables by

$$y_1 = x_1, y_2 = x_2^{-1}x_1, y_3 = x_3^{-1}x_1, \ldots, y_\ell = x_\ell^{-1}x_1.$$

We find that (4.1) equals the integral, over $y_1 \in G'_{\gamma'}(F_v)\backslash G'(F_v)$ and $(y_2, \ldots, y_\ell)$ in $(G'(F_v))^{\ell-1}$, of

$$f_1(y_1^{-1}\gamma' y_1 y_2^{-1})f_2(y_2 y_3^{-1})\cdots f_\ell(y_\ell)v_M(y_1, y_1y_2^{-1}, \ldots, y_1y_\ell^{-1}). \tag{4.2}$$

We intend to extract two applications of the equality of (4.1) with the integral of (4.2). The first applies to nonArchimedean fields.Suppose that v is nonArchimedean and that f_1 is invariant under an open compact subgroup κ of $G'(F_v)$. Choose each of the functions $f_2, \ldots, f_\ell$ to be the characteristic function of κ divided by the volume of κ. Then the integral of (4.2) equals

$$\int\limits_{G'_{\gamma'}(F_v)\backslash G'(F_v)} f_1(y_1^{-1}\gamma' y_1)v_M(y_1, \ldots, y_1)dy_1.$$

Since $G' = G_\theta$, Lemma 8.3 of [1(e)] tells us that

$$v_M(y_1, \ldots, y_1) = v_{M'}(y_1).$$

But

$$f' = f_1 * f_2 * \cdots * f_\ell = f_1,$$

so that (4.1) equals

$$\int_{G'_{\gamma'}(F_v)\backslash G'(F_v)} f'(y^{-1}\gamma' y)v_{M'}(y)dy.$$

By definition, $J_{M'}(\gamma', f')$ is the product of this expression with $|D^{G'}(\gamma')|_v^{\frac{1}{2}}$. By Lemma 1.1, we have

$$|D^G(\gamma)|_v^{\frac{1}{2}} = |\ell|_v^{\frac{1}{2}\dim G_\gamma}|D^{G'}(\gamma')|_v^{\frac{1}{2}}.$$

It follows that for f of the special form described above,

$$J_M(\gamma, f) = |\ell|_v^{\frac{1}{2}\dim G_\gamma} J_{M'}(\gamma', f'). \tag{4.3}$$

It is really the invariant distribution $I_M(\gamma, f)$ that we want to study. However, the following lemma is an easy consequence of the definitions. It was established as Lemma 2.1 in [1(g)].

LEMMA 4.1: *Suppose that v is an unramified (finite) place for G, and that $f \in \mathcal{H}(G(F_v))$ is K_v-bi-invariant. Then*

$$I_M(\gamma, f) = J_M(\gamma, f), \qquad \gamma \in M(F_v). \quad \blacksquare$$

LEMMA 4.2: *Suppose that v is an unramified (finite) place at which G splits, and that $f \in \mathcal{H}(G(F_v))$ is K_v-bi-invariant. Then*

$$I_M(\gamma, f) = I_M^{\mathcal{E}}(\gamma, f) \qquad \gamma \in M(F_v).$$

Proof. As above, we embed G' diagonally in G^0. The dependence of $I_M(\gamma, f)$ and $I_M^{\mathcal{E}}(\gamma, f)$ on f is only through the function $\operatorname{tr}\pi(f)$, with

$$\pi^0 = \underbrace{\pi' \otimes \cdots \otimes \pi'}_{\ell}, \qquad \pi' \in \Pi_{\mathrm{temp}}(G'(F_v)).$$

We can therefore assume that

$$f \longleftrightarrow (f_1, \chi, \dots, \chi),$$

where f_1 is a K'_v-bi-invariant function in $\mathcal{H}(G'(F_v))$, and χ equals the characteristic function of K'_v. We can also assume that

$$\gamma = (\gamma', 1, \dots, 1), \qquad \gamma' \in M'(F_v).$$

Suppose that γ' is G'-regular. Then

$$I_M(\gamma, f) = J_M(\gamma, f) = |\ell|_v^{n/2} J_{M'}(\gamma', f') = |\ell|_v^{n/2} I_{M'}(\gamma', f'),$$

by Lemma 4.1, applied to both G and G', and (4.3) with $\kappa = K'_v$. Since

$$|\ell|_v^{n/2} I_{M'}(\gamma', f') = I_M^{\mathcal{E}}(\gamma, f)$$

if γ' is G-regular, the lemma holds in this case. It then follows from Lemma 3.6 that the formula

$$I_M(\gamma, f) = I_M^{\mathcal{E}}(\gamma, f)$$

holds for any $\gamma \in M(F_v)$. ∎

While we are at it, we shall record a weaker version of Lemma 4.2 that holds if v is assumed only to be unramified. It is not related to convolution, but follows directly from a recent result of Kottwitz.

LEMMA 4.3: *Suppose that v is an unramified (finite) place for G, and that f is the characteristic function of the subset $K_v \rtimes \theta$ of $G(F_v)$. Then*

$$I_M(\gamma, f) = I_M^{\mathcal{E}}(\gamma, f), \qquad \gamma \in M(F_v).$$

Proof. Let f' be the characteristic function of K'_v in $G'(F_v)$. Its image in $\mathcal{I}(G'(F_v))$ coincides with that of f. Suppose that $\gamma \in M(F_v)$ is G-regular. In [29(c)] Kottwitz has shown that

$$\int_{G_\gamma(F_v)\backslash G^0(F_v)} f(x^{-1}\gamma x) v_M(x)\,dx = \int_{G'_{\gamma'}(F_v)\backslash G'(F_v)} f'(x^{-1}\gamma' x) v_{M'}(x)\,dx.$$

Since $|D^G(\gamma)|_v^{\frac{1}{2}}$ equals $|\ell|_v^{n/2} |D^{G'}(\gamma')|_v^{\frac{1}{2}}$ this implies that

$$J_M(\gamma, f) = |\ell|_v^{n/2} J_{M'}(\gamma', f').$$

Lemma 4.1, applied to both G and G', then tells us that

$$I_M(\gamma, f) = |\ell|_v^{n/2} I_{M'}(\gamma', f').$$

Since γ is G-regular, we have

$$|\ell|_v^{n/2} I_{M'}(\gamma', f') = I_M^{\mathcal{E}}(\gamma, f),$$

so the lemma holds in this case. It then follows from Lemma 3.6 that the formula

$$I_M(\gamma, f) = I_M^{\mathcal{E}}(\gamma, f)$$

holds for any $\gamma \in M(F_v)$. ∎

Next, we will take v to be Archimedean. We shall show that $I_M^{\mathcal{E}}(\gamma)$ satisfies the same differential equations as $I_M(\gamma)$.

LEMMA 4.4: *Suppose that v is any Archimedean place of F. Then*

$$(3.5)^{\mathcal{E}} \qquad I_M^{\mathcal{E}}(\gamma, zf) = \sum_{L\in\mathcal{L}(M)} \partial_M^L(\gamma, z_L) I_L^{\mathcal{E}}(\gamma, f)$$

for $z \in \mathcal{Z}(G(F_v))$, $f \in \mathcal{H}(G(F_v))$ and $\gamma \in M(F_v) \cap G_{\mathrm{reg}}$.

Proof. Let H^0 be any θ-stable maximal torus in M^0 which is defined over F_v, and set

$$T = T_0 \rtimes \theta,$$

where T_0 is the centralizer of θ in H^0. Notice that T_0 contains A_M. It is actually enough to prove the lemma with γ in $T_{\mathrm{reg}}(F_v) = T(F_v) \cap G_{\mathrm{reg}}$. To see this, note that for general γ, the distribution $I_M^{\mathcal{E}}(\gamma, f)$ depends only on the $M^0(F_v)$-orbit of γ. The differential operators $\partial_M^L(\gamma, z_L)$ are obtained from the differential equation (3.5). They too depend only on the $M^0(F_v)$-orbit of γ. But any regular $M^0(F_v)$-orbit contains an element of the form $\gamma^0 \rtimes \theta$, where γ^0 is a regular element in $G^0(F_v)$ which commutes with θ ([11(a)], Proposition 2.10). Letting H^0 be the centralizer of γ^0 in G^0, we see that it is indeed enough to take $\gamma \in T_{\mathrm{reg}}(F_v)$.

We shall regard $G^0(F_v)$ as a real Lie group. Let $\mathfrak{g}_v$ be the complexification of its Lie algebra. Then θ defines a linear automorphism of $\mathfrak{g}_v$, which we shall also denote by θ. Its fixed point set is $\mathfrak{g}'_v$, the complexified Lie algebra of the real Lie group $G'(F_v)$. Since $\mathfrak{g}_v$ and $\mathfrak{g}'_v$ are complex rather than real Lie algebras, there is a canonical isomorphism

$$\mathfrak{g}_v \xrightarrow{\sim} \underbrace{\mathfrak{g}'_v \oplus \cdots \oplus \mathfrak{g}'_v}_{\ell}$$

in which θ acts on the right by the standard permutation. (See for example §2 of [11(a)].) Moreover,

$$T(F_v) = \{(t, \ldots, t) \rtimes \theta : t \in T'(F_v)\},$$

where $T'(F_v)$ is a Cartan subgroup of $G'(F_v)$ which contains $A_{M'}(F_v)$. Thus, the triple $(\mathfrak{g}_v, T(F_v), A_M(F_v))$ is no different in general than it is in the special case that G splits at v. But according to Corollary 12.3 of [1(e)], the differential operators $\partial_M^L(\gamma, z)$ depend only on this triple. It is therefore enough to prove the lemma under the assumption that G splits at v.

Assuming that G splits at v, we adopt the earlier notation of this section. Let $\mathcal{Z}_v$ and $\mathcal{Z}'_v$ be the centers of universal enveloping algebras of $\mathfrak{g}_v$ and $\mathfrak{g}'_v$

respectively. Then

$$\mathcal{Z}_v = \underbrace{\mathcal{Z}'_v \otimes \cdots \otimes \mathcal{Z}'_v}_{\ell}.$$

If

$$z = z_1 \otimes \cdots \otimes z_\ell, \qquad z_i \in \mathcal{Z}'_v,$$

set

$$z' = z_1 \cdots z_l.$$

Then

$$(zf)' = z'f', \qquad f \in \mathcal{H}(G(F_v)).$$

Take

$$f \longleftrightarrow (f_1, \ldots, f_\ell), \qquad f_i \in \mathcal{H}(G'(F_v)),$$

$$\gamma = (\gamma', 1, \ldots, 1) \rtimes \theta, \qquad \gamma' \in T'_{\mathrm{reg}}(F_v), \tag{4.4}$$

and

$$z = (z', 1, \ldots, 1), \qquad z' \in \mathcal{Z}'_v. \tag{4.5}$$

Exploiting the equality of (4.1) with the integral of (4.2), we find that $J_M(\gamma, zf)$ equals the product of $|D^G(\gamma)|_v^{\frac{1}{2}}$ with the integral over $y_1 \in T'(F_v)\backslash G'(F_v)$ and $(y_2, \ldots, y_\ell)$ in $(G'(F_v))^{\ell-1}$ of

$$(z'f_1)(y_1^{-1}\gamma' y_1 y_2^{-1}) f_2(y_2 y_3^{-1}) \cdots f_\ell(y_\ell) v_M(y_1, y_1 y_2^{-1}, \ldots, y_1 y_\ell^{-1}).$$

Let $f_2, \ldots, f_\ell$ all approach the Dirac distribution at 1 on $G'(F_v)$. Then $J_M(\gamma, zf)$ approaches

$$\begin{aligned} &|D^G(\gamma)|_v^{\frac{1}{2}} \int_{T'(F_v)\backslash G'(F_v)} (z'f_1)(y_1^{-1}\gamma' y_1) v_M(y_1, \ldots, y_1) dy_1 \\ &= |\ell|_v^{n/2} |D^{G'}(\gamma')|_v^{\frac{1}{2}} \int_{T'(F_v)\backslash G'(F_v)} (z'f_1)(y_1^{-1}\gamma' y_1) v_{M'}(y_1) dy_1 \\ &= |\ell|_v^{n/2} J_{M'}(\gamma', z'f_1). \end{aligned}$$

The differential equation (3.5) actually arose from a similar equation

$$J_M(\gamma, zf) = \sum_{L \in \mathcal{L}(M)} \partial_M^L(\gamma, z_L) J_L(\gamma, f)$$

for the weighted orbital integrals (Proposition 11.1 of [1(e)]). Consider each side of this equation with f being as above. Then the left-hand side approaches the product of $|\ell|_v^{n/2}$ with

$$J_{M'}(\gamma', z'f_1) = \sum_{L\in\mathcal{L}(M)} \partial_{M'}^{L'}(\gamma', z'_{L'})J_{L'}(\gamma', f_1),$$

while the right-hand side approaches the product of $|\ell|_v^{n/2}$ with

$$\sum_{L\in\mathcal{L}(M)} \partial_M^L(\gamma, z_L)J_{L'}(\gamma', f_1).$$

Assume inductively that $\partial_{M'}^{L'}(\gamma', z'_{L'})$ equals $\partial_M^L(\gamma, z)$ for any $L \in \mathcal{L}(M)$ with $L \neq G$. Then

$$\partial_{M'}^{G'}(\gamma', z')J_{G'}(\gamma', f_1) = \partial_M^G(\gamma, z)J_{G'}(\gamma', f_1).$$

It follows that $\partial_{M'}^{L'}(\gamma', z'_{L'}) = \partial_M^L(\gamma, z_L)$ for $L = G$ and hence for all L.

Now, suppose that f is an arbitrary element in $\mathcal{H}(G(F_v))$. Since γ is G-regular, we have

$$\begin{aligned} I_M^{\mathcal{E}}(\gamma, zf) &= |\ell|_v^{n/2} I_{M'}(\gamma', (zf)') \\ &= |\ell|_v^{n/2} I_{M'}(\gamma', z'f') \\ &= \sum_{L\in\mathcal{L}(M)} \partial_{M'}^{L'}(\gamma', z'_{L'})|\ell|_v^{n/2} I_{L'}(\gamma', f') \\ &= \sum_{L\in\mathcal{L}(M)} \partial_M^L(\gamma, z_L) I_L^{\mathcal{E}}(\gamma, f). \end{aligned}$$

This is the required differential equation. We have proved it only for γ and z of the form (4.4) and (4.5). However, this suffices, since for general γ and z each side of the equation depends only on γ' and z'. ∎

5. Statement of Theorem A

We shall first discuss the global implications of the definitions in §3. Let S_{ram} be the finite set of all valuations of F which are either Archimedean or ramified for G. Suppose that S is a finite set which contains S_{ram}. By multiplying any function $f \in \mathcal{H}(G(F_S))$ by the characteristic function of

$$\Big(\prod_{v \notin S} K_v\Big) \rtimes \theta,$$

we obtain an embedding of $\mathcal{H}(G(F_S))$ into $\mathcal{H}(G(\mathbf{A}))$. The geometric side of the trace formula for G is an expansion

$$I(f) = \sum_{M \in \mathcal{L}} |W_0^M||W_0^G|^{-1} \sum_{\gamma \in (M(F))_{M,S}} a^M(S,\gamma) I_M(\gamma, f), \tag{5.1}$$

in terms of the distributions discussed in §3. Here, $(M(F))_{M,S}$ consists of what in general were called (M,S)-equivalence classes in $M(F)$, but which in the present case are just the $M^0(F)$-orbits in $M(F)$. Also, $a^M(S,\gamma)$ is a certain constant whose dependence on γ is essentially through the unipotent part. More precisely, suppose that $\gamma = \sigma u$ is the Jordan decomposition of γ. Set $i^M(S,\sigma)$ equal to 1 if σ is F-elliptic in M, and if in addition, the $M^0(F_v)$-orbit of σ meets $(K_v \cap M^0(F_v)) \rtimes \theta$ for every valuation v outside of S. Otherwise, set $i^M(S,\sigma)$ equal to 0. Given the special nature of G, it follows without difficulty from (3.2) of [1(h)] that

$$a^M(S,\gamma) = i^M(S,\sigma) a^{M_\sigma}(S,u). \tag{5.2}$$

Define

$$I^{\mathcal{E}}(f) = \hat{I}(f'), \qquad f \in \mathcal{H}(G(\mathbf{A})). \tag{5.3}$$

If $\gamma = \sigma u$ is an element in $M(F)$, and S is a large finite set, define

$$a^{M,\mathcal{E}}(S,\gamma) = a^{M'}(S,\gamma'). \tag{5.4}$$

By comparing the characteristic polynomials of σ and σ', it is easy to see that $i^M(S,\sigma)$ equals $i^{M'}(S,\sigma')$. Applying (5.2) to M', we obtain

$$a^{M,\mathcal{E}}(S,\gamma) = i^M(S,\sigma) a^{M_\sigma,\mathcal{E}}(S,u). \tag{5.2$^{\mathcal{E}}$}$$

PROPOSITION 5.1: *We have*

$$I^{\mathcal{E}}(f) = \sum_{M \in \mathcal{L}} |W_0^M||W_0^G|^{-1} \sum_{\gamma \in (M(F))_{M,S}} a^{M,\mathcal{E}}(S,\gamma) I_M^{\mathcal{E}}(\gamma, f). \tag{5.1$^{\mathcal{E}}$}$$

Proof. Applying (5.1) to G', we see that

$$I^{\mathcal{E}}(f) = \sum_{L \in \mathcal{L}'} |W_0^L||W_0^{G'}|^{-1} \sum_{\zeta \in (L(F))_{L,S}} a^L(S,\zeta)\hat{I}_L(\zeta, f').$$

The distribution

$$\hat{I}_L(\zeta, f'), \qquad L \in \mathcal{L}', \ \zeta \in L(F),$$

has a global vanishing property. According to Proposition 8.1 of [1(h)], it equals 0 unless $L = M'$ and $\zeta = \gamma'$ for some $M \in \mathcal{L}$ and $\gamma \in M(F)$. Since $\gamma \to \gamma'$ is an injection of $(M(F))_{M,S}$ into $(M'(F))_{M',S}$, we obtain

$$I^{\mathcal{E}}(f) = \sum_{M \in \mathcal{L}} |W_0^M||W_0^G|^{-1} \sum_{\gamma \in (M(F))_{M,S}} a^{M'}(S,\gamma')\hat{I}_{M'}(\gamma', f').$$

Applying Corollary 3.4 and the definition (5.4), we see that this equals

$$\sum_{M \in \mathcal{L}} |W_0^M||W_0^G|^{-1} \sum_{\gamma \in (M(F))_{M,S}} a^{M,\mathcal{E}}(S,\gamma)I_M^{\mathcal{E}}(\gamma, f),$$

as required. ∎

We have not actually described the role in (5.1) and $(5.1)^{\mathcal{E}}$ of the finite set S of valuations of F. If f is a given function in $\mathcal{H}(G(\mathbf{A}))$, let $V = V(f)$ be the smallest set of valuations which contains S_{ram} and such that f belongs to $\mathcal{H}(G(F_V))$. A precise assertion is that (5.1) and $(5.1)^{\mathcal{E}}$ hold for any S which is suitably large in a sense that depends only on $\mathrm{supp}(f)$ and $V(f)$. (As usual, $\mathrm{supp}(f)$ denotes the support of f.) In addition, the sums over γ in (5.1) and $(5.1)^{\mathcal{E}}$ can both be taken over a finite set, that again depends only on $\mathrm{supp}(f)$ and $V(f)$. This follows from Theorem 3.3 of [1(h)], applied to both G and G'.

THEOREM A: *(i) Suppose that S is any finite set of valuations which contains S_{ram}. Then*

$$I_M^{\mathcal{E}}(\gamma, f) = I_M(\gamma, f) \qquad \gamma \in M(F_S), f \in \mathcal{H}(G(F_S)).$$

(ii) Suppose that γ is an element in $M(F)$. Then

$$a^{M,\mathcal{E}}(S,\gamma) = a^M(S,\gamma)$$

for any suitably large finite set S.

This theorem, which consists of a local assertion and a global assertion, is one of the two main results of Chapter 2. It implies a term by term identification of the geometric sides of the trace formulas of G and G'. The correspondence between automorphic representations will come from the resulting equality of spectral sides.

Theorem A will be proved together with a dual result (Theorem B) which we will announce presently. The process will take up the remainder of Chapter 2. We begin by making an induction hypothesis that will remain in force until the end of Chapter 2. We assume that Theorem A holds if G is replaced by any G_1 with $\dim G_1 < \dim G$, where G_1 is a product of varieties each satisfying the same conditions as G. In particular, the theorem holds if G is replaced by any Levi subset $L \in \mathcal{L}$ with $L \neq G$. More generally, suppose that $M \in \mathcal{L}$ and that σ is a semisimple element in $M(F)$. Then M_σ satisfies the same assumptions as G. Moreover, $\dim M_\sigma < \dim G$ unless $M = G$, $\ell = 1$, and σ belongs to $A_G(F)$.

The induction hypothesis has some immediate consequences. Let S be a finite set of valuations which contains S_{ram}, and consider a Levi subset $M_1 \in \mathcal{L}$ with $M_1 \subsetneq M$. If γ belongs to $M_1(F_S) \cap G_{\mathrm{reg}}$, we have

$$I_M^{\mathcal{E}}(\gamma, f) - I_M(\gamma, f) = \sum_{L \in \mathcal{L}(M_1)} d_{M_1}^G(M, L)(\hat{I}_{M_1}^{L,\mathcal{E}}(\gamma, f_L) - \hat{I}_{M_1}^L(\gamma, f_L))$$

by (3.3) and $(3.3)^{\mathcal{E}}$. Remember that the constant $d_{M_1}^G(M, L)$ vanishes unless the map

$$\mathfrak{a}_{M_1}^M \oplus \mathfrak{a}_{M_1}^L \to \mathfrak{a}_{M_1}^G$$

is an isomorphism. Since $M_1 \subsetneq M$, the constant will vanish if $L = G$. However, if $L \neq G$, the local part of the induction hypothesis tells us that

$$\hat{I}_{M_1}^{L,\mathcal{E}}(\gamma, f_L) - \hat{I}_{M_1}^L(\gamma, f_L) = 0.$$

We conclude that

$$I_M^{\mathcal{E}}(\gamma, f) - I_M(\gamma, f) = 0, \qquad \gamma \in M_1(F_S) \cap G_{\mathrm{reg}}. \tag{5.5}$$

Next, take S to be a disjoint union of S_0 and S_1, where S_0 contains S_{ram} and S_1 consists of one unramified valuation. Suppose that $f = f_0 f_1$ and $\gamma = \gamma_0 \gamma_1$ are corresponding decompositions. Then the difference between $I_M^{\mathcal{E}}(\gamma, f)$ and $I_M(\gamma, f)$ equals

$$\sum_{L_0, L_1 \in \mathcal{L}(M)} d_M^G(L_0, L_1)\big(\hat{I}_M^{L_0,\mathcal{E}}(\gamma_0, f_{0,L_0})\hat{I}_M^{L_1,\mathcal{E}}(\gamma_1, f_{1,L_1}) - \hat{I}_M^{L_0}(\gamma_0, f_{0,L_0})\hat{I}_M^{L_1}(\gamma_1, f_{1,L_1})\big),$$

by (3.4) and $(3.4)^{\mathcal{E}}$. We shall see in a moment that the local assertion (i) of Theorem A implies the equality of $I_M^{\mathcal{E}}(\gamma_1, f_1)$ and $I_M(\gamma_1, f_1)$. Our induction hypothesis then allows us to write

$$\hat{I}_M^{L_0,\mathcal{E}}(\gamma_0, f_{0,L_0})\hat{I}_M^{L_1,\mathcal{E}}(\gamma_1, f_{1,L_1}) - \hat{I}_M^{L_0}(\gamma_0, f_{0,L_0})\hat{I}_M^{L_1}(\gamma_1, f_{1,L_1}) = 0,$$

if neither L_0 nor L_1 equals G. On the other hand, if one of the Levi subsets L_0 or L_1 equals G, the constant $d_M^G(L_0, L_1)$ will vanish unless the other one equals M. According to the definitions in §7 of [1(g)],

$$d_M^G(G, M) = d_M^G(M, G) = 1.$$

We conclude that

$$\begin{aligned} &I_M^{\mathcal{E}}(\gamma, f) - I_M(\gamma, f) \\ (5.6) \qquad &= \sum\nolimits_{i=0}^{1} (I_M^{\mathcal{E}}(\gamma_i, f_i) - I_M(\gamma_i, f_i)) \prod_{j \neq i} \hat{I}_M^M(\gamma_j, f_{j,M}). \end{aligned}$$

(Of course there is only one factor in the product on the right.) Notice that Theorem A(i) implies the vanishing of the left-hand side of (5.6) as well as the summand with $i = 0$ on the right. It therefore also implies the equality of $I_M^{\mathcal{E}}(\gamma_1, f_1)$ and $I_M(\gamma_1, f_1)$, as we claimed above.

The induction hypothesis also has a global consequence. Given $M \in \mathcal{L}$, take an element $\gamma = \sigma u$ in $M(F)$. In the case that $M = G$, assume that σ does not belong to $A_G(F)$. Then $\dim M_\sigma < \dim G$, so we can apply the global part of the induction hypothesis to M_σ. If S is a suitably large finite set of valuations, we conclude from (5.2) and $(5.2)^{\mathcal{E}}$ that

$$a^{M,\mathcal{E}}(S, \gamma) = a^M(S, \gamma).$$

Thus the global assertion of the theorem follows in most cases from the induction hypothesis. From (5.1), $(5.1)^{\mathcal{E}}$ and (3.9) we obtain the following lemma.

LEMMA 5.2: *The distribution*

$$I^{\mathcal{E}}(f) - I(f), \qquad f \in \mathcal{H}(G(F_S)),$$

is the sum of

$$\sum_{M \in \mathcal{L}} |W_0^M||W_0^G|^{-1} \sum_{\gamma \in (M(F))_{M,S}} a^M(S, \gamma)(I_M^{\mathcal{E}}(\gamma, f) - I_M(\gamma, f))$$

and

$$\sum_{\xi \in A_G(F)} \sum_{u \in (\mathcal{U}_G(F))_{G,S}} (a^{G,\mathcal{E}}(S, u) - a^G(S, u)) I_G(\xi u, f).$$

(By definition, $\mathcal{U}_G$ is empty unless $\ell = 1$. In other words, the second term vanishes unless $G = G^0$.) ∎

6. Comparison of $I_M^{\mathcal{E}}(\gamma, f)$ and $I_M(\gamma, f)$

In this paragraph we shall derive some consequences of the local assertion (i) of Theorem A. The assertion applies only if every valuation outside S is unramified for G. It would be natural to consider more general finite sets S. For example, if v is any valuation of F, we could ask whether the distributions $I_M^{\mathcal{E}}(\gamma_v)$ and $I_M(\gamma_v)$ are equal. The next theorem provides a partial answer.

THEOREM 6.1: *Fix an element $M \in \mathcal{L}$ and a finite set S of valuations with the closure property. In the special case that $S \supset S_{\mathrm{ram}}$, we suppose that*

$$I_L^{\mathcal{E}}(\gamma, f) = I_L(\gamma, f), \qquad \gamma \in L(F_S),\ f \in \mathcal{H}(G(F_S)),$$

for any $L \in \mathcal{L}(M)$. Then there are unique constants

$$\varepsilon_L(S) = \varepsilon_L^G(S), \qquad L \in \mathcal{L}(M),$$

such that

$$I_M^{\mathcal{E}}(\gamma, f) = \sum_{L \in \mathcal{L}(M)} \hat{I}_M^L(\gamma, \varepsilon_L(S) f_L),\ \gamma \in M(F_S),\ f \in \mathcal{H}(G(F_S)). \tag{6.1}$$

The constants have the descent property

$$\varepsilon_M(S) = \sum_{L \in \mathcal{L}(M_1)} d_{M_1}^G(M, L)\varepsilon_{M_1}^L(S), \qquad M_1 \subset M, \tag{6.2}$$

and the splitting property

$$\varepsilon_M(S) = \sum_{L_1, L_2 \in \mathcal{L}(M)} d_M^G(L_1, L_2)\varepsilon_M^{L_1}(S)\varepsilon_M^{L_2}(S), \quad S = S_1 \cup S_2. \tag{6.3}$$

Proof If $M = G$, the theorem holds with

$$\varepsilon_G(S) = 1.$$

Fix $M \neq G$, and assume inductively that the theorem is valid whenever M is replaced by any element $L \in \mathcal{L}(M)$ with $L \neq M$. In particular, we assume that the constants

$$\varepsilon_L(S), \qquad L \supsetneq M,$$

have all been defined. The main step is the following lemma.

LEMMA 6.2: *The function*

$$\varepsilon_M(\gamma, f) = I_M^{\mathcal{E}}(\gamma, f) - \sum_{L \supsetneq M} \hat{I}_M^L(\gamma, \varepsilon_L(S) f_L),\ \gamma \in M(F_S), f \in \mathcal{H}(G(F_S)),$$

has descent and splitting properties which are identical to (3.3) and (3.4).

Proof. These properties will hold for $\varepsilon_M(\gamma, f)$ essentially because they hold for $I_M^{\mathcal{E}}(\gamma, f)$, $\hat{I}_M^L(\gamma, f_L)$ and $\varepsilon_L(S)$. Let us verify the descent property. Take M_1 and γ as in (3.3). By (3.3) and $(3.3)^{\mathcal{E}}$ we may express $\varepsilon_M(\gamma, f)$ as the difference between

$$\sum_{L \in \mathcal{L}(M_1)} d_{M_1}^G(M, L) \hat{I}_{M_1}^{L,\mathcal{E}}(\gamma, f_L)$$

and

$$\sum_{L \supsetneq M} \sum_{L_1 \in \mathcal{L}^L(M_1)} d_{M_1}^L(M, L_1) \hat{I}_{M_1}^{L_1}(\gamma, f_{L_1}) \varepsilon_L(S). \tag{6.4}$$

Consider the expression (6.4). Since we need only consider terms for which $d_{M_1}^L(M, L_1) \neq 0$, we may write (6.4) as

$$\sum_{L_1 \supsetneq M_1} \sum_{L \in \mathcal{L}(L_1)} d_{M_1}^L(M, L_1) \hat{I}_{M_1}^{L_1}(\gamma, f_{L_1}) \varepsilon_L(S).$$

The element L in the sum will be strictly larger than M. Therefore our induction assumption implies that $\varepsilon_L(S)$ satisfies the descent property (6.2) of the theorem. Combining this with a formal property ([1(g)], formula (7.1)) of the constants $d_{M_1}^G(\cdot, \cdot)$ we obtain

$$\sum_{L \in \mathcal{L}(L_1)} d_{M_1}^L(M, L_1) \varepsilon_L(S) = \sum_{L \in \mathcal{L}(L_1)} d_{M_1}^G(M, L) \varepsilon_{L_1}^L(S).$$

Consequently, (6.4) equals

$$\sum_{L_1 \supsetneq M_1} \sum_{L \in \mathcal{L}(L_1)} d_{M_1}^G(M, L) \hat{I}_{M_1}^{L_1}(\gamma, \varepsilon_{L_1}^L(S) f_{L_1}).$$

We have shown that $\varepsilon_M(\gamma, f)$ equals

$$\sum_{L \in \mathcal{L}(M_1)} d_{M_1}^G(M, L) \Big(\hat{I}_{M_1}^{L,\mathcal{E}}(\gamma, f_L) - \sum_{\{L_1 : M_1 \subsetneq L_1 \subset L\}} \hat{I}_{M_1}^{L_1}(\gamma, \varepsilon_{L_1}^L(S) f_{L_1} \Big).$$

We obtain

$$\varepsilon_M(\gamma, f) = \sum_{L \in \mathcal{L}(M_1)} d_{M_1}^G(M, L) \hat{\varepsilon}_{M_1}^L(\gamma, f_L), \tag{6.2*}$$

the required descent property.

For the splitting property, we take $\gamma = \gamma_1 \gamma_2$ and $f = f_1 f_2$ as in (3.4). It is proved in much the same way. One applies the splitting properties (3.4),

$(3.4)^{\mathcal{E}}$ and (6.3) (with M replaced by $L \supsetneq M$) to the formula for $\varepsilon_M(\gamma, f)$. We shall skip the details. The final result is

$$(6.3^*) \quad \varepsilon_M(\gamma, f) = \sum_{L_1, L_2 \in \mathcal{L}(M)} d_M^G(L_1, L_2) \hat{\varepsilon}_M^{L_1}(\gamma_1, f_{1,L_1}) \hat{\varepsilon}_M^{L_2}(\gamma_2, f_{2,L_2}). \quad \blacksquare$$

Remark. Lemma 6.2 is proved under the given assumption of Theorem 6.1 This is actually slightly stronger than what we used to prove the lemma. The formulas (6.2*) and (6.3*) hold if we only assume that

$$I_L^{\mathcal{E}}(\gamma, f) = I_L(\gamma, f), \quad \gamma \in L(F_S), f \in \mathcal{H}(G(F_S)), S \supset S_{\text{ram}},$$

for elements $L \in \mathcal{L}(M)$ with $L \neq M$.

We can now prove Theorem 6.1. It is obvious that the constant $\varepsilon_M(S)$ is uniquely determined by the required condition, so we have only to prove its existence. We shall do so by decreasing induction on the number of valuations in S. If S contains S_{ram}, the theorem holds with $\varepsilon_M(S) = 0$, by hypothesis. Assume inductively that the theorem holds for a given set S. In particular, we assume that $\varepsilon_M(S)$ is defined. The required condition is just

$$\varepsilon_M(\gamma, f) = \varepsilon_M(S) \hat{I}_M^M(\gamma, f_M), \qquad \gamma \in M(F_S).$$

Now, suppose that S is a disjoint union of S_1 and S_2. We shall show that the theorem holds for S_1 and S_2.

If $\gamma = \gamma_1 \gamma_2$ and $f = f_1 f_2$, the splitting property (6.3*) allows us to express $\varepsilon_M(\gamma, f)$ as

$$\varepsilon_M(\gamma_1, f_1) \hat{I}_M^M(\gamma_2, f_{2,M}) + \hat{I}_M^M(\gamma_1, f_{1,M}) \varepsilon_M(\gamma_2, f_2) + c_0 \hat{I}_M^M(\gamma_1, f_{1,M}) \hat{I}_M^M(\gamma_2, f_{2,M}),$$

where

$$c_0 = \sum_{\substack{L_1, L_2 \in \mathcal{L}(M) \\ L_1, L_2 \neq G}} d_M^G(L_1, L_2) \varepsilon_M^{L_1}(S_1) \varepsilon_M^{L_2}(S_2).$$

Fix γ_2 and f_2 so that $\hat{I}_M^M(\gamma_2, f_{2,M}) \neq 0$. Let γ_1 be any element such that $\hat{I}_M^M(\gamma_1, f_{1,M_1})$ vanishes. Then

$$\varepsilon_M(\gamma, f) = \varepsilon_M(S) \hat{I}_M^M(\gamma_1, f_{1,M}) \hat{I}_M^M(\gamma_2, f_{2,M}) = 0.$$

This implies that

$$\varepsilon_M(\gamma_1, f_1) \hat{I}_M^M(\gamma_2, f_{2,M}) = 0,$$

and that $\varepsilon_M(\gamma_1, f_1)$ vanishes. It then follows for any γ_1 that

$$\varepsilon_M(\gamma_1, f_1) = \varepsilon_M(S_1, \gamma_1) \hat{I}_M^M(\gamma_1, f_{1,M}),$$

for some function $\varepsilon_M(S_1, \gamma_1)$. Similarly,

$$\varepsilon_M(\gamma_2, f_2) = \varepsilon_M(S_2, \gamma_2)\hat{I}_M^M(\gamma_2, f_{2,M}),$$

for some function $\varepsilon_M(S_2, \gamma_2)$. Substituting back into the original expression, we see that

$$\varepsilon_M(S_1, \gamma_1) + \varepsilon_M(S_2, \gamma_2) + c_0 = \varepsilon_M(S).$$

It follows that $\varepsilon_M(S, \gamma_1)$ and $\varepsilon_M(S, \gamma_2)$ do not depend on γ_1 and γ_2. We have shown that if $i = 1, 2$, there is a constant $\varepsilon_M(S_i)$ such that

$$\varepsilon_M(\gamma_i, f_i) = \varepsilon_M(S_i)\hat{I}_M^M(\gamma_i, f_{i,M}).$$

This completes the inductive definition of the constants $\varepsilon_M(S)$.

We have the two supplementary properties to check. However, these follow immediately from (6.2*) and (6.3*). The proof of the theorem is therefore complete. ∎

COROLLARY 6.3: *Suppose that S either contains S_{ram} or consists of one unramified valuation. Then*

$$\varepsilon_M(S) = \begin{cases} 1, & M = G \\ 0, & M \neq G. \end{cases} \blacksquare$$

It seems likely that Corollary 6.3 is true for arbitrary S. We shall investigate this question only in the case that $\ell = 1$.

PROPOSITION 6.4: *Suppose that $\ell = 1$. As in Theorem 6.1, assume that*

$$I_L^{\mathcal{E}}(\gamma, f) = I_L(\gamma, f), \qquad \gamma \in L(F_S), f \in \mathcal{H}(G(F_S)),$$

if $L \in \mathcal{L}(M)$ and $S \supset S_{\text{ram}}$. Then

$$\varepsilon_M(S) = \begin{cases} 1, & M = G \\ 0, & M \neq G \end{cases}$$

for any finite set S of valuations with the closure property.

Proof. By Theorem 6.1, we know that the constants $\varepsilon_L(S)$ exist. The proposition is trivial if $M = G$, so we shall fix $M \subsetneq G$. We may assume inductively that $\varepsilon_M^L(S) = 0$ if $M \subsetneq L \subsetneq G$. It follows from the descent property (6.2) that $\varepsilon_M(S) = 0$ unless M is minimal. Moreover, from the splitting property (6.3), we see that

$$\varepsilon_M(S) = \sum_{v \in S} \varepsilon_M(v).$$

It is therefore enough to show that each number $\varepsilon_M(v)$ vanishes.

We are assuming that $\ell = 1$. Consequently,

$$G(F) = A^*(F),$$

where A is a simple algebra of degree n over F. For each v, A has an invariant i_v which is an element in $\mathbf{Q}/\mathbf{Z}$ whose order d_v divides n. The constant $\varepsilon_M(v)$ depends only on the pair $(G(F_v), M(F_v))$. We may as well fix n and assume that M is minimal over F_v. Then

$$\varepsilon(i_v) = \varepsilon_M(v)$$

is a complex number which depends only on the element $i_v \in \mathbf{Z}/n\mathbf{Z}$. There is a (unique) simple algebra A over F attached to any finite set

$$\{i_v \in \mathbf{Z}/n\mathbf{Z} : v \in S\}$$

such that

$$\sum i_v = 0.$$

We know that

$$\sum_{v \in S} \varepsilon(i_v) = \varepsilon(S) = 0,$$

if $S \supset S_{\mathrm{ram}}$. It follows easily from this that all of the constants $\varepsilon(i_v)$ vanish. ∎

7. Comparison of germs

Our induction assumption of the last section leads us to define a certain subspace of $\mathcal{H}(G(F_S))$. If $\ell = 1$, G is just the group of units of a central simple algebra. In this case, let S_G denote the set of finite places at which G does not split. If $\ell \neq 1$, simply take S_G to be empty. Define $\mathcal{H}(G(F_S))^0$ to be the subspace of $\mathcal{H}(G(F_S))$ spanned by functions

$$f = \prod_{v \in S} f_v, \qquad f_v \in \mathcal{H}(G(F_v)),$$

which satisfy the following condition. For each $v \in S_G \cap S$, the orbital integral of f_v vanishes at any element

$$\gamma_v = \xi_v u_v, \qquad \xi_v \in A_G(F_v), u_v \in \mathcal{U}_G(F_v),$$

such that $u_v \neq 1$. Orbital integrals are of course invariant, and they define distributions on the space $\mathcal{I}(G(F_S))$. We can therefore define a subspace $\mathcal{I}(G(F_S))^0$ of $\mathcal{I}(G(F_S))$ in the same way. It is clear that we can also define further spaces $\mathcal{H}(G(\mathbf{A}))^0$ and $\mathcal{I}(G(\mathbf{A}))^0$.

Suppose that v is a nonArchimedean valuation of F. The purpose of this section is to show that if f belongs to $\mathcal{H}(G(F_v))^0$, then $\mathcal{I}_M^{\mathcal{E}}(\gamma, f)$ and $\mathcal{I}_M(\gamma, f)$ have the same germ expansions. In order to exploit our induction hypothesis, we shall first show that Theorem A implies an identity of germs.

The germs for $G(F_v)$ and $G'(F_v)$ belong to different equivalence classes, but it turns out that they can be compared directly. Choose a semisimple element σ in $M(F_v)$, and consider the germ expansion about $\tau = \sigma'$ for

$$\hat{I}_{M'}(\zeta, f'), \qquad f \in \mathcal{H}(G(F_v)).$$

Any Levi subgroup in $\mathcal{L}(M')$ equals L', for a unique element $L \in \mathcal{L}(M)$. Consequently

$$\hat{I}_{M'}(\zeta, f') \overset{(M'_\tau, \tau)}{\sim} \sum_{L \in \mathcal{L}(M)} \sum_{\eta \in \tau(\mathcal{U}_{L'_\tau}(F_v))} g_{M'}^{L'}(\zeta, \eta) \hat{I}_{L'}(\eta, f'),$$

for $\zeta \in \tau M'_\tau(F_v) \cap G'_{\text{reg}}$. The vanishing formula (3.7) tells us that $\hat{I}_{L'}(\eta, f') = 0$ unless $\eta = \delta'$ for some $\delta \in \sigma(\mathcal{U}_{L_\sigma}(F_v))$. Therefore

$$\hat{I}_{M'}(\zeta, f') \overset{(M'_\tau, \tau)}{\sim} \sum_{L \in \mathcal{L}(M)} \sum_{\delta \in \sigma(\mathcal{U}_{L_\sigma}(F_v))} g_{M'}^{L'}(\zeta, \delta') \hat{I}_{L'}(\delta', f').$$

But by (3.7), the function $I_{M'}(\zeta, f')$ vanishes unless $\zeta = \gamma'$ for some $\gamma \in M(F_v) \cap G_{\text{reg}}$. We claim that for each L and δ, there is a function

$g_{M'}^{L'}(\zeta,\delta')$ within the (M',τ)-equivalence class which has the same property. We can certainly assume inductively that this is true if $L\neq G$. Fix $\delta_1\in\sigma(\mathcal{U}_G(F_v))$, and choose $f_1\in\mathcal{H}(G(F_v))$ so that

$$I_G(\delta,f_1)=\begin{cases}\Lambda^G(\delta_1) & , \quad \delta=\delta_1,\\ 0 & , \quad \text{otherwise.}\end{cases}$$

Then by (1.4),

$$\hat{I}_{G'}(\delta',f_1')=\begin{cases}1, & \delta=\delta_1,\\ 0, & \text{otherwise.}\end{cases}$$

Substituting f_1 into the expansion above, we justify the claim. Now the orbital integral of a function on $M'(F_v)$ which vanishes if $\zeta\neq\gamma'$ is equal to an orbital integral in γ of a function on $M(F_v)$ (Proposition I.3.1). Each germ

$$g_{M'}^{L'}(\gamma',\delta'),\qquad \gamma\in\sigma M_\sigma(F_v)\cap G_{\text{reg}},$$

may therefore be regarded as an (M,σ)-equivalence class. It is in this sense that we can compare the germs for G and G'. The expansion above becomes

$$\hat{I}_M(\gamma',f')\overset{(M,\sigma)}{\sim}\sum_{L\in\mathcal{L}(M)}\ \sum_{\delta\in\sigma(\mathcal{U}_{L_\sigma}(F_v))} g_{M'}^{L'}(\gamma',\delta')\hat{I}_{L'}(\delta',f'),$$

for $\gamma\in\sigma M_\sigma(F_v)\cap G_{\text{reg}}$. If we apply (3.8) to the left-hand side, we can rewrite this as

$$\text{(7.1)}\qquad I_M^{\mathcal{E}}(\gamma,f)\overset{(M,\sigma)}{\sim}|\ell|_v^{n/2}\sum_{L\in\mathcal{L}(M)}\ \sum_{\delta\in\sigma(\mathcal{U}_{L_\sigma}(F_v))} g_{M'}^{L'}(\gamma',\delta')\hat{I}_{L'}(\delta',f').$$

PROPOSITION 7.1: *Suppose that $\ell=1$ and that Theorem A holds for G. Then for each nonArchimedean valuation v of F, and each $u\in(\mathcal{U}_G(F_v))$,*

$$g_{M'}^{G'}(\gamma',u')\overset{(M,1)}{\sim}e^G(u)g_M^G(\gamma,u),\qquad \gamma\in M(F_v)\cap G_{\text{reg}}.$$

Proof. By hypothesis G and M satisfy the conditions of Proposition 6.4. Combining this proposition with Theorem 6.1, we obtain

$$I_M^{\mathcal{E}}(\gamma,f)=I_M(\gamma,f),\qquad \gamma\in M(F_v),f\in\mathcal{H}(G(F_v)).$$

Moreover, by Corollary 3.3,

$$\hat{I}_{L'}(u',f')=e^L(u)^{-1}I_L^{\mathcal{E}}(u,f)=e^L(u)^{-1}I_L(u,f).$$

It follows from (7.1) that

$$I_M(\gamma,f)\overset{(M,1)}{\sim}\sum_{L\in\mathcal{L}(M)}\ \sum_{u\in(\mathcal{U}_L(F_v))} g_{M'}^{L'}(\gamma',u')e^L(u)^{-1}I_L(u,f).$$

On the other hand, applying the original expansion (3.6), we have

$$I_M(\gamma, f) \overset{(M,1)}{\sim} \sum_{L \in \mathcal{L}(M)} \sum_{u \in (\mathcal{U}_L(F_v))} g_M^L(\gamma, u) I_L(u, f).$$

We may assume inductively that

$$e^L(u)^{-1} f_{M'}^{L'}(\gamma', u') \overset{(M,1)}{\sim} g_M^L(\gamma, u),$$

if $L \neq G$. It follows that

$$\sum_{u \in (\mathcal{U}_G(F_v))} (e^G(u)^{-1} g_{M'}^{G'}(\gamma', u') - g_M^G(\gamma, u)) I_G(u, f) \overset{(M,1)}{\sim} 0.$$

This is a formula in the space of $(M, 1)$-equivalence classes of germs of functions. Since it is valid for any f, we obtain

$$e^G(u)^{-1} g_{M'}^{G'}(\gamma', u') - g_M^G(\gamma, u) \overset{(M,1)}{\sim} 0, \quad u \in (\mathcal{U}_G(F_v)).$$

This is the required formula. ∎

We are carrying the induction hypothesis that Theorem A holds if G is replaced by a proper Levi subset. In Proposition 7.3 we shall combine this with the last lemma to deduce the equality of most of the germs. However, there is one pair of germs which we can compare without recourse to Theorem A (and the global methods its proof entails).

LEMMA 7.2: *Suppose that $\ell = 1$ and that v is a nonArchimedean valuation of F. Then*

$$g_{M'}^{G'}(\gamma', 1) \overset{(M,1)}{\sim} e_v g_M^G(\gamma, 1), \qquad \gamma \in M(F_v) \cap G_{\mathrm{reg}}.$$

Proof. Since $\ell = 1$, G is the multiplicative group of a central simple algebra. In this case the local correspondence is an injection $\pi \to \pi'$ from $\Pi_{\mathrm{temp}}(G(F_v))$ into $\Pi_{\mathrm{temp}}(G'(F_v))$ such that

$$\mathrm{tr}\, \pi(f) = e_v f'(\pi'), \qquad f \in \mathcal{H}(G(F_v)),$$

and

$$\Theta_\pi(\gamma) = e_v \Theta_{\pi'}(\gamma'), \qquad \gamma \in G(F_v)_{\mathrm{reg}}.$$

Any supercuspidal representation in $\Pi(G'(F_v))$ is of the form π', for a (unique) supercuspidal representation π in $\Pi(G(F_v))$. This follows from the character identity above and an easy argument based on Casselman's theorem [10(c)]. Fix such a pair π and π'.

Let f be a matrix coefficient of the contragredient $\tilde{\pi}$ such that $\operatorname{tr}\pi(f) \neq 0$. Since π is supercuspidal, f belongs to $\mathcal{H}(G(F_v))$. The main result of [1(i)] asserts that for γ in $M(F_v) \cap G_{\text{reg}}$,

$$I_M(\gamma, f) =$$
$$(-1)^{\dim(A_M/A_G)} \cdot \operatorname{vol}(G_\gamma(F_v)/A_M(F_v))^{-1} \cdot \operatorname{tr}\pi(f) \cdot |D_M^G(\gamma)|_v^{\frac{1}{2}} \Theta_\pi(\gamma).$$

Notice that if γ is not F_v-elliptic in M, the F_v-split component of G_γ is larger than A_M, and the right-hand side vanishes. The function f' is a priori only an element in $\mathcal{I}(G'(F_v))$, but we can clearly represent it as a matrix coefficient of π'. Since

$$(-1)^{\dim(A_M/A_G)} \cdot \operatorname{vol}(G'_{\gamma'}(F_v)/A_{M'}(F_v))^{-1} \cdot \operatorname{tr}\pi(f) \cdot |D^G(\gamma)|_v^{\frac{1}{2}} \Theta_\pi(\gamma)$$

equals

$$(-1)^{\dim(A_{M'}/A_{G'})} \cdot \operatorname{vol}(G'_{\gamma'}(F_v)/A_{M'}(F_v))^{-1} \cdot \operatorname{tr}\pi'(f') \cdot |D^{G'}(\gamma')|_v^{\frac{1}{2}} \Theta_{\pi'}(\gamma'),$$

for any G-regular element $\gamma \in M(F_v)$, we see that

$$I_M(\gamma, f) = I_{M'}(\gamma', f').$$

But for any such γ, $I_M(\gamma', f')$ equals $I_M^{\mathcal{E}}(\gamma, f)$. It follows from Lemma 3.6 that $I_M(\gamma, f)$ equals $I_M^{\mathcal{E}}(\gamma, f)$ for *any* element γ in $M(F_v)$. We shall use this fact with $\gamma = 1$. In this case we obtain

$$I_M(1, f) = I_M^{\mathcal{E}}(1, f) = e_v I_{M'}(1, f'),$$

from Corollary 3.3.

We shall also need to know that if u is a unipotent element in $M(F_v)$ which is not equal to 1, then $I_M(u, f) = 0$. Since G comes from a central simple algebra, u can be represented as an induced unipotent conjugacy class

$$u_1^M, \qquad u_1 \in (\mathcal{U}_{M_1}(F_v)),$$

where M_1 is a proper Levi subgroup of M. (We can in fact assume that $u_1 = 1$.) The descent formula in Corollary 8.2 of [1(g)] then applies. We obtain

$$I_M(u, f) = I_M(u_1^M, f) = \sum_{L \in \mathcal{L}(M_1)} d_{M_1}^G(M, L) \hat{I}_{M_1}^L(u_1, f_L).$$

But f is a supercusp form on $G(F_v)$, so that $f_L = 0$ for any proper Levi subgroup L of G. If $L = G$, the constant $d_{M_1}^G(M, L)$ is equal to 0. Consequently, $I_M(u, f)$ vanishes, as required. An identical argument applied to G' leads to the vanishing of $I_{M'}(u', f')$.

Take γ to be a G-regular element in $M(F_v)$ which is close to 1. Then

$$I_M(\gamma, f) \overset{(M,1)}{\sim} \sum_{L\in\mathcal{L}(M)} \sum_{u\in(\mathcal{U}_L(F_v))} g_M^L(\gamma, u) I_L(u, f).$$

But from (7.1) we also have

$$I_{M'}(\gamma', f') \overset{(M,1)}{\sim} \sum_{L\in\mathcal{L}(M)} \sum_{u\in(\mathcal{U}_L(F_v))} g_{M'}^{L'}(\gamma', u') I_{L'}(u', f').$$

We have seen that the left-hand sides are equal. Substituting the formulas we have proved into the resulting equality of right-hand sides, we obtain

$$\sum_{L\in\mathcal{L}(M)} e_v g_M^L(\gamma, 1) I_{L'}(1, f') \overset{(M,1)}{\sim} \sum_{L\in\mathcal{L}(M)} g_{M'}^{L'}(\gamma', 1) I_{L'}(1, f').$$

Assume inductively that

$$e_v g_M^L(\gamma, 1) \overset{(M,1)}{\sim} g_{M'}^{L'}(\gamma', 1),$$

if $L \neq G$. It then follows that

$$e_v g_M^G(\gamma, 1) I_{G'}(1, f') \overset{(M,1)}{\sim} g_{M'}^{G'}(\gamma', 1) I_{G'}(1, f').$$

Since $I_{G'}(1, f') \neq 0$, this gives the lemma. ∎

PROPOSITION 7.3: *Suppose that v is a nonArchimedean place of F, and that σ is a semisimple element in $M(F_v)$. Assume that $\mathfrak{a}_{M_\sigma} = \mathfrak{a}_M$. Then*

$$(3.6)^{\mathcal{E}} \qquad I_M^{\mathcal{E}}(\gamma, f) \overset{(M,\sigma)}{\sim} \sum_{L\in\mathcal{L}(M)} \sum_{\delta\in\sigma(\mathcal{U}_{L_\sigma}(F_v))} g_M^L(\gamma, \delta) I_L^{\mathcal{E}}(\delta, f),$$

for $\gamma \in \sigma M_\sigma(F_v) \cap G_{\mathrm{reg}}$ and $f \in \mathcal{H}(G(F_v))^0$.

Proof. It is known that the germs depend only on the unipotent part of δ. More precisely, suppose that

$$\delta = \sigma u, \qquad u \in \mathcal{U}_{L_\sigma}(F_v),$$

and

$$\gamma = \sigma\mu, \qquad \mu \in M_\sigma(F_v).$$

Then by Lemma 9.2 of [1(e)],

$$(7.2) \qquad g_M^L(\gamma, \delta) = \begin{cases} g_{M_\sigma}^{L_\sigma}(\mu, u) & , \text{ if } \mathfrak{a}_{L_\sigma} = \mathfrak{a}_L, \\ 0 & , \text{ otherwise.} \end{cases}$$

This formula will allow us to limit our consideration to varieties of dimension smaller than G, where we can apply the induction hypothesis of §5.

According to (7.1), $I_M^{\mathcal{E}}(\gamma, f)$ is (M,σ)-equivalent to the sum over $L \in \mathcal{L}(M)$ of

$$(7.3) \qquad |\ell|_v^{n/2} \sum_{\delta \in \sigma(\mathcal{U}_{L_\sigma}(F_v))} g_{M'}^{L'}(\gamma', \delta') \hat{I}_{L'}(\delta', f').$$

Set $\tau = \sigma'$. Applying (7.2) to (G', τ), and taking account of (1.1), we see that

$$g_{M'}^{L'}(\gamma', \delta') = g_{M'}^{L'}(\tau\mu_\tau^\ell, \tau u_\tau^\ell) = \begin{cases} g_{M'_\tau}^{L'_\tau}(\mu_\tau^\ell, u_\tau^\ell) & , \quad \text{if } \mathfrak{a}_{L_\sigma} = \mathfrak{a}_L, \\ 0 & , \quad \text{otherwise,} \end{cases}$$

for γ and δ as in (7.2). In particular, we need only sum (7.3) over elements $L \in \mathcal{L}(M)$ with $\mathfrak{a}_{L_\sigma} = \mathfrak{a}_L$. Take such an L, and suppose in addition that $L_\sigma \neq G$. Then by our induction hypothesis, Theorem A holds for L_σ. Applying Lemma 7.1 to L_σ, we see that

$$g_{M'_\tau}^{L'_\tau}(\mu_\tau^\ell, u_\tau^\ell) \overset{(M_\sigma,1)}{\sim} e^{L_\sigma}(u^\ell) g_{M_\sigma}^{L_\sigma}(\mu^\ell, u^\ell).$$

There is a homogeneity property of germs (Proposition 10.2 of [1(e)]) which allows us to express $g_{M_\sigma}^{L_\sigma}(\mu^\ell, u^\ell)$ in terms of a certain sum over $\mathcal{L}^{L_\sigma}(M_\sigma)$.But any group in this set equals $L_{1\sigma}$, for a unique $L_1 \in \mathcal{L}^L(M)$. The homogeneity property then asserts the equality of $g_{M_\sigma}^{L_\sigma}(\mu^\ell, u^\ell)$ with

$$|\ell|_v^d \sum_{L_1 \in \mathcal{L}^L(M)} \sum_{u_1 \in (\mathcal{U}_{L_{1\sigma}}(F_v))} g_{M_\sigma}^{L_{1\sigma}}(\mu, u_1) c_{L_{1\sigma}}^{L_\sigma}(u_1, \ell) \left[u_1^{L_\sigma} : u\right],$$

where $c_{L_{1\sigma}}^{L_\sigma}(u_1, \ell)$ follows the notation of §3,

$$d = \frac{1}{2}(\dim L_{\sigma u} - n),$$

and $\left[u_1^{L_\sigma} : u\right]$ equals 0 or 1, depending on whether the induced conjugacy class $u_1^{L_\sigma}$ equals u or not. Suppose that $\left[u_1^{L_\sigma} : u\right] = 1$, and set

$$\delta_1 = \sigma u_1.$$

Then

$$d = \frac{1}{2} \dim L_{1,\delta_1} - \frac{1}{2} n.$$

Since u and u^ℓ represent the same unipotent conjugacy class in $(\mathcal{U}_{L_\sigma}(F_v))$, we have

$$e^{L_\sigma}(u^\ell) = e^{L_\sigma}(u) = e^{L_{1\sigma}}(u_1) = e^{L_1}(\delta_1),$$

so that

$$|\ell|_v^{n/2} |\ell|_v^d e^{L_\sigma}(u^\ell) = \Lambda^{L_1}(\delta_1).$$

Moreover, Lemma 3.5 tells us that

$$c_{L_{1\sigma}}^{L_\sigma}(u_1,\ell) = c_{L_1}^{L}(\sigma u_1,\ell) = c_{L_1}^{L}(\delta_1,\ell).$$

Finally, by (7.2) we have

$$g_{M_\sigma}^{L_{1\sigma}}(\mu,u_1) = g_M^{L_1}(\gamma,\delta_1).$$

Gathering these facts together, we see that (7.3) equals the expression

$$\sum_{L_1\in\mathcal{L}^L(M)} \sum_{\delta_1\in\sigma(\mathcal{U}_{L_{1\sigma}}(F_v))} \Lambda^{L_1}(\delta_1) g_M^{L_1}(\gamma,\delta_1) c_{L_1}^{L}(\delta_1,\ell)\hat{I}_{L'}((\delta_1^L)',f'). \tag{7.4}$$

The equality of (7.3) and (7.4) was established for any $L \in \mathcal{L}(M)$ with $\mathfrak{a}_{L_\sigma} = \mathfrak{a}_L$ and $L_\sigma \neq G$. Suppose L is such that $\mathfrak{a}_{L_\sigma} \neq \mathfrak{a}_L$. It follows from (7.2) and Lemma 3.5 that

$$g_{M'}^{L'}(\gamma',\delta') = 0,$$

and

$$g_M^{L_1}(\gamma,\delta_1)c_{L_1}^{L}(\delta_1,\ell) = 0.$$

Consequently, (7.3) and (7.4) both vanish, and in particular remain equal to each other. The only other case is when L_σ equals G. Assume this is so. Then $L = G$ and σ is central. This implies $\ell = 1$ and $G = G^0$. The expression (7.3) then equals

$$\sum_{\delta\in\sigma(\mathcal{U}_G(F_v))} g_{M'}^{G'}(\gamma',\delta')\hat{I}_{G'}(\delta',f'). \tag{7.3*}$$

Since $\ell = 1$, $c_{L_1}^{L}(\delta_1,\ell)$ comes from a constant (G, M) family, and vanishes unless $L_1 = L = G$. Consequently (7.4) equals

$$\sum_{\delta\in\sigma(\mathcal{U}_G(F_v))} e^G(\delta) g_M^G(\gamma,\delta)\hat{I}_{G'}(\delta',f'). \tag{7.4*}$$

We are assuming that f belongs to $\mathcal{H}(G(F_v))^0$. If G splits at v, this poses no restriction on f. But then $G = G'$ and (7.3*) and (7.4*) are certainly equal. If G does not split at v,

$$\hat{I}_{G'}(\delta',f') = 0$$

unless $\delta = \sigma$. Since we are assuming σ is central,

$$g_{M'}^{G'}(\delta',\sigma') = g_{M'}^{G'}(\gamma',1) = e_v g_M^G(\gamma,1) = e^G(\sigma)g_M^G(\gamma,\sigma),$$

by Lemma 7.2. It follows that (7.3*) equals (7.4*) in this case as well.

We have shown that (7.3) equals (7.4) for any $L \in \mathcal{L}(M)$. In other words, $I_M^{\mathcal{E}}(\gamma, f)$ is (M, σ)-equivalent to the sum over $L \in \mathcal{L}(M)$ of (7.4). Interchange the sums over L and L_1 in the resulting expression. By Corollary 3.2,

$$\Lambda^{L_1}(\delta_1) \sum_{L \in \mathcal{L}(L_1)} c_{L_1}^L(\delta_1, \ell) \hat{I}_{L'}((\delta_1^L)', f') = I_{L_1}^{\mathcal{E}}(\delta_1, f).$$

Therefore (7.1) becomes

$$I_M^{\mathcal{E}}(\gamma, f) \overset{(M,\sigma)}{\sim} \sum_{L_1 \in \mathcal{L}(M)} \sum_{\delta_1 \in \sigma(\mathcal{U}_{L_{1\sigma}}(F_v))} g_M^{L_1}(\gamma, \delta_1) I_{L_1}^{\mathcal{E}}(\delta_1, f),$$

which is the required formula. ∎

8. The distributions $I_M(\pi, X)$ and $I^{\mathcal{E}}_M(\pi, X)$

We shall now direct our attention to the other side of the trace formula. The spectral side is similar to the geometric side, in that its terms can be separated into local and global constituents. We shall discuss the local properties in this paragraph.

As usual, S denotes a finite set of valuations of F with the closure property and M is an element in $\mathcal{L}$. The local constituents of the spectral side are related to the distributions

$$I_M(\pi, X, f), \qquad \pi \in \Pi^+(M(F_S)),\ X \in \mathfrak{a}_{M,S}, f \in \mathcal{H}(G(F_S)),$$

introduced in [1(g), §3]. These distributions are also defined for standard representations $\rho \in \Sigma^+(M(F_S))$. The two are connected by an expansion formula

$$I_M(\pi, X, f) =$$

$$\sum_P \omega_P \sum_L \sum_\rho \int_{\varepsilon_P + i\mathfrak{a}^*_{M,S}/i\mathfrak{a}^*_{L,S}} r^L_M(\pi_\lambda, \rho_\lambda) I_L(\rho_\lambda, h_L(X), f) e^{-\lambda(X)} d\lambda, \tag{8.1}$$

where P, L and ρ are summed over $\mathcal{P}(M)$, $\mathcal{L}(M)$ and $\Sigma^+(M(F_S))$ respectively ([1(g), (3.2)]). For each P, ε_P is a small point in $(\mathfrak{a}^*_P)^+$, and

$$\omega_P = \mathrm{vol}\{X \in \mathfrak{a}^+_P : \|X\| \leq 1\} \cdot \mathrm{vol}\{X \in \mathfrak{a}_P : \|X\| \leq 1\}^{-1},$$

while for any L, $h_L(X)$ denotes the projection of X onto $\mathfrak{a}_L$. The function $r^L_M(\pi_\lambda, \rho_\lambda)$ was introduced in §6 of [1(f)]. It is obtained from a certain (G, M) family built out of the local normalizing factors. It is a rational function of the variables

$$\{\lambda(\alpha^\vee), q_v^{-\lambda(\alpha^\vee)}\}, \tag{8.2}$$

in which α ranges over the roots of (G, A_M), and v ranges over the discrete valuations in S with residue degree q_v. In the special case that π is unitary, the formula (8.1) simplifies somewhat to

$$I_M(\pi, X, f) =$$

$$\sum_L \sum_\rho \int_{\varepsilon_M + i\mathfrak{a}^*_{M,S}/i\mathfrak{a}^*_{L,S}} r^L_M(\pi_\lambda, \rho_\lambda) I_L(\rho_\lambda, h_L(X), f) e^{-\lambda(X)} d\lambda,$$

where ε_M is a small regular point in $\mathfrak{a}^*_M$.

The lattice $X(M')_F$ has a quotient

$$\mathcal{C}_M = X(M')_F / \ell X(M')_F$$

of order $\ell^{\dim A_M}$. Note that there are natural embeddings

$$\mathcal{C}_L \subset \mathcal{C}_M$$

for the elements $L \in \mathcal{L}(M)$. We shall fix a primitive Grössencharacter $\eta_{E/F}$ associated to E/F by class field theory. Then for each $\xi \in \mathcal{C}_M$,

$$\eta_{E/F}(\xi(m')), \qquad m' \in M'(F_S),$$

is a character of $M'(F_S)$. There is an action

$$(\xi\rho')(m') = \eta_{E/F}(\xi(m'))\rho'(m'), \quad \xi \in \mathcal{C}_M, \rho' \in \Sigma(M'(F_S))$$

of $\mathcal{C}_M$ on $\Sigma(M'(F_S))$. There is a similar action on $\Pi(M'(F_S))$ and also an adjoint action of $\mathcal{C}_M$ on $\mathcal{I}(M'(F_S))$. If ξ belongs to the subgroup $\mathcal{C}_G$ of $\mathcal{C}_M$, it follows easily from (1.5) that

$$\hat{I}_{M'}(\xi\rho', X', f') = \hat{I}_{M'}(\rho', X', \xi f') = \hat{I}_{M'}(\rho', X', f'),$$

for any (ρ', X') in $\Sigma(M'(F_S)) \times \mathfrak{a}_{M',S}$.

As we noted in §1, the local correspondence gives us a map $\rho \to \Sigma'(\rho)$ from $\Sigma^+(M(F_S))$ onto a collection of finite disjoint subsets of $\Sigma(M'(F_S))$ such that

$$\operatorname{tr}\rho(h) = e_S h'(\rho'), \qquad \rho \in \Sigma^+(M(F_S)), \rho' \in \Sigma'(\rho),$$

for any function $h \in \mathcal{H}(M(F_S))$. We also have a map $\pi \longrightarrow \Pi'(\pi)$ from $\Pi^+(M(F_S))$ onto a collection of finite disjoint subsets of $\Pi(M'(F_S))$. However, for nontempered π this map does not give a simple character identity unless h is in the unramified Hecke algebra. Suppose that $\rho \in \Sigma^+(M(F_S))$. Take any $\rho' \in \Sigma'(\rho)$ and define

$$I_M^{\mathcal{E}}(\rho, X, f) = e_S \sum_{\xi \in \mathcal{C}_M} \hat{I}_{M'}(\xi\rho', X', f').$$

LEMMA 8.1: *As the notation suggests, $I_M^{\mathcal{E}}(\rho, X, f)$ depends only on ρ, and not on the element $\rho' \in \Sigma'(\rho)$.*

Proof. The distribution $I_{M'}(\rho', X', f')$ is left unchanged if ρ' is transformed by an element in $\mathcal{C}_G$. It will be convenient to write

$$I_M^{\mathcal{E}}(\rho, X, f) = e_S \ell^{\dim A_G} \sum_{\xi \in \mathcal{C}_M/\mathcal{C}_G} \hat{I}_{M'}(\xi\rho', X', f').$$

We can use a splitting formula (Proposition 9.4 of [1(g)]) to reduce the lemma to the case that S contains one element v. For suppose that S is a disjoint union of two sets S_1 and S_2 which both have the closure property, and that $\rho = \rho_1 \otimes \rho_2$ and $f = f_1 f_2$ are corresponding decompositions. Suppose that $\rho' = \rho'_1 \otimes \rho'_2$ is any representation in $\Sigma'(\rho)$. For a given point

$$(X_1, X_2) \in (\mathfrak{a}_{M,S_1} \oplus \mathfrak{a}_{M,S_2}),$$

the splitting formula expresses the Fourier transform

$$\int \sum_{\xi \in \mathcal{C}_M/\mathcal{C}_G} \hat{I}_{M'}(\xi\rho'_{\Lambda'}, X'_1 + X'_2, f')e^{-\Lambda'(X'_1, X'_2)}d\Lambda',$$

with

$$\Lambda \in (i\mathfrak{a}^*_{M,S_1} \oplus i\mathfrak{a}^*_{M,S_2})/i\mathfrak{a}^*_{M,S},$$

as the sum over $L_1, L_2 \in \mathcal{L}(M)$ of the product of $d^G_M(L_1, L_2)$ with

$$\sum_{\xi \in \mathcal{C}_M/\mathcal{C}_G} \hat{I}^{L'_1}_{M'}(\xi\rho'_1, X'_1, f'_{1,L'_1})\hat{I}^{L'_2}_{M'}(\xi\rho'_2, X'_2, f'_{2,L'_2}).$$

A given summand will vanish unless the map

$$\mathfrak{a}^{L_1}_M \oplus \mathfrak{a}^{L_2}_M \to \mathfrak{a}^G_M$$

is an isomorphism. For any such L_1 and L_2, the natural map

$$\mathcal{C}_M/\mathcal{C}_G \to (\mathcal{C}_M/\mathcal{C}_{L_1}) \oplus (\mathcal{C}_M/\mathcal{C}_{L_2})$$

is an isomorphism. This is a consequence of Lemma 10.1 of [1(h)]. Therefore, the lemma will follow for S if it can be established for S_1 and S_2.

We may therefore assume that $S = \{v\}$. Choose a minimal element $M_1 \in \mathcal{L}^M$ for which ρ is induced from a representation $\rho_1 \in \Sigma^+(M_1(F_v))$. Then ρ' will be induced from a representation $\rho'_1 \in \Sigma(M'_1(F_v))$. Suppose that X_1 is a point in $\mathfrak{a}_{M_1,S}$ whose projection onto $\mathfrak{a}_{M,S}$ equals X. A formula of descent (Corollary 8.5 of [1(g)]) expresses the Fourier transform

$$\int_{i\mathfrak{a}^*_{M_1,S}/i\mathfrak{a}^*_{M,S}} \sum_{\xi \in \mathcal{C}_M/\mathcal{C}_G} \hat{I}_{M'}(\xi\rho'_{\Lambda'}, X', f')e^{-\Lambda'(X'_1)}d\Lambda'$$

as the sum over $L \in \mathcal{L}(M)$ of the product of $d^G_{M_1}(M, L)$ with the function

$$\sum_{\xi \in \mathcal{C}_M/\mathcal{C}_G} \hat{I}^{L'}_{M'_1}(\xi\rho'_1, X'_1, f'_{L'}). \tag{8.3}$$

A given summand will vanish unless the map

$$\mathfrak{a}^M_{M_1} \oplus \mathfrak{a}^L_{M_1} \to \mathfrak{a}^G_{M_1}$$

is an isomorphism. For any such L, the natural map

$$\mathcal{C}_M/\mathcal{C}_G \to \mathcal{C}_{M_1}/\mathcal{C}_L$$

is an isomorphism. Therefore the function (8.3) equals

$$\sum_{\xi \in \mathcal{C}_{M_1}/\mathcal{C}_L} \hat{I}^{L'}_{M'_1}(\xi\rho'_1, X'_1, f'_{L'}).$$

Given our choice of M_1, the theory of the local lifting (Proposition I.6.7) tells us that

$$\Sigma'(\rho_1) = \{\xi\rho'_1 : \xi \in \mathcal{C}_{M_1}\}.$$

It follows that (8.3) depends only on ρ_1. Consequently, the original function depends only on ρ, as required. ∎

For each $L \in \mathcal{L}(M)$, $\rho \in \Sigma^+(M(F_S))$ and $\lambda \in \mathfrak{a}^*_{M,\mathbf{C}}$, there is an induced representation ρ^L_λ of $L(F_S)$. If λ is in general position, ρ^L_λ belongs to $\Sigma^+(L(F_S))$. As in (8.1), we will often drop the superscript L, so that

$$I^{\mathcal{E}}_L(\rho_\lambda, X, f) = I^{\mathcal{E}}_L(\rho^L_\lambda, X, f).$$

If $\pi \in \Pi^+(M(F_S))$, we define

$$I^{\mathcal{E}}_M(\pi, X, f) =$$

$$\text{(8.4)} \quad \sum_P \omega_P \sum_L \sum_\rho \int_{\epsilon_P + i\mathfrak{a}^*_{M,S}/i\mathfrak{a}^*_{L,S}} r^L_M(\pi_\lambda, \rho_\lambda) I^{\mathcal{E}}_L(\rho_\lambda, h_L(X), f) e^{-\lambda(X)} d\lambda.$$

To describe the local constituents of the trace formula one changes notation slightly. If π stands for a representation in $\Pi^+(M(F_S)^1)$, let us agree to identify π with an orbit $\{\pi_\lambda : \lambda \in \mathfrak{a}^*_{M,\mathbf{C}}\}$ of $\mathfrak{a}^*_{M,\mathbf{C}}$ in $\Pi^+(M(F_S))$. Usually π will be unitary, in which case we will identify it with the smaller orbit $\{\pi_\lambda : \lambda \in i\mathfrak{a}^*_M\}$ of $i\mathfrak{a}^*_M$ in $\Pi^+_{\text{unit}}(M(F_S))$. We shall also adopt these conventions, sometimes without comment, for representations in $\Pi^+(M(\mathbf{A})^1)$ and $\Pi^+_{\text{unit}}(M(\mathbf{A})^1)$. If π belongs to $\Pi^+_{\text{unit}}(M(F_S)^1)$, we set

$$I_M(\pi, f) = I_M(\pi_\lambda, 0, f)$$

and

$$I^{\mathcal{E}}_M(\pi, f) = I^{\mathcal{E}}_M(\pi_\lambda, 0, f).$$

These expressions are independent of λ. The former describes the local spectral terms of the trace formula of G. The latter is closely related to the analogous terms for G'. Both expressions are independent of S if S is suitably large, and so may be defined for $\pi \in \Pi^+_{\text{unit}}(M(\mathbf{A})^1)$.

As in [1(f), §5], let $\Delta(\pi, \rho)$ and $\Gamma(\rho, \pi)$ be the constants which describe the transformation formulas between standard characters and irreducible characters. That is,

$$\mathrm{tr}(\rho) = \sum_{\pi \in \Pi^+(M(F_S))} \Gamma(\rho, \pi)\, \mathrm{tr}(\pi), \qquad \rho \in \Sigma^+(M(F_S)),$$

and

$$\mathrm{tr}(\pi) = \sum_{\rho \in \Sigma^+(M(F_S))} \Delta(\pi, \rho)\, \mathrm{tr}(\rho), \qquad \pi \in \Pi^+(M(F_S)).$$

Now suppose that $\tau \in \Pi(M'(F_S))$ and $\rho \in \Sigma^+(M(F_S))$. The constants above are not immediately defined, since the representations are for two separate groups. However, we shall set

$$\Delta(\tau, \rho) = e_S \sum_{\rho' \in \Sigma'(\rho)} \Delta(\tau, \rho').$$

For each $\pi \in \Pi^+(M(F_S))$ we then define

$$\delta(\tau, \pi) = \sum_{\rho \in \Sigma^+(M(F_S))} \Delta(\tau, \rho)\Gamma(\rho, \pi).$$

If $G = G'$, we have

$$\delta(\tau, \pi) = \begin{cases} 1, & \tau = \pi \\ 0, & \text{otherwise,} \end{cases}$$

but in general the situation is more complicated. Observe that

$$\delta(\tau, \pi) = \prod_{v \in S} \delta(\tau_v, \pi_v) \tag{8.5}$$

if $\tau = \bigotimes_{v \in S} \tau_v$ and $\pi = \bigotimes_{v \in S} \pi_v$.

PROPOSITION 8.2: *We have*

$$h'(\tau) = \sum_{\pi \in \Pi^+(M(F_S))} \delta(\tau, \pi)\, \mathrm{tr}\, \pi(h)$$

for any $h \in \mathcal{H}(M(F_S))$ *and* $\tau \in \Pi(M'(F_S))$.

Proof. The sets $\Sigma'(\rho)$ are all disjoint. Recall that $h'(\rho')$ equals $e_S \operatorname{tr} \rho(h)$ if ρ' belongs to $\Sigma'(\rho)$, and vanishes if ρ' belongs to no such set. Therefore

$$\begin{aligned} h'(\tau) &= \sum_{\rho' \in \Sigma(M'(F_S))} \Delta(\tau, \rho') h'(\rho') \\ &= \sum_{\rho \in \Sigma^+(M(F_S))} \Delta(\tau, \rho) \operatorname{tr} \rho(h) \\ &= \sum_{\rho \in \Sigma^+(M(F_S))} \sum_{\pi \in \Pi^+(M(F_S))} \Delta(\tau, \rho) \Gamma(\rho, \pi) \operatorname{tr} \pi(h) \\ &= \sum_{\pi} \delta(\tau, \pi) \operatorname{tr} \pi(h), \end{aligned}$$

as required. ∎

COROLLARY 8.3: *Suppose that S consists of one unramified place v and that $\pi \in \Pi^+(M(F_v))$ is unramified. Then for any $\tau \in \Pi(M'(F_S))$,*

$$\delta(\tau, \pi) = \begin{cases} 1, & \tau \in \Pi'(\pi), \\ 0, & \textit{otherwise.} \end{cases}$$

Proof. Take h to be an arbitary function in $\mathcal{H}(M(F_v))$ which is bi-invariant under $K_v \cap M^0(F_v)$. Since v is unramified, $e_v = 1$. The fundamental lemma (Theorem I.4.5) tells us that $h'(\tau) = \operatorname{tr} \pi(h)$ for any $\tau \in \Pi'(\pi)$. The corollary then follows from the proposition. ∎

Now suppose that $\tau = \bigotimes_v \tau_v$ and $\pi = \bigotimes_v \pi_v$ are representations in $\Pi(M'(\mathbf{A}))$ and $\Pi^+(M(\mathbf{A}))$ respectively. Define

$$\delta(\tau, \pi) = \prod_v \delta(\tau_v, \pi_v).$$

By the corollary, almost all the terms in the product are either 0 or 1, so the product can be taken over a finite set. The adèlic formulation is therefore included in the previous definitions, and satisfies all the formulas above. In particular,

$$h'(\tau) = \sum_{\pi \in \Pi(M(\mathbf{A}))} \delta(\tau, \pi) \operatorname{tr} \pi(h), \quad h \in \mathcal{H}(M(\mathbf{A})), \ \tau \in \Pi(M'(\mathbf{A})).$$

Suppose instead that we take τ and π to be representations in $\Pi(M'(\mathbf{A})^1)$ and $\Pi^+(M(\mathbf{A})^1)$ respectively. As we have agreed, we may identify these representations with orbits $\{\tau_\eta\}$ and $\{\pi_\lambda\}$ in $\Pi(M'(\mathbf{A}))$ and $\Pi^+(M(\mathbf{A}))$. Then in this situation, we define

$$\delta(\tau, \pi) = \sum_{\lambda \in \mathfrak{a}^*_{M,\mathbf{C}}} \delta(\tau_\eta, \pi_\lambda).$$

There can be at most one nonzero summand on the right, and its value is independent of η.

9. Statement of Theorem B

In this paragraph we shall describe the global constituents $a^M(\pi)$ of the spectral side. We shall then state Theorem B which, together with the dual Theorem A, is the main result of Chapter 2.

Let I be the distribution defined by (5.1). The spectral side of the trace formula is a sum

$$I(f) = \sum_{t \geq 0} I_t(f),$$

where

$$I_t(f) = \sum_{M \in \mathcal{L}} |W_0^M||W_0^G|^{-1} \int_{\Pi(M,t)} a^M(\pi) I_M(\pi, f) d\pi. \tag{9.1}$$

In particular, it is an expansion of $I(f)$ in terms of the distributions

$$I_M(\pi, f) = I_M(\pi_\lambda, 0, f)$$

discussed in §8. The variable t, which ranges over the nonnegative real numbers, is required for convergence. We shall recall in a moment how it is used to keep track of the size of Archimedean infinitesimal characters. We shall then briefly review the definitions of $a^M(\pi)$ and $\Pi(M,t)$ from [1(h)].

Let S_∞ denote the set of Archimedean valuations of F, and set $F_\infty = F_{S_\infty}$. Then $\mathrm{GL}(n, F_\infty)$ can be regarded as a real Lie group. Let $\mathfrak{h}'_{\mathbf{C}}$ denote the standard Cartan subalgebra of its complex Lie algebra. Let $\mathfrak{h}' \subset \mathfrak{h}'_{\mathbf{C}}$ be the real form of $\mathfrak{h}'_{\mathbf{C}}$ associated to the split real form of $\mathrm{GL}(n, F_\infty)$. Then $\mathfrak{h}'$ is invariant under the complex Weyl group W' of $\mathrm{GL}(n, F_\infty)$. Set

$$\mathfrak{h} = \underbrace{\mathfrak{h}' \oplus \cdots \oplus \mathfrak{h}'}_{\ell}.$$

By means of the inner twist η, we can identify $\mathfrak{h}_{\mathbf{C}}$ with a Cartan subalgebra of the complex Lie algebra of the real Lie group $G^0(F_\infty)$. Then $\mathfrak{h}$ is invariant under the complex Weyl group W of $G^0(F_\infty)$. It contains each of the real vector spaces $\mathfrak{a}_M$. It is convenient to fix a Euclidean norm $\|\cdot\|$ on $\mathfrak{h}$ which is invariant under W. We shall also write $\|\cdot\|$ for the dual Hermitian norm on $\mathfrak{h}^*_{\mathbf{C}}$. To any representation $\pi \in \Pi^+(M(\mathbf{A}))$, $M \in \mathcal{L}$, we can associate the induced representation π^G of $G^+(\mathbf{A})$. Let ν_π be the infinitesimal character of its Archimedean constituent. It is a W-orbit in $\mathfrak{h}^*_{\mathbf{C}}$ which meets $(\mathfrak{h}')^*_{\mathbf{C}}$. We shall be more concerned with the case that π is a representation in $\Pi^+(M(\mathbf{A})^1)$. Then ν_π is a priori determined only as an orbit of $\mathfrak{a}^*_{M,\mathbf{C}}$ in $\mathfrak{h}^*_{\mathbf{C}}$. However, this orbit has a unique point of smallest norm in $\mathfrak{h}^*_{\mathbf{C}}$ (up to translation by W), and it is this point which we will denote by ν_π. If $t \geq 0$, define $\Pi^+(M(\mathbf{A})^1, t)$ to be the set of representations $\pi \in \Pi^+(M(\mathbf{A})^1)$ such that

$$\| \operatorname{Im}(\nu_\pi) \| = t,$$

where $\operatorname{Im}(\nu_\pi)$ is the imaginary part of ν_π relative to the real form $\mathfrak{h}^*$ of $\mathfrak{h}^*_{\mathbf{C}}$.

The global constituents of (9.1) are defined in terms of a function

$$a_{\text{disc}}(\pi) = a^G_{\text{disc}}(\pi), \qquad \pi \in \Pi^+(G(\mathbf{A})^1, t).$$

It in turn is defined by rewriting the expression

$$I_{\text{disc},t}(f) = \tag{9.2}$$
$$\sum_{L \in \mathcal{L}} |W_0^L||W_0^G|^{-1} \sum_{s \in W(\mathfrak{a}_L)_{\text{reg}}} |\det(s-1)_{\mathfrak{a}^G_L}|^{-1} \operatorname{tr}(M(s,0)\rho_{Q,t}(0,f))$$

as

$$\sum_{\pi \in \Pi^+(G(\mathbf{A})^1, t)} a^G_{\text{disc}}(\pi) f_G(\pi), \tag{9.3}$$

a linear combination of characters. The terms in (9.2) are as in [1(h), §4]. In particular, Q is any element in $\mathcal{P}(L)$, and $\rho_{Q,t}$ is the induced representation of $G^+(\mathbf{A})^1$ obtained from the subrepresentation of $M^+(\mathbf{A})^1$ on $L^2(M^0(F)\backslash M^0(\mathbf{A})^1)$ which decomposes into a discrete sum of elements in $\Pi^+(M(\mathbf{A})^1, t)$. Moreover, $M(s,0)$ is the global intertwining operator associated to an element in

$$W(\mathfrak{a}_L)_{\text{reg}} = \{s \in W(\mathfrak{a}_L) : \det(s-1)_{\mathfrak{a}^G_L} \neq 0\}.$$

(Here $W(\mathfrak{a}_L)$ denotes the Weyl group of $\mathfrak{a}_L$.) For any function $f \in \mathcal{H}(G(\mathbf{A}))$, the sum in (9.3) can be taken over a finite subset of $\Pi^+_{\text{unit}}(G(\mathbf{A})^1, t)$, and it is understood that

$$f_G(\pi) = I_G(\pi, f) = I_G(\pi_\lambda, 0, f).$$

Suppose that $M_1 \in \mathcal{L}$. As in [1(h)], we write $\Pi_{\text{disc}}(M_1, t)$ for the subset of $\Pi^+_{\text{unit}}(M_1(\mathbf{A})^1, t)$ consisting of irreducible constituents of induced representations

$$\sigma^{M_1}_\lambda, \qquad L \in \mathcal{L}^{M_1}, \sigma \in \Pi^+_{\text{unit}}(L(\mathbf{A})^1, t), \lambda \in i\mathfrak{a}^*_L/i\mathfrak{a}^*_{M_1},$$

in which σ_λ satisfies the following two conditions.

(i) $a^L_{\text{disc}}(\sigma) \neq 0$.

(ii) There is an element $s \in W^{M_1}(\mathfrak{a}_L)_{\text{reg}}$ such that $s\sigma_\lambda = \sigma_\lambda$.

Then for any M, $\Pi(M,t)$ is the disjoint union over $M_1 \in \mathcal{L}^M$ of the sets

$$\Pi_{M_1}(M,t) = \{\pi = \pi_{1,\lambda} : \pi_1 \in \Pi_{\text{disc}}(M_1,t), \lambda \in i\mathfrak{a}^*_{M_1}/i\mathfrak{a}^*_M\}.$$

The global datum in (9.1) is the function

$$a^M(\pi) = a^{M_1}_{\text{disc}}(\pi_1)\ r^M_{M_1}(\pi_{1,\lambda}), \qquad \pi \in \Pi_{M_1}(M,t),$$

also introduced in [1(h), §4]. It can be defined for any representation

$$\pi = \pi_{1,\lambda}, \qquad \pi_1 \in \Pi^+(M_1(\mathbf{A})^1), \lambda \in i\mathfrak{a}^*_{M_1}/i\mathfrak{a}^*_M,$$

but it vanishes unless π belongs to $\Pi_{M_1}(M,t)$ for some t. The function $r^M_{M_1}(\pi_{1,\lambda})$ is obtained from a (G, M_1) family which is built out of the global normalizing factors. We shall discuss it in more detail in §11. Finally, the measure in (9.1) is given by

$$d\pi = d\pi_{1,\lambda} = |W^{M_1}_0||W^M_0|^{-1}d\lambda, \qquad \pi \in \Pi_{M_1}(M,t).$$

In our notation $\{\pi_{1,\lambda}\}$ stands for the orbit of $i\mathfrak{a}^*_{M_1}/i\mathfrak{a}^*_M$ in

$$\Pi^+_{\text{unit}}(M_1(\mathbf{A}) \cap M(\mathbf{A})^1)$$

associated to a given $\pi_1 \in \Pi_{\text{disc}}(M_1,t)$, but we shall often identify $\pi = \pi_{1,\lambda}$ with the induced representation $\pi^M_{1,\lambda}$ in $\Pi^+_{\text{unit}}(M(\mathbf{A})^1)$. It is in this sense that the distribution $I_M(\pi, f)$ in (9.1) is defined. We should perhaps emphasize that the function

$$I_M(\pi, f) = I_M(\pi^M_{1,\lambda}, f), \qquad \lambda \in i\mathfrak{a}^*_{M_1}/i\mathfrak{a}^*_M,$$

is rapidly decreasing. It in fact extends to a meromorphic function in the complex domain which is rapidly decreasing on cylinders, as one sees directly from the definition [1(g)] of the distribution. This property is implicit in the formula (9.1) (as well as (8.1)), and will be used later without comment.

The integral over $\Pi(M,t)$ in (9.1) converges absolutely. So does the sum $\sum_t I_t(f)$. However, it is not known that the two converge together as a double integral over (t,π). It is because of this difficulty that we introduced the sum over t in the first place. However, it does not seem unreasonable from an aesthetic standpoint that we should be forced to keep track of Archimedean infinitesimal characters.

A similar expansion of course holds for G'. However, we would like to define functions which we can compare directly with $a^M(\pi)$.

LEMMA 9.1: *If π_1 is any representation in $\Pi^+(M_1(\mathbf{A})^1)$, the series*

$$\sum_{\tau_1 \in \Pi(M'_1(\mathbf{A})^1)} a^{M'_1}_{\text{disc}}(\tau_1)\delta(\tau_1, \pi_1)$$

can be summed over a finite set.

Proof. Lift π_1 to a representation in $\Pi^+(M_1(\mathbf{A}))$, and choose a finite set S of valuations outside of which G and π_1 are unramified. Let $\pi_1 = \pi_{1,S} \otimes \pi_1^S$ be the decomposition of π_1 corresponding to

$$M_1(\mathbf{A}) = M_1(F_S)\Big(\prod_{v \notin S} M_1(F_v)\Big).$$

Consider the representations $\tau_1 = \tau_{1,S} \otimes \tau_1^S$ in $\Pi(M_1'(\mathbf{A}))$ such that the number

$$\delta(\tau_1, \pi_1) = \delta(\tau_{1,S}, \pi_{1,S}) \prod_{v \notin S} \delta(\tau_{1,v}, \pi_{1,v})$$

does not vanish. By Corollary 8.3, τ_1^S is unramified, and it is clear that there are only finitely many choices for $\tau_{1,S}$. It follows from Lemma 4.2 of [1(h)] that there are only finitely many such τ_1 with $a_{\text{disc}}^{M_1}(\tau_1) \neq 0$. Therefore, there are only finitely many nonzero summands in the series. ∎

Define

$$a_{\text{disc}}^{M_1,\mathcal{E}}(\pi_1) = \ell^{-\dim A_{M_1}} \sum_{\tau_1 \in \Pi(M_1'(\mathbf{A})^1)} a_{\text{disc}}^{M_1'}(\tau_1)\delta(\tau_1, \pi_1). \tag{9.4}$$

Then if

$$\pi = \pi_{1,\lambda}, \qquad \pi_1 \in \Pi^+(M_1(\mathbf{A})^1), \quad \lambda \in \mathfrak{a}^*_{M_1,\mathbf{C}}/\mathfrak{a}^*_{M,\mathbf{C}},$$

we define

$$a^{M,\mathcal{E}}(\pi) = a_{\text{disc}}^{M_1,\mathcal{E}}(\pi_1)\, r_{M_1}^M(\pi_{1,\lambda}).$$

The function $r_{M_1}^M(\pi_{1,\lambda})$ is obtained from global normalizing factors, and is well behaved only when π_1 is automorphic. Therefore, it is not *a priori* clear that the definition of $a^{M,\mathcal{E}}(\pi)$ makes sense. This will follow from the induction hypothesis introduced below.

THEOREM B: *(i) Suppose that S is a finite set of valuations which contains S_{ram}. Then*

$$I_M^{\mathcal{E}}(\pi, f) = I_M(\pi, f), \qquad \pi \in \Pi^+_{\text{unit}}(M(\mathbf{A})^1), f \in \mathcal{H}(G(F_S)).$$

(ii) For any given

$$\pi = \pi_{1,\lambda}, \qquad \pi_1 \in \Pi^+(M_1(\mathbf{A})^1), \lambda \in \mathfrak{a}^*_{M_1,\mathbf{C}}/\mathfrak{a}^*_{M,\mathbf{C}},$$

we have

$$a^{M,\mathcal{E}}(\pi) = a^M(\pi).$$

This theorem, which consists of a local assertion and a global assertion, is the second main result of Chapter 2. It will imply a term by term identification of the spectral sides of the trace formulas of G and G'. It is the second assertion (ii) which will allow us to deduce the correspondence between automorphic representations.

Theorem B will be proved in conjunction with Theorem A. As we shall see in the next section, the local assertion (i) can be proved from our induction assumption of §5. However, the global assertion (ii) requires its own induction assumption. We assume that for any $M_1 \in \mathcal{L}$, with $M_1 \neq G$, that

$$a_{\mathrm{disc}}^{M_1,\mathcal{E}}(\pi_1) = a_{\mathrm{disc}}^{M_1}(\pi_1), \qquad \pi_1 \in \Pi(M_1(\mathbf{A})^1).$$

Then $a_{\mathrm{disc}}^{M_1,\mathcal{E}}(\pi_1)$ vanishes unless π_1 belongs to $\Pi_{\mathrm{disc}}(M_1,t)$ for some t. This means, in particular, that π_1 must be unitary. But if π_1 is unitary, and $M \supset M_1$,

$$\pi = \pi_{1,\lambda}, \qquad \lambda \in i\mathfrak{a}_{M_1}^*/i\mathfrak{a}_M^*,$$

is well defined. Moreover, if π_1 belongs to $\Pi_{\mathrm{disc}}(M_1,t)$, the function $r_{M_1}^M(\pi_{1,\lambda})$ is defined. It follows that the function $a^{M,\mathcal{E}}(\pi)$ is well defined and that

$$a^{M,\mathcal{E}}(\pi) = a^M(\pi)$$

whenever $M_1 \neq G$.

10. Comparison of $I_M^{\mathcal{E}}(\pi, X, f)$ and $I_M(\pi, X, f)$

We shall establish the local assertion (i) of Theorem B. We are actually going to show that Theorem A(i) implies the equality of the distributions $I_M(\pi, X, f)$ and $I_M^{\mathcal{E}}(\pi, X, f)$ described in §8. We will use the constructions of [1(g)], which were designed for this purpose.

Fix a finite set S of valuations of F with the closure property. In [1(f), §11] and [1(g),§4], we defined function spaces

$$\tilde{\mathcal{H}}_{ac}(G(F_S)) \supset \mathcal{H}_{ac}(G(F_S)) \supset \mathcal{H}(G(F_S))$$

and

$$\tilde{\mathcal{I}}_{ac}(G(F_S)) \supset \mathcal{I}_{ac}(G(F_S)) \supset \mathcal{I}(G(F_S)).$$

The definitions were set up so that the spaces in the second row become the images under invariant Fourier transform of the corresponding spaces in the first row. We shall not describe them further, except to say that those of the second row consist of functions $\phi(\pi, X)$ on $\Pi_{\text{temp}}(G(F_S)) \times \mathfrak{a}_{G,S}$ with different conditions on the second variable. The conditions on $\mathcal{I}(G(F_S))$ require that $\phi(\pi, X)$ be smooth and compactly supported in X. For $\mathcal{I}_{ac}(G(F_S))$ the compactness of support is relaxed, and for $\tilde{\mathcal{I}}_{ac}(G(F_S))$ the smoothness condition is also weakened. All the invariant distributions on $\mathcal{H}(G(F_S))$ that we have described extend naturally to linear forms on $\tilde{\mathcal{H}}_{ac}(G(F_S))$. Their Fourier transforms therefore extend to $\tilde{\mathcal{I}}_{ac}(G(F_S))$.

In [1(g), §4] we also defined maps θ_M^L and ${}^c\theta_M^L$ from $\tilde{\mathcal{H}}_{ac}(L(F_S))$ to $\tilde{\mathcal{I}}_{ac}(M(F_S))$, for every pair $M \subset L$ of Levi subsets in $\mathcal{L}$. These maps satisfy

$$\sum_{L \in \mathcal{L}(M)} \hat{\theta}_M^L({}^c\theta_L(f)) = \sum_{L \in \mathcal{L}(M)} {}^c\hat{\theta}_M^L(\theta_L(f)) = 0, \tag{10.1}$$

$$I_M(\gamma, f) = \sum_{L \in \mathcal{L}(M)} {}^c\hat{I}_M^L(\gamma, \theta_L(f)), \tag{10.2}$$

and

$${}^cI_M(\gamma, f) = \sum_{L \in \mathcal{L}(M)} \hat{I}_M^L(\gamma, {}^c\theta_L(f)), \tag{10.3}$$

for $\gamma \in M(F_S)$ and $f \in \tilde{\mathcal{H}}_{ac}(G(F_S))$. Here ${}^cI_M^L(\gamma)$ is an invariant distribution on $\tilde{\mathcal{H}}_{ac}(L(F_S))$ which depends only on the $M^0(F_S)$-orbit of γ. (As usual, we have suppressed the superscript if it is G.) The key feature of ${}^cI_M$ is a property of compact support. If f actually belongs to $\mathcal{H}(G(F_S))$

then ${}^cI_M(\gamma, f)$ is compactly supported as a function of γ in the space of $M^0(F_S)$-orbits in $M(F_S)$ ([1(g), Lemma 4.4]).

We have similar objects for G', of course, and we can pull these back to G. Define

$$^cI_M^{\mathcal{E}}(\gamma, f) = |\ell|_S^{n/2}\, {}^c\hat{I}_{M'}(\gamma', f'), \qquad f \in \tilde{\mathcal{H}}_{ac}(G(F_S)),$$

for any G-regular element $\gamma \in M(F_S)$. If $f \in \tilde{\mathcal{H}}_{ac}(G(F_S))$, $\pi \in \Pi^+_{\text{temp}}(M(F_S))$ and $X \in \mathfrak{a}_{M,S}$, we also set

$$\theta_M^{\mathcal{E}}(f, \pi, X) = e_S \sum_{\xi \in \mathcal{C}_M} \hat{\theta}_{M'}(f', \xi\pi', X')$$

and

$$^c\theta_M^{\mathcal{E}}(f, \pi, X) = e_S \sum_{\xi \in \mathcal{C}_M} {}^c\hat{\theta}_{M'}(f', \xi\pi', X'),$$

where π' is any representation in $\Pi'(\pi)$.

LEMMA 10.1: *As the notation suggests, $\theta_M^{\mathcal{E}}(f, \pi, X)$ and ${}^c\theta_M^{\mathcal{E}}(f, \pi, X)$ are independent of the choice of $\pi' \in \Pi'(\pi)$. The functions $\theta_M^{\mathcal{E}}(f)$ and ${}^c\theta_M^{\mathcal{E}}(f)$ of (π, X) defined by these expressions both belong to $\tilde{\mathcal{I}}_{ac}(M(F_S))$. Moreover, we have*

$$(10.1)^{\mathcal{E}} \qquad \sum_{L \in \mathcal{L}(M)} \hat{\theta}_M^{L,\mathcal{E}}({}^c\theta_L^{\mathcal{E}}(f)) = \sum_{L \in \mathcal{L}(M)} {}^c\hat{\theta}_M^{L,\mathcal{E}}(\theta_L^{\mathcal{E}}(f)) = 0,$$

$$(10.2)^{\mathcal{E}} \qquad I_M^{\mathcal{E}}(\gamma, f) = \sum_{L \in \mathcal{L}(M)} {}^c\hat{I}_M^{L,\mathcal{E}}(\gamma, \theta_L^{\mathcal{E}}(f)),$$

and

$$(10.3)^{\mathcal{E}} \qquad {}^cI_M^{\mathcal{E}}(\gamma, f) = \sum_{L \in \mathcal{L}(M)} \hat{I}_M^{L,\mathcal{E}}(\gamma, {}^c\theta_L^{\mathcal{E}}(f)).$$

Proof. According to Lemma 4.7 of [1(g)],

$$\hat{\theta}_{M'}(f', \pi', X') = \sum_{P \in \mathcal{P}(M)} \omega_P(X) e^{-\nu'_P(X')} \hat{I}_{M'}(\pi'_{\nu'_P}, X', f'),$$

for any $\pi \in \Pi^+_{\text{temp}}(M'(F_S))$. Here

$$\omega_P(X) = \text{vol}(\mathfrak{a}_P^+ \cap B)\, \text{vol}(B)^{-1},$$

where B is a small ball in $\mathfrak{a}_P$ centered at the origin, while ν_P stands for any point in the chamber $(\mathfrak{a}_P^*)^+$ which is far from the walls. It follows from Lemma 8.1 that $\theta_M^{\mathcal{E}}(f, \pi, X)$ is independent of $\pi' \in \Pi'(\pi)$. Moreover, Proposition 10.3 of [1(g)] implies that $\hat{\theta}_{M'}(f', \pi', X')$ vanishes if π' does not belong to a set $\Pi'(\pi)$. It follows from the definitions that $\theta_M^{\mathcal{E}}(f)$ belongs to $\tilde{\mathcal{I}}_{ac}(M(F_S))$. We have thus established the two required properties of $\theta_M^{\mathcal{E}}(f)$. To see that they also hold for ${}^c\theta_M^{\mathcal{E}}(f)$, we must first make an observation.

For any function $f \in \tilde{\mathcal{H}}_{ac}(G(F_S))$, we have

$$\begin{aligned}\theta_G^{\mathcal{E}}(f)'(\pi', X') &= e_S \ell^{-\dim(A_G)} \theta_G^{\mathcal{E}}(f, \pi, X) \\ &= \ell^{-\dim(A_G)} \sum_{\xi \in \mathcal{C}_G} \hat{\theta}_{G'}(f', \xi\pi', X') \\ &= \hat{\theta}_{G'}(f', \pi', X'),\end{aligned}$$

by (1.7) and the definitions above. In other words

$$\theta_G^{\mathcal{E}}(f)' = \hat{\theta}_{G'}(f').$$

This formula is rather trivial, for the maps are defined in [1(g)] so that

$$\theta_G(f) = {}^c\theta_G(f) = f_G.$$

If G is replaced by an arbitrary element $L \in \mathcal{L}$, the corresponding formula does not hold. However, suppose that I' is an invariant distribution on $\tilde{\mathcal{H}}_{ac}(L'(F_S))$ which is supported on characters, and annihilates any function which vanishes on the K_S-bi-invariant set $L'(F_S)^L$. Fourier inversion on the finite abelian group

$$L'(F_S)/L'(F_S)^L$$

then yields the partial result

$$\hat{I}'(\theta_L^{\mathcal{E}}(f)') = \hat{I}'(\hat{\theta}_{L'}(f')).$$

The distribution

$$I'(h) = \sum_{\xi \in \mathcal{C}_M} {}^c\theta_{M'}^{L'}(h, \xi\pi', X'), \qquad h \in \tilde{\mathcal{H}}_{ac}(L'(F_S)),$$

satisfies the two conditions above. Consequently,

$$\begin{aligned}&\sum_{L \in \mathcal{L}(M)} {}^c\hat{\theta}_M^{L,\mathcal{E}}(\theta_L^{\mathcal{E}}(f), \pi, X) \\ &= \sum_L e_S \sum_{\xi \in \mathcal{C}_M} {}^c\hat{\theta}_{M'}^{L'}(\theta_L^{\mathcal{E}}(f)', \xi\pi', X') \\ &= e_S \sum_{\xi \in \mathcal{C}_M} \sum_{L \in \mathcal{L}(M)} {}^c\hat{\theta}_{M'}^{L'}(\hat{\theta}_{L'}(f'), \xi\pi', X').\end{aligned}$$

This vanishes by (10.1), applied to G'. Since $\theta_G^{\mathcal{E}}(f)$ equals f_G, we obtain

$$^c\theta_M^{\mathcal{E}}(f,\pi,X) = -\sum_{\{L\in\mathcal{L}(M):L\neq G\}} {}^c\hat{\theta}_M^{L,\mathcal{E}}(\theta_L^{\mathcal{E}}(f),\pi,X).$$

It follows inductively that $^c\theta_M^{\mathcal{E}}(f,\pi,X)$ is independent of $\pi'\in\Pi'(\pi)$, and that $^c\theta_M^{\mathcal{E}}(f)$ belongs to $\tilde{\mathcal{I}}_{ac}(M(F_S))$. One half of the required formula $(10.1^{\mathcal{E}})$ is also an immediate consequence of this identity.

The remaining assertions of the lemma follow by similar arguments. For if I' is an invariant distribution on $\tilde{\mathcal{H}}_{ac}(L'(F_S))$ which satisfies the two given conditions, we can also establish a formula

$$\hat{I}'({}^c\theta_L^{\mathcal{E}}(f)') = \hat{I}'({}^c\hat{\theta}_{L'}(f')),$$

as above. This holds in particular if

$$I'(h), \qquad h\in\tilde{\mathcal{H}}_{ac}(L'(F_S)),$$

is one of the distributions $\sum_{\xi\in\mathcal{C}_M}\theta_{M'}^{L'}(h,\xi\pi',X')$ or $I_{M'}^{L'}(\gamma',h)$. The other half of $(10.1)^{\mathcal{E}}$, as well as $(10.2)^{\mathcal{E}}$ and $(10.3)^{\mathcal{E}}$, follows without difficulty. ∎

THEOREM 10.2: *Fix an element $M\in\mathcal{L}$ and a finite set S of valuations with the closure property. In the special case that $S\supset S_{\text{ram}}$, assume that*

$$I_L^{\mathcal{E}}(\gamma,f) = I_L(\gamma,f), \qquad f\in\mathcal{H}(G(F_S)),$$

for each $L\in\mathcal{L}(M)$ and $\gamma\in L(F_S)$. Then for any $f\in\tilde{\mathcal{H}}_{ac}(G(F_S))$ and $X\in\mathfrak{a}_{M,S}$, we have

(a) $$\theta_M^{\mathcal{E}}(f) = \theta_M(f),$$

(b) $$^c\theta_M^{\mathcal{E}}(f) = {}^c\theta_M(f),$$

(c) $$I_M^{\mathcal{E}}(\rho,X,f) = I_M(\rho,X,f), \qquad \rho\in\Sigma^+(M(F_S)),$$

and

(d) $$I_M^{\mathcal{E}}(\pi,X,f) = I_M(\pi,X,f), \qquad \pi\in\Pi^+(M(F_S)).$$

Proof. According to the induction assumption of §5,

$$I_L^{L_1,\mathcal{E}}(\gamma) = I_L^{L_1}(\gamma), \qquad \gamma\in L(F_S),$$

if $L\subset L_1\subsetneq G$ and $S\supset S_{\text{ram}}$. We may therefore assume inductively that the four required formulas of this theorem hold if G is replaced by any such L_1. We shall also assume inductively that the four formulas hold for G, but with M replaced by any Levi subset $L\supsetneq M$.

It suffices to prove the theorem for a fixed function $f \in \mathcal{H}(G(F_S))$. For the restriction of a given function in $\tilde{\mathcal{H}}_{ac}(G(F_S))$ to any fixed set

$$G(F_S)^Z = \{x \in G(F_S) : H_G(x) = Z\}, \qquad Z \in \mathfrak{a}_{G,S},$$

coincides with that of some function in $\mathcal{H}(G(F_S))$. Suppose that $\gamma \in M(F_S)$ is G-regular. The given hypothesis permits us to use Theorem 6.1, and in particular the expansion (6.1). Combined with the descent properties (3.3) and (6.2), this becomes

$$I_M^{\mathcal{E}}(\gamma, f) = \sum_{L_1 \in \mathcal{L}(M)} \varepsilon_M^{L_1}(S) I_{L_1}(\gamma, f). \tag{10.4}$$

Anticipating a similar formula for ${}^cI_M^{\mathcal{E}}(\gamma, f)$, let us consider the expression

$${}^cI_M^{\mathcal{E}}(\gamma, f) - \sum_{L_1 \in \mathcal{L}(M)} \varepsilon_M^{L_1}(S)\, {}^cI_{L_1}(\gamma, f). \tag{10.5}$$

By (10.3) and $(10.3)^{\mathcal{E}}$, we can write this as the sum of

$$\hat{I}_M^M(\gamma, {}^c\theta_M^{\mathcal{E}}(f) - {}^c\theta_M(f)) \tag{10.6}$$

and

$$\sum_{L \supsetneq M} \Big\{ \hat{I}_M^{L,\mathcal{E}}(\gamma, {}^c\theta_L^{\mathcal{E}}(f)) - \sum_{L_1 \in \mathcal{L}^L(M)} \varepsilon_M^{L_1}(S) \hat{I}_{L_1}^L(\gamma, {}^c\theta_L(f)) \Big\}.$$

Now if $L \neq M$, we have

$$\sum_{L_1 \in \mathcal{L}^L(M)} \varepsilon_M^{L_1}(S) \hat{I}_{L_1}^L(\gamma, {}^c\theta_L(f)) = \hat{I}_M^{L,\mathcal{E}}(\gamma, {}^c\theta_L(f)) = \hat{I}_M^{L,\mathcal{E}}(\gamma, {}^c\theta_L^{\mathcal{E}}(f)),$$

by our induction hypothesis and (10.4) (with G replaced by L). Therefore the second expression vanishes, and (10.5) equals (10.6). Since f belongs to $\mathcal{H}(G(F_S))$, the expression (10.5) has bounded support as a function of γ in the space of $M^0(F_S)$-orbits in $M(F_S)$. The same is therefore true of (10.6). For a given $X \in \mathfrak{a}_{M,S}$, (10.6) is the orbital integral in

$$\{\gamma \in M(F_S) : H_M(\gamma) = X\}$$

of a function defined on

$$M(F_S)^X = \{x \in M(F_S) : H_M(x) = X\}.$$

The tempered characters of this function are just

$${}^c\theta_M^{\mathcal{E}}(f, \pi, X) - {}^c\theta_M(f, \pi, X), \qquad \pi \in \Pi_{\text{temp}}(M(F_S)).$$

It follows that the difference is compactly supported in X.

In [1(g), §5] we defined a meromorphic function

$$^c\theta_M(f,\pi_\lambda)=\int_{\mathfrak{a}_{M,S}} {}^c\theta_M(f,\pi,X)e^{\lambda(X)}dX, \qquad \lambda\in\mathfrak{a}^*_{M,\mathbf{C}},$$

for each $\pi\in\Pi^+_{\text{temp}}(M(F_S))$. The definition extends readily to standard representations $\rho\in\Sigma^+(M(F_S))$ by analytic continuation. Define

$$^cI_M(\rho,X,f)=\sum_{L\in\mathcal{L}(M)}\hat{I}^L_M(\rho,X,{}^c\theta_L(f)), \quad \rho\in\Sigma^+(M(F_S)).$$

Then Proposition 5.4 of [1(g)] asserts that

$$(10.7)\quad {}^cI_M(\rho,X,f)=\lim_{\beta}\sum_{P\in\mathcal{P}(M)}\omega_P\int_{\varepsilon_P+i\mathfrak{a}^*_{M,S}}\hat{\beta}(\lambda){}^c\theta_M(f,\rho_\lambda)e^{-\lambda(X)}d\lambda,$$

where β stands for a test function in $C^\infty_c(\mathfrak{a}_{M,S})$ which approaches the Dirac measure at the origin, and $X\in\mathfrak{a}_{M,S}$ is any point at which the left-hand side is smooth. (If $\mathfrak{a}_{M,S}$ is discrete, X can be any point and β may be removed from the formula.) Thus, $I_M(\rho,X,f)$ may be computed inductively from the function ${}^c\theta_M(f,\rho_\lambda)$.

In a similar fashion, we define

$$^c\theta^{\mathcal{E}}_M(f,\pi_\lambda)=\int_{\mathfrak{a}_{M,S}} {}^c\theta^{\mathcal{E}}_M(f,\pi,X)e^{\lambda(X)}dX, \qquad \lambda\in\mathfrak{a}^*_{M,\mathbf{C}},$$

for each $\pi\in\Pi^+_{\text{temp}}(M(F_S))$. If π' is any representation in $\Pi'(\pi)$,

$$\begin{aligned}
{}^c\theta^{\mathcal{E}}_M(f,\pi_\lambda)&=\int_{\mathfrak{a}_{M,S}}\sum_{\xi\in C_M}e_S{}^c\hat{\theta}_M(f',\xi\pi',X')e^{\lambda(X)}dX\\
&=\ell^{-\dim A_M}\int_{\mathfrak{a}_{M',S}}\sum_{\xi}e_S{}^c\hat{\theta}_M(f',\xi\pi',X')e^{\lambda'(X')}dX'\\
&=\ell^{-\dim A_M}\sum_{\xi}e_S{}^c\hat{\theta}_M(f',\xi\pi'_{\lambda'}),
\end{aligned}$$

since dX equals $\ell^{-\dim A_M}dX'$. These formulas again extend by analytic continuation to representations $\rho\in\Sigma^+(M(F_S))$. We can also define

$$^cI^{\mathcal{E}}_M(\rho,X,f)=\sum_{L\in\mathcal{L}(M)}\hat{I}^{L,\mathcal{E}}_M(\rho,X,{}^c\theta^{\mathcal{E}}_L(f)).$$

Then

$$\begin{aligned}{}^{c}I_M^{\mathcal{E}}(\rho, X, f) &= \sum_{L\in\mathcal{L}(M)} \sum_{\xi\in\mathcal{C}_M} e_S \hat{I}_{M'}^{L'}(\xi\rho', X', {}^{c}\theta_L^{\mathcal{E}}(f)') \\ &= \sum_{\xi\in\mathcal{C}_M} e_S \sum_{L\in\mathcal{L}(M)} \hat{I}_{M'}^{L'}(\xi\rho', X', {}^{c}\hat{\theta}_{L'}(f')) \\ &= \sum_{\xi\in\mathcal{C}_M} e_S {}^{c}\hat{I}_{M'}(\xi\rho', X', f').\end{aligned}$$

(See the proof of Lemma 10.1.) To this last expression we apply (10.7), (with f, ρ, and X replaced by f', $\xi\rho'$ and X'). We obtain

$$\lim_{\beta} \sum_{P\in P(M)} \omega_P \int_{\varepsilon_P + i\mathfrak{a}^*_{M,S}} \hat{\beta}(\lambda') \sum_{\xi\in\mathcal{C}_M} e_S {}^{c}\hat{\theta}_M(f', \xi\rho'_{\lambda'}) e^{-\lambda'(X')} d\lambda'.$$

Finally, we substitute the formula above for ${}^{c}\theta_M^{\mathcal{E}}(f, \rho_\lambda)$. Remembering that $d\lambda'$ equals $\ell^{-\dim(A_M)} d\lambda$, we see that

$$(10.7)^{\mathcal{E}} \quad {}^{c}I_M^{\mathcal{E}}(\rho, X, f) = \lim_{\beta} \sum_{P\in\mathcal{P}(M)} \omega_P \int_{\varepsilon_P + i\mathfrak{a}^*_{M,S}} \hat{\beta}(\lambda) {}^{c}\theta_M^{\mathcal{E}}(f, \rho_\lambda) e^{-\lambda(X)} d\lambda,$$

for any smooth point X.

We shall apply (10.7) with $\rho = \pi_\mu$, where $\pi \in \Pi^+_{\text{temp}}(M(F_S))$ and $\mu \in \mathfrak{a}^*_M$. We may as well take μ to be in general position. Then the contours of integration on the right-hand side of (10.7) can all be deformed to $i\mathfrak{a}^*_{M,S}$. Consequently,

$$\begin{aligned}{}^{c}I_M(\pi_\mu, X, f) &= \lim_{\beta} \int_{i\mathfrak{a}^*_{M,S}} \hat{\beta}(\lambda) {}^{c}\theta_M(f, \pi_{\mu+\lambda}) e^{-\lambda(X)} d\lambda \\ &= e^{\mu(X)} \lim_{\beta} \int_{\mu + i\mathfrak{a}^*_{M,S}} \hat{\beta}(\lambda - \mu) {}^{c}\theta_M(f, \pi_\lambda) e^{-\lambda(X)} d\lambda.\end{aligned}$$

Remember that β is to approach the Dirac measure at the origin. But

$$\lambda \to \hat{\beta}(\lambda - \mu)$$

is the Fourier–Laplace transform of a function

$$X \to e^{-\mu(X)} \beta(X)$$

which also approaches the Dirac measure at the origin. We may therefore replace $\hat{\beta}(\lambda - \mu)$ by $\hat{\beta}(\lambda)$. We obtain

$$e^{-\mu(X)} {}^{c}I_M(\pi_\mu, X, f) = \lim_{\beta} \int_{\mu + i\mathfrak{a}^*_{M,S}} \hat{\beta}(\lambda) {}^{c}\theta_M(f, \pi_\lambda) e^{-\lambda(X)} d\lambda.$$

A similar formula arises from $(10.7)^{\mathcal{E}}$. Taking the difference of the two, we see that

$$e^{-\mu(X)}({}^cI_M^{\mathcal{E}}(\pi_\mu, X, f) - {}^cI_M(\pi_\mu, X, f)) \tag{10.8}$$

equals

$$\lim_{\beta} \int_{\mu+i\mathfrak{a}^*_{M,S}} \hat{\beta}(\lambda)({}^c\theta_M^{\mathcal{E}}(f, \pi_\lambda) - {}^c\theta_M(f, \pi_\lambda))e^{-\lambda(X)}d\lambda.$$

Now

$$\lambda \to {}^c\theta_M^{\mathcal{E}}(f, \pi_\lambda) - {}^c\theta_M(f, \pi_\lambda)$$

is the Fourier transform of the compactly supported function

$$X \to {}^c\theta_M^{\mathcal{E}}(f, \pi, X) - {}^c\theta_M(f, \pi, X),$$

and is therefore entire. Consequently, the integral over $\mu + i\mathfrak{a}^*_{M,S}$ can be deformed to any other translate of $i\mathfrak{a}^*_{M,S}$. The outcome of this discussion is that (10.8) is independent of μ. At least this is the case for almost all μ and X. But there are formulas in [1(g), §4] which allow us to express the value of (10.8) at any μ and X in terms of its values at nearby points in general position. It follows that (10.8) is independent of μ without restriction.

According to Lemma 4.5 of [1(g)],

$$\sum_{P \in \mathcal{P}(M)} \omega_P(X)e^{-\nu_P(X)}\, {}^cI_M(\pi_{\nu_P}, X, f) = 0.$$

(The notation $\omega_P(X)$ and ν_P was described at the beginning of the proof of Lemma 10.1.) Applying the same formula to G', we obtain

$$\sum_{P \in \mathcal{P}(M)} \omega_P(X)e^{-\nu_P(X)}\, {}^cI_M^{\mathcal{E}}(\pi_{\nu_P}, X, f) = 0.$$

Since (10.8) is independent of μ, we can express its value at any μ as the sum over $P \in \mathcal{P}(M)$ of the product of $\omega_P(X)$ with its value at ν_P. Consequently, (10.8) vanishes for any μ. We have therefore established that

$${}^cI_M^{\mathcal{E}}(\pi_\mu, X, f) = {}^cI_M(\pi_\mu, X, f),$$

for any $\pi \in \Pi^+_{\mathrm{temp}}(M(F_S))$ and $\mu \in \mathfrak{a}^*_M$. The next step is to set $\mu = 0$. For there is another result (Lemma 4.7 of [1(g)]) which asserts that

$${}^cI_M(\pi, X, f) = {}^c\theta_M(f, \pi, X), \qquad \pi \in \Pi^+_{\mathrm{temp}}(M(F_S)).$$

The same result applied to G' yields

$${}^cI_M^{\mathcal{E}}(\pi, X, f) = {}^c\theta_M^{\mathcal{E}}(f, \pi, X).$$

Combining the three formulas, we see at last that ${}^c\theta^{\mathcal{E}}_M(f)$ equals ${}^c\theta_M(f)$. This is the second assertion of the theorem. The first assertion is the equality of $\theta^{\mathcal{E}}_M(f)$ and $\theta_M(f)$. This follows from the second assertion, our induction hypothesis, and the formulas (10.1) and $(10.1)^{\mathcal{E}}$.

The third assertion of the theorem will follow from a comparison of (10.7) with $(10.7)^{\mathcal{E}}$. Our definitions, together with the second assertion of the theorem (which we have already proved), imply that

$${}^c\theta^{\mathcal{E}}_M(f,\pi_\lambda) = {}^c\theta_M(f,\pi_\lambda), \qquad \pi \in \Pi^+_{\text{temp}}(M(F_S)),\ \lambda \in \mathfrak{a}^*_{M,\mathbf{C}}.$$

But by analytic continuation, this formula remains valid if π is replaced by the standard representation ρ. Therefore, the right-hand sides of (10.7) and $(10.7)^{\mathcal{E}}$ are equal. Consider, then, the resulting equality

$${}^cI^{\mathcal{E}}_M(\rho,X,f) = {}^cI_M(\rho,X,f)$$

of left-hand sides. By our induction assumption and the second assertion of the theorem, we have

$$\hat{I}^{L,\mathcal{E}}_M(\rho,X,{}^c\theta^{\mathcal{E}}_L(f)) = \hat{I}^L_M(\rho,X,{}^c\theta_L(f))$$

for any $L \in \mathcal{L}(M)$ with $L \neq G$. We therefore obtain

$$I^{\mathcal{E}}_M(\rho,X,f) = I_M(\rho,X,f),$$

the third assertion of the theorem.

Finally, suppose that π is any representation in $\Pi^+(M(F_S))$. We defined $I^{\mathcal{E}}_M(\pi,X,f)$ by (8.4), an expansion in the distributions

$$I^{\mathcal{E}}_L(\rho_\lambda,h_L(X),f), \quad \rho \in \Sigma^+(M(F_S)),\ \lambda \in \varepsilon_P + i\mathfrak{a}^*_{M,S},\ L \in \mathcal{L}(M).$$

We also noted that $I_M(\pi,X,f)$ satisfies (8.1), an identical expansion in the distributions

$$I_L(\rho_\lambda,h_L(X),f).$$

The fourth and final assertion of the theorem then follows from the third assertion, with M replaced by L. ∎

COROLLARY 10.3: *Under the assumptions of the theorem, we have*

$${}^cI^{\mathcal{E}}_M(\gamma,f) = \sum_{L\in\mathcal{L}(M)} \varepsilon^L_M(S)\,{}^cI_L(\gamma,f), \qquad f \in \tilde{\mathcal{H}}_{ac}(G(F_S)),$$

for any G-regular element $\gamma \in M(F_S)$.

Proof. In the proof of the theorem we established the equality of (10.5) with (10.6). But by the second assertion of the theorem, (10.6) vanishes. The corollary follows. ∎

We shall now use Theorem 10.2 to prove the local assertion (i) of Theorem B. We are not at liberty to assume the equality of $I_L^{\mathcal{E}}(\gamma)$ and $I_L(\gamma)$. However, we are carrying our original induction assumption from §5. Consequently, if $L \subset L_1 \subsetneq G$, we may assume that $I_L^{L_1,\mathcal{E}}(\gamma)$ and $I_L^{L_1}(\gamma)$ are equal.

Now, suppose that $\pi \in \Pi_{\mathrm{unit}}^+(M(F_S))$. We must show that

$$I_M^{\mathcal{E}}(\pi, 0, f) = I_M(\pi, 0, f), \qquad f \in \mathcal{H}(G(F_S)).$$

As we noted near the end of the proof above, these distributions have identical expansions in terms of the distributions associated to the standard representations $\rho \in \Sigma^+(M(F_S))$. Moreover, only those ρ with $\Delta(\pi, \rho) \neq 0$ can occur in the expansions. Since π is unitary, any such ρ must have a unitary central character. It is therefore sufficient to establish the formula

$$(10.9) \qquad I_L^{\mathcal{E}}(\rho_\lambda, h_L(X), f) = I_L(\rho_\lambda, h_L(X), f), \quad L \in \mathcal{L}(M),\ X \in \mathfrak{a}_M,$$

for any such ρ and any point $\lambda \in \mathfrak{a}_{M,\mathbb{C}}^*$ with small real part.

We will use the splitting and descent formulas for $I_L(\rho_\lambda, h_L(X), f)$ and $I_L^{\mathcal{E}}(\rho_\lambda, h_L(X), f)$. (See the proof of Lemma 8.1.) By the splitting property, we need only establish (10.9) under the assumption that S contains one valuation v. Suppose that this is the case. Since the central character is unitary, ρ is either tempered or induced from a proper parabolic subset. If ρ is tempered, we have

$$I_L^{\mathcal{E}}(\rho_\lambda, h_L(X), f) = I_L(\rho_\lambda, h_L(X), f) = \begin{cases} 0 & , \quad L \neq G, \\ f_G(\rho_\lambda^G, h_G(X)) & , \quad L = G, \end{cases}$$

as may be seen from the proof of [1(g), Lemma 3.1]. In the other case

$$\rho = \rho_1^M, \qquad M_1 \subsetneq M,\ \rho_1 \in \Sigma^+(M_1(F_v)),$$

and we can make use of the descent property of each side of (10.9). We find that we need only establish the formula

$$\hat{I}_{M_1}^{L_1,\mathcal{E}}(\rho_{1,\lambda}, X_1, f_{L_1}) = \hat{I}_{M_1}^{L_1}(\rho_{1,\lambda}, X_1, f_{L_1}), \quad L_1 \in \mathcal{L}(M_1),\ X_1 \in \mathfrak{a}_{M_1},$$

with $L_1 \neq G$. Since we are assuming the equality of $I_L^{L_1,\mathcal{E}}(\gamma)$ and $I_L^{L_1}(\gamma)$, the formula follows from Theorem 10.2, with (G, M) replaced by (L_1, M_1).

This completes the verification of (10.9) and therefore of Theorem B(i).

11. More on normalizing factors

This section is a digression. We shall discuss some further questions related to the comparison of normalizing factors.

Suppose that $\pi \in \Pi_{\mathrm{disc}}(M,t)$. Then π can be identified with an orbit $\{\pi_\lambda\}$ of $i\mathfrak{a}_M^*$ in $\Pi_{\mathrm{unit}}^+(M(\mathbf{A}))$. The global normalizing factors are functions

$$r_{P_1|P_2}(\pi_\lambda) = \prod_{\alpha \in \Sigma_{P_1} \cap \Sigma_{\overline{P}_2}} r_\alpha(\pi, \lambda(\alpha^\vee)), \qquad P_1, P_2 \in \mathcal{P}(M),$$

which are meromorphic in $\lambda \in \mathfrak{a}_{M,\mathbf{C}}^*$ and which are regular for imaginary λ. They satisfy

$$(11.1) \qquad r_{P_1|P_3}(\pi_\lambda) = r_{P_1|P_2}(\pi_\lambda)\, r_{P_2|P_3}(\pi_\lambda).$$

(See [1(h), §4].) The global factors $r_\alpha(\quad)$ can be obtained from the local normalizing factors defined in §2. In fact, if π is identified with a unitary representation $\bigotimes_v \pi_v$ of $M(\mathbf{A})$, they are related by an infinite product

$$r_\alpha(\pi, \lambda(\alpha^\vee)) = \prod_v r_\alpha(\pi_v, \lambda(\alpha^\vee)),$$

which converges in some right half plane. Suppose that $\pi' = \bigotimes_v \pi'_v$ is a representation in $\Pi'(\pi)$. The formula (2.3) for the local normalizing factors can be written

$$r_\alpha(\pi_v, s) = \lambda_{\alpha,v} \prod_{\xi \in \mathcal{C}_M/\mathcal{C}_{M_\alpha}} r_{\alpha'}(\xi\pi'_v, s).$$

Here $M_\alpha \supset M$ is the Levi subset such that

$$\mathfrak{a}_{M_\alpha} = \{ H \in \mathfrak{a}_M : \alpha(H) = 0 \}.$$

Since

$$\prod_v \lambda_{\alpha,v} = 1,$$

we obtain

$$(11.2) \qquad r_\alpha(\pi, s) = \prod_{\xi \in \mathcal{C}_M/\mathcal{C}_{M_\alpha}} r_{\alpha'}(\xi\pi', s).$$

In particular, the expression on the right depends only on π and not π'. Observe that if the representation $\pi' \in \Pi'(\pi)$ is not automorphic, the constituents $r_{\alpha'}(\xi\pi', s)$ may not be defined for all s. They are defined in general only for s in some right half plane.

The functions which occur in the trace formula are built out of the (G, M) family

$$(11.3)\quad r_P(\nu, \pi_\lambda, P_0) = r_{P|P_0}(\pi_\lambda)^{-1} r_{P|P_0}(\pi_{\lambda+\nu}), \qquad P \in \mathcal{P}(M),\ \nu \in i\mathfrak{a}_M^*.$$

The associated functions $r_M^L(\pi_\lambda)$, $L \in \mathcal{L}(M)$, are analytic in $\lambda \in i\mathfrak{a}_M^*$, and are independent of P_0.

LEMMA 11.1: . *For each $L \in \mathcal{L}(M)$ we have*

$$r_M^L(\pi_\lambda) = \ell^{-\dim(A_M/A_L)} \sum_{\xi \in \mathcal{C}_M/\mathcal{C}_L} r_{M'}^{L'}(\xi\pi'_{\lambda'}),$$

where π' is any representation in $\Pi'(\pi)$.

Proof. Since the (G, M) family (11.3) is defined as a product of functions indexed by roots, we can apply Lemma 7.1 of [1(c)]. We obtain

$$r_M^L(\pi_\lambda) = \sum_{\Phi} \mathrm{vol}(\mathfrak{a}_M^L/\mathbf{Z}(\Phi^\vee)) \prod_{\alpha\in\Phi} r_\alpha(\pi, \lambda(\alpha^\vee))^{-1} r_\alpha^{(1)}(\pi, \lambda(\alpha^\vee)),$$

where Φ is taken over all subsets of the roots of (G, A_M) for which

$$\Phi^\vee = \{\alpha^\vee : \alpha \in \Phi\}$$

is a basis of $\mathfrak{a}_M^L$, and $r_\alpha^{(1)}$ denotes the derivative with respect to the second variable. Observe that the map $X \to X'$ from $\mathfrak{a}_M^L$ to $\mathfrak{a}_{M'}^{L'}$ sends any "co-root" $\alpha^\vee$ to the corresponding "co-root" $(\alpha')^\vee$. Since the map expands volume by a factor $\ell^{\dim(A_M/A_L)}$, relative to our fixed measure on $\mathfrak{a}_M^L \cong \mathfrak{a}_{M'}^{L'}$, we have

$$\mathrm{vol}(\mathfrak{a}_M^L/\mathbf{Z}(\Phi^\vee)) = \ell^{-\dim(A_M/A_L)} \mathrm{vol}(\mathfrak{a}_{M'}^{L'}/\mathbf{Z}((\Phi')^\vee))$$

for any Φ. Notice also that there is a canonical isomorphism

$$\mathcal{C}_M/\mathcal{C}_L \xrightarrow{\sim} \bigoplus_{\alpha\in\Phi} \mathcal{C}_M/\mathcal{C}_{M_\alpha}.$$

Combined with (11.2) and the fact $\lambda(\alpha^\vee) = \lambda'((\alpha')^\vee)$, this gives

$$\prod_{\alpha\in\Phi} r_\alpha(\pi, \lambda(\alpha^\vee))^{-1} r_\alpha^{(1)}(\pi, \lambda(\alpha^\vee))$$
$$= \sum_{\xi\in\mathcal{C}_M/\mathcal{C}_L} \prod_{\alpha'\in\Phi'} r_{\alpha'}(\xi\pi', \lambda'((\alpha')^\vee))^{-1} r_{\alpha'}^{(1)}(\xi\pi', \lambda'(\alpha')^\vee)).$$

Applying Lemma 7.1 of [1(c)] to the function

$$r_{M'}^{L'}(\xi\pi'_{\lambda'}),$$

we obtain the required formula. ∎

It is actually a slight variant of Lemma 11.1 that we will need to use. If τ is any representation in $\Pi_{\mathrm{disc}}(M', t)$, set

$$r_{P_1|P_2}(\check{\tau}_\lambda) = \prod_{\alpha \in \Sigma_{P_1} \cap \Sigma_{\overline{P}_2}} \prod_{\xi \in \mathcal{C}_M / \mathcal{C}_{M_\alpha}} r_{\alpha'}(\xi\tau, \lambda'((\alpha')^\vee)),$$

and let

$$r_P(\nu, \check{\tau}_\lambda, P_0) = r_{P|P_0}(\check{\tau}_\lambda)^{-1} r_{P|P_0}(\check{\tau}_{\lambda+\nu}), \qquad P \in \mathcal{P}(M),\ \nu \in i\mathfrak{a}_M^*,$$

be the associated (G, M) family. The proof of the lemma gives

COROLLARY 11.2: .

$$r_M^L(\check{\tau}_\lambda) = \ell^{-\dim(A_M/A_L)} \sum_{\xi \in \mathcal{C}_M/\mathcal{C}_L} r_{M'}^{L'}(\xi\tau_{\lambda'}),$$

for each $L \in \mathcal{L}(M)$. ∎

Suppose that π is as above, and that

$$\pi_\lambda = \bigotimes_v \pi_{v,\lambda}, \qquad \pi_v \in \Pi_{\mathrm{unit}}^+(M(F_v)).$$

We can write

$$r_{P_1|P_2}(\pi_\lambda) = \prod_v r_{P_1|P_2}(\pi_{v,\lambda}),$$

the right hand side being defined by analytic continuation. The local factors do not satisfy the product formula (11.1). However, it is important to consider certain quotients of local factors which do satisfy this formula. The functions $r_M^L(\pi_\lambda, \rho_\lambda)$ discussed in §8 arise from examples of this sort. More examples are provided by coupling representations of G and G'.

Suppose that $\tau = \bigotimes_v \tau_v$ and $\pi = \bigotimes_v \pi_v$ are arbitrary representations in $\Pi(M'(\mathbf{A}))$ and $\Pi^+(M(\mathbf{A}))$ respectively. Assume that the number

$$\delta(\tau, \pi) = \prod_v \delta(\pi_v, \tau_v)$$

defined in §8 does not vanish. For each root α of (G, A_M) and each v, set

$$\widetilde{r}_\alpha(\check{\tau}_v, \pi_v, s) = \Big(\lambda_{\alpha,v} \prod_{\xi \in \mathcal{C}_M/\mathcal{C}_{M_\alpha}} r_{\alpha'}(\xi\tau_v, s)\Big)^{-1} r_\alpha(\pi_v, s).$$

This function is constructed out of local L-functions. If τ_v and π_v are unramified, the representation τ_v must belong to $\Pi'(\pi_v)$. It follows from the product formula above that in this case

$$\widetilde{r}_\alpha(\check{\tau}_v, \pi_v, s) = 1.$$

We can therefore define

$$\widetilde{r}_\alpha(\check{\tau},\pi,s)=\prod_v \widetilde{r}_\alpha(\check{\tau}_v,\pi_v,s)$$

as a product over a finite set S. We can then set

$$\widetilde{r}_{P_1|P_2}(\check{\tau}_\lambda,\pi_\lambda)=\prod_{\alpha\in\Sigma_{P_1}\cap\Sigma_{\overline{P}_2}} \widetilde{r}_\alpha(\check{\tau},\pi,\lambda(\alpha^\vee)).$$

Now for any given v,

$$\delta(\tau_v,\pi_v)=\sum_{\rho_v\in\Sigma^+(M(F_v))}\ \sum_{\rho'_v\in\Sigma'(\rho_v)} e_v\Delta(\tau_v,\rho'_v)\,\Gamma(\rho_v,\pi_v)\neq 0,$$

so we may choose ρ_v and ρ'_v so that $\Delta(\tau_v,\rho'_v)\,\Gamma(\rho_v,\pi_v)$ does not vanish. Since

$$r_\alpha(\rho_v,s)=\lambda_{\alpha,v}\sum_{\xi\in\mathcal{C}_M/\mathcal{C}_{M_\alpha}} r_{\alpha'}(\xi\rho'_v,s),$$

we can write $\widetilde{r}_\alpha(\check{\tau}_v,\pi_v,s)$ as the product of

$$\left(\prod_{\xi\in\mathcal{C}_M/\mathcal{C}_{M_\alpha}} r_{\alpha'}(\xi\tau_v,s)\ r_{\alpha'}(\xi\rho'_v,s)^{-1}\right)^{-1}$$

and

$$r_\alpha(\rho_v,s)^{-1}\,r_\alpha(\pi_v,s).$$

By Lemma 5.2 of [1(f)], these are rational functions of s if v is Archimedean, and of q_v^{-s} if v is non-Archimedean of residual order q_v. The same is therefore true of $\widetilde{r}_\alpha(\check{\tau}_v,\pi_v,s)$. Another consequence of the lemma is the formula

$$\widetilde{r}_\alpha(\check{\tau}_v,\pi_v,s)\widetilde{r}_{-\alpha}(\check{\tau}_v,\pi_v,s)=1.$$

In other words,

$$\widetilde{r}_\alpha(\check{\tau},\pi,s)\,\widetilde{r}_{-\alpha}(\check{\tau},\pi,s)=1.$$

From this we see easily that

$$\widetilde{r}_{P_1|P_3}(\check{\tau}_\lambda,\pi_\lambda)=\widetilde{r}_{P_1|P_2}(\check{\tau}_\lambda,\pi_\lambda)\,\widetilde{r}_{P_2|P_3}(\check{\tau}_\lambda,\pi_\lambda),$$

for any P_1, P_2 and P_3 in $\mathcal{P}(M)$.

We define a (G,M) family

$$\widetilde{r}_P(\nu,\check{\tau}_\lambda,\pi_\lambda,P_0)=\widetilde{r}_{P|P_0}(\check{\tau}_\lambda,\pi_\lambda)^{-1}\,\widetilde{r}_{P|P_0}(\check{\tau}_{\lambda+\nu},\pi_{\lambda+\nu}),\quad P\in\mathcal{P}(M),\ \nu\in i\mathfrak{a}_M^*.$$

LEMMA 11.3: . (a) *Take τ and π as above. Then for each $L \in \mathcal{L}(M)$, $\widetilde{r}_M^L(\check{\tau}_\lambda, \pi_\lambda)$ is a rational function of the variables (8.2) which is independent of P_0. (In (8.2) S is understood to be any finite set of valuations outside of which τ and π are unramified.)*

(b) *Suppose in addition that $\tau \in \Pi_{\text{disc}}(M', t)$ and $\pi \in \Pi_{\text{disc}}(M, t)$. Then*

$$\widetilde{r}_{P_1|P_2}(\check{\tau}_\lambda, \pi_\lambda) = r_{P_1|P_2}(\check{\tau}_\lambda)^{-1} r_{P_1|P_2}(\pi_\lambda).$$

In particular, for each $L \in \mathcal{L}(M)$ the function $\widetilde{r}_M^L(\check{\tau}_\lambda, \pi_\lambda)$ is regular for $\lambda \in i\mathfrak{a}_M^*$. Moreover,

$$r_M^L(\pi_\lambda) = \sum_{L_1 \in \mathcal{L}^L(M)} r_M^{L_1}(\check{\tau}_\lambda)\, \widetilde{r}_{L_1}^L(\check{\tau}_\lambda, \pi_\lambda).$$

(As we have done before, we have written τ_λ and π_λ when in the last formula we really mean the induced representations $\tau_\lambda^{L_1'}$ and $\pi_\lambda^{L_1}$.)

Proof. Everything but the last assertion of the lemma follows from the discussion above. Notice that under the conditions of (b) we have a decomposition

$$r_P(\nu, \pi_\lambda, P_0) = \widetilde{r}_P(\nu, \check{\tau}_\lambda, \pi_\lambda, P_0)\, r_P(\nu, \check{\tau}_\lambda, P_0)$$

into a product of (G, M) families. The last assertion therefore follows from Lemma 6.5 of [1(b)]. ∎

It is clear that there are other (G, M) families which are similar to those just discussed. For example, suppose that $\rho = \bigotimes_v \rho_v$ is a standard representation in $\Sigma^+(M(\mathbf{A}))$ such that the number

$$\Delta(\tau, \rho) = \prod_v \Delta(\tau_v, \rho_v)$$

does not vanish. Then by replacing π by ρ in the discussion above, we can define a (G, M) family

$$\widetilde{r}_P(\nu, \check{\tau}_\lambda, \rho_\lambda, P_0) = \widetilde{r}_{P|P_0}(\check{\tau}_\lambda, \rho_\lambda)^{-1}\, \widetilde{r}_{P|P_0}(\check{\tau}_{\lambda+\nu}, \rho_{\lambda+\nu}).$$

For another example, let σ and τ be representations in $\Pi(M'(\mathbf{A}))$ such that for some $\pi \in \Pi(M(\mathbf{A}))$, the numbers $\delta(\sigma, \pi)$ and $\delta(\tau, \pi)$ are nonzero. Then

$$\widetilde{r}_P(\nu, \check{\sigma}_\lambda, \check{\tau}_\lambda, P_0) = \widetilde{r}_P(\nu, \check{\sigma}_\lambda, \pi_\lambda, P_0)\, \widetilde{r}_P(\nu, \check{\tau}_\lambda, \pi_\lambda, P_0)^{-1}$$

is a (G, M) family which is independent of π. It satisfies an obvious analogue of Lemma 11.3. Notice that Lemma 6.5 of [1(b)] provides additional expansion formulas

$$\tilde{r}_M^L(\check{\tau}_\lambda, \rho_\lambda) = \sum_{L_1 \in \mathcal{L}^L(M)} \tilde{r}_M^{L_1}(\check{\tau}_\lambda, \pi_\lambda)\tilde{r}_{L_1}^L(\pi_\lambda, \rho_\lambda) \tag{11.4}$$

and

$$\tilde{r}_M^L(\check{\sigma}_\lambda, \pi_\lambda) = \sum_{L_1 \in \mathcal{L}^L(M)} \tilde{r}_M^{L_1}(\check{\sigma}_\lambda, \check{\tau}_\lambda)\tilde{r}_{L_1}^L(\check{\tau}_\lambda, \pi_\lambda). \tag{11.5}$$

Finally, suppose that τ, ρ and π are *arbitrary* representations in $\Pi(M'(\mathbf{A}))$, $\Sigma^+(M(\mathbf{A}))$ and $\Pi^+(M(\mathbf{A}))$ respectively. The functions $r_M^L(\pi_\lambda, \rho_\lambda)$ used in §8 are obtained from the (G, M) family

$$r_P(\nu, \pi_\lambda, \rho_\lambda, P_0) = \Delta(\pi, \rho)\tilde{r}_P(\nu, \pi_\lambda, \rho_\lambda, P_0), \qquad P \in \mathcal{P}(M),\ \nu \in i\mathfrak{a}_M^*.$$

In a similar fashion, we define

$$r_P(\nu, \check{\tau}_\lambda, \pi_\lambda, P_0) = \delta(\tau, \pi)\tilde{r}_P(\nu, \check{\tau}_\lambda, \pi_\lambda, P_0)$$

and

$$r_P(\nu, \check{\tau}_\lambda, \rho_\lambda, P_0) = \Delta(\tau, \rho)\tilde{r}_P(\nu, \check{\tau}_\lambda, \rho_\lambda, P_0)$$

for $P \in \mathcal{P}(M)$ and $\nu \in i\mathfrak{a}_M^*$. These two new (G, M) families (as well as the earlier ones) satisfy versions of Lemma 11.1. We shall comment explicitly only on the case of the latter one.

LEMMA 11.4: . *For each $L \in \mathcal{L}(M)$ we have*

$$r_M^L(\check{\tau}_\lambda, \rho_\lambda) = \ell^{-\dim(A_M/A_L)} \sum_{\xi \in \mathcal{C}_M/\mathcal{C}_L} \sum_{\rho' \in \Sigma'(\rho)} r_{M'}^{L'}(\xi\tau_{\lambda'}, \xi\rho'_{\lambda'}).$$

Proof. By definition,

$$r_M^L(\check{\tau}_\lambda, \rho_\lambda) = \Delta(\tau, \rho)\,\tilde{r}_M^L(\check{\tau}_\lambda, \rho_\lambda),$$

Arguing as in Lemma 11.1, we see that

$$\tilde{r}_M^L(\check{\tau}_\lambda, \rho_\lambda) = \ell^{-\dim(A_M/A_L)} \sum_{\xi \in \mathcal{C}_M/\mathcal{C}_L} \tilde{r}_M^L(\xi\tau_{\lambda'}, \xi\rho'_{\lambda'}),$$

for any $\rho' \in \Sigma'(\rho)$ with $\Delta(\tau, \rho') \neq 0$. In particular, the expression on the right is independent of ρ'. Moreover,

$$\Delta(\tau, \rho) = \sum_{\rho' \in \Sigma'(\rho)} \Delta(\tau, \rho'),$$

since $e_S = 1$ for any large finite set S of valuations. Therefore

$$r_M^L(\check{\tau}_\lambda, \rho_\lambda) = \ell^{-\dim(A_M/A_L)} \sum_{\xi \in \mathcal{C}_M/\mathcal{C}_L} \sum_{\rho' \in \Sigma'(\rho)} \Delta(\tau, \rho')\,\tilde{r}_{M'}^{L'}(\xi\tau_{\lambda'}, \xi\rho'_{\lambda'}).$$

Since $\Delta(\tau,\rho')$ equals $\Delta(\xi\tau,\xi\rho')$, this becomes

$$\ell^{-\dim(A_M/A_L)}\sum_{\xi}\sum_{\rho'} r_{M'}^{L'}(\xi\tau_{\lambda'},\xi\rho'_{\lambda'}),$$

as required. ∎

12. A formula for $I_t^{\mathcal{E}}(f)$

We return to our discussion of Theorem B. We established the local assertion (i) in §10, so we can concentrate on the global assertion (ii). Our goal for this section is to obtain an expansion for $I^{\mathcal{E}}(f)$ which is dual to that of Proposition 5.1. However, we must first establish an inversion formula for $I_M^{\mathcal{E}}(\pi, f)$.

LEMMA 12.1: . *Suppose that* $\tau \in \Pi_{\mathrm{unit}}(M'(\mathbf{A})^1)$. *Then the distribution*

$$I_M^{\mathcal{E}}(\check{\tau}, f) = \sum_{\xi \in \mathcal{C}_M} \hat{I}_{M'}(\xi\tau, f')$$

equals

$$\sum_{L \in \mathcal{L}(M)} \sum_{\pi \in \Pi^+(M(\mathbf{A})^1)} \int_{\varepsilon_M + i\mathfrak{a}_M^*/i\mathfrak{a}_L^*} r_M^L(\check{\tau}_\lambda, \pi_\lambda)\, I_L^{\mathcal{E}}(\pi_\lambda, f)\, d\lambda, \tag{12.1}$$

where ε_M *is a small point in general position in* $\mathfrak{a}_M^*$.

Remark. Only those π with $\delta(\tau, \pi) \neq 0$ will contribute to (12.1). Since τ is assumed to be unitary, any such π will have a unitary central character. Consequently, π may be identified with an orbit $\{\pi_\lambda\}$ of $i\mathfrak{a}_M^*$ in $\Pi^+(M(\mathbf{A}))$.

Proof. For the proof, it is convenient to re-label the summation index L in (12.1) by L_1. We can then insert the expression

$$\sum_{Q \subset \mathcal{P}(L_1)} \omega_Q \sum_{L \in \mathcal{L}(L_1)} \sum_{\rho \in \Sigma^+(M(\mathbf{A})^1)} \int_{\varepsilon_Q + i\mathfrak{a}_{L_1}^*/i\mathfrak{a}_L^*} r_{L_1}^L(\pi_{\lambda+\mu}, \rho_{\lambda+\mu}) I_L^{\mathcal{E}}(\rho_{\lambda+\mu}, f)\, d\mu$$

for $I_{L_1}^{\mathcal{E}}(\pi_\lambda, f)$ into the formula. The point ε_Q can be taken to be small relative to the point ε_M. Since λ belongs to $\varepsilon_M + i\mathfrak{a}_M^*$, we can deform the contours of integration in μ from $\varepsilon_Q + i\mathfrak{a}_{L_1}^*/i\mathfrak{a}_L^*$ to $i\mathfrak{a}_{L_1}^*/i\mathfrak{a}_L^*$. Therefore (12.1) equals the sum over elements $L_1, L \in \mathcal{L}(M)$, with $L_1 \subset L$, of

$$\sum_{\pi \in \Pi^+(M(\mathbf{A})^1)} \sum_{\rho \in \Sigma^+(M(\mathbf{A})^1)} \int_{\varepsilon_M + i\mathfrak{a}_M^*/i\mathfrak{a}_L^*} r_M^{L_1}(\check{\tau}_\lambda, \pi_\lambda) r_{L_1}^L(\pi_\lambda, \rho_\lambda) I_L^{\mathcal{E}}(\rho_\lambda, f)\, d\lambda.$$

We shall take the sums over L_1 and π inside the integral. Their contribution will be given by

$$\sum_{\pi} \sum_{\{L_1 : M \subset L_1 \subset L\}} r_M^{L_1}(\check{\tau}_\lambda, \pi_\lambda)\, r_{L_1}^L(\pi_\lambda, \rho_\lambda).$$

This expression comes from a product of two (G, M) families. In fact, by (11.4), it equals

$$\tilde{r}_M^L(\check{\tau}_\lambda, \rho_\lambda)\Big(\sum_{\pi\in\Pi^+(M(\mathbf{A})^1)} \delta(\tau,\pi)\Delta(\pi,\rho)\Big).$$

But

$$\sum_{\pi\in\Pi^+(M(\mathbf{A})^1)} \delta(\tau,\pi)\Delta(\pi,\rho) = \sum_{\rho_1\in\Sigma^+(M(\mathbf{A})^1)} \Delta(\tau,\rho_1) \sum_{\pi\in\Pi^+(M(\mathbf{A})^1)} \Gamma(\rho_1,\pi)\Delta(\pi,\rho) = \Delta(\tau,\rho).$$

Since the product of this with $\tilde{r}_M^L(\check{\tau}_\lambda, \rho_\lambda)$ equals $r_M^L(\check{\tau}_\lambda, \rho_\lambda)$, (12.1) can be written as

$$\sum_{L\in\mathcal{L}(M)} \sum_{\rho\in\Sigma^+(M(\mathbf{A})^1)} \int_{\varepsilon_M + i\mathfrak{a}_M^*/i\mathfrak{a}_L^*} r_M^L(\check{\tau}_\lambda, \rho_\lambda) I_L^{\mathcal{E}}(\rho_\lambda, f)\, d\lambda. \tag{12.2}$$

Now, consider the expression

$$\sum_{\rho\in\Sigma^+(M(\mathbf{A})^1)} r_M^L(\check{\tau}_\lambda, \rho_\lambda) I_L^{\mathcal{E}}(\rho_\lambda, f).$$

By Lemma 11.4, this equals

$$\ell^{-\dim(A_M/A_L)} \sum_{\rho} \sum_{\rho'\in\Sigma'(\rho)} \sum_{\xi\in\mathcal{C}_M/\mathcal{C}_L} r_{M'}^{L'}(\xi\tau_{\lambda'}, \xi\rho'_{\lambda'}) I_L^{\mathcal{E}}(\rho_\lambda, f).$$

But Lemma 8.1 permits us to write

$$I_L^{\mathcal{E}}(\rho_\lambda, f) = \sum_{\zeta\in\mathcal{C}_L} \hat{I}_{L'}(\zeta\xi\rho'_{\lambda'}, f'),$$

so the expression equals

$$\ell^{-\dim(A_M/A_L)} \sum_{\rho} \sum_{\rho'\in\Sigma'(\rho)} \sum_{\xi\in\mathcal{C}_M} r_{M'}^{L'}(\xi\tau_{\lambda'}, \xi\rho'_{\lambda'}) \hat{I}_{L'}(\xi\rho'_{\lambda'}, f').$$

If $\rho' \in \Sigma(M'(\mathbf{A})^1)$ does not belong to one of the sets $\Sigma'(\rho)$,

$$I_{L'}(\xi\rho'_{\lambda'}, f') = 0,$$

by Proposition 10.3 of [1(g)]. We may therefore sum over all ρ' in $\Sigma(M'(\mathbf{A})^1)$. The expression equals

$$\ell^{-\dim(A_M/A_L)} \sum_{\rho'} \sum_{\xi\in\mathcal{C}_M} r_{M'}^{L'}(\xi\tau_{\lambda'}, \xi\rho'_{\lambda'}) \hat{I}_{L'}(\xi\rho'_{\lambda'}, f')$$

$$= \ell^{-\dim(A_M/A_L)} \sum_{\rho'} \sum_{\xi} r_{M'}^{L'}(\xi\tau_{\lambda'}, \rho'_{\lambda'}) \hat{I}_{L'}(\rho'_{\lambda'}, f').$$

We substitute this expression into (12.2). Since $d\lambda$ equals $\ell^{\dim(A_M/A_L)}d\lambda'$, we obtain

$$\sum_{\xi\in\mathcal{C}_M}\sum_{L\in\mathcal{L}(M)}\sum_{\rho'\in\Sigma(M'(\mathbf{A})^1)}\int_{\varepsilon_M+i\mathfrak{a}_M^*/i\mathfrak{a}_L^*} r_{M'}^{L'}(\xi\tau_{\lambda'},\rho'_{\lambda'})\hat{I}_{L'}(\rho'_{\lambda'},f')\,d\lambda.$$

But τ is unitary, so we can apply the formula quoted at the beginning of §8. It follows that (12.2) equals

$$\sum_{\xi\in\mathcal{C}_M}\hat{I}_{M'}(\xi\tau,f'),$$

as required. ∎

In §5 we defined $I^{\mathcal{E}}$ as the pullback to G of the distribution I on G'. Applying (9.1) to G', we obtain

$$I^{\mathcal{E}}(f)=\sum_{t\geq 0}I_t^{\mathcal{E}}(f),$$

where

$$I_t^{\mathcal{E}}(f)=\hat{I}_t(f')=\sum_{L\in\mathcal{L}'}|W_0^L||W_0^{G'}|^{-1}\int_{\Pi(L,t)}a^L(\tau)\hat{I}_L(\tau,f')\,d\tau.$$

According to Proposition 8.2 of [1(h)], $\hat{I}_L(\tau,f')$ vanishes unless L is the image of an element in $\mathcal{L}$. Therefore,

$$I_t^{\mathcal{E}}(f)=\sum_{M\in\mathcal{L}}|W_0^M||W_0^G|^{-1}\int_{\Pi(M',t)}a^{M'}(\tau)\hat{I}_{M'}(\tau,f')\,d\tau. \tag{12.3}$$

We would like to transform this into a formula involving the functions $a^{M,\mathcal{E}}$ and $I_M^{\mathcal{E}}$.

For convenience we shall define $\Pi_{\rm disc}^{\mathcal{E}}(M_1,t)$ to be the subset of $\Pi^+(M_1(\mathbf{A})^1,t)$ consisting of irreducible constituents of induced representations

$$\sigma_\lambda^{M_1},\qquad L\in\mathcal{L}^{M_1},\ \sigma\in\Pi^+(L(\mathbf{A})^1,t),\ \lambda\in\mathfrak{a}_L^*/i\mathfrak{a}_{M_1}^*,$$

in which σ_λ satisfies the following two conditions.

(i) $a_{\rm disc}^{L,\mathcal{E}}(\sigma)\neq 0$.

(ii) There is an element $s\in W^{M_1}(\mathfrak{a}_L)_{\rm reg}$ such that $s\sigma_\lambda=\sigma_\lambda$.

But for the superscript $\mathcal{E}$, this definition is identical to that of $\Pi_{\rm disc}(M_1,t)$. It follows from our induction assumption that $\Pi_{\rm disc}^{\mathcal{E}}(M_1,t)$ equals $\Pi_{\rm disc}(M_1,t)$ if $M_1\neq G$. Copying the definition of $\Pi(M,t)$, we take $\Pi^{\mathcal{E}}(M,t)$ to be the disjoint union over $M_1\in\mathcal{L}^M$ of the sets

$$\Pi_{M_1}^{\mathcal{E}}(M,t)=\{\pi=\pi_{1,\lambda}\,:\,\pi_1\in\Pi_{\rm disc}^{\mathcal{E}}(M_1,t),\ \lambda\in i\mathfrak{a}_{M_1}^*/i\mathfrak{a}_M^*\}.$$

Proposition 12.2: . *Suppose that* $t \geq 0$ *and* $f \in \mathcal{H}(G(\mathbf{A}))$. *Then*

$$(9.1)^{\mathcal{E}} \qquad I_t^{\mathcal{E}}(f) = \sum_{M \in \mathcal{L}} |W_0^M||W_0^G|^{-1} \int_{\Pi^{\mathcal{E}}(M,t)} a^{M,\mathcal{E}}(\pi) I_M^{\mathcal{E}}(\pi, f)\, d\pi.$$

Proof. It follows from the formulas above that $I_t^{\mathcal{E}}(f)$ equals

$$\sum_{\{M_1, M \in \mathcal{L} : M \supset M_1\}} |W_0^{M_1}||W_0^G|^{-1} \sum_{\tau \in \Pi_{\mathrm{disc}}(M_1',t)} \int_{ia_{M_1}^*/ia_M^*} a_{\mathrm{disc}}^{M_1'}(\tau) r_{M_1'}^{M'}(\tau_{\lambda'}) \hat{I}_{M'}(\tau_{\lambda'}, f')\, d\lambda'.$$

Replace each τ by $\xi\tau$, $\xi \in \mathcal{C}_{M_1}$, and then sum over ξ. Since ξ permutes $\Pi_{\mathrm{disc}}(M_1', t)$, this will still equal $I_t^{\mathcal{E}}(f)$ provided that we multiply by $\ell^{-\dim(A_{M_1})}$. According to Lemma 4.3 of [1(h)],

$$a_{\mathrm{disc}}^{M_1'}(\xi\tau) = a_{\mathrm{disc}}^{M_1'}(\tau).$$

Consequently $I_t^{\mathcal{E}}(f)$ equals the sum over $\{M_1 \subset M\}$ of the product of

$$(12.4) \qquad \ell^{-\dim(A_{M_1})}|W_0^{M_1}||W_0^G|^{-1}$$

with

$$\sum_{\tau} \int_{ia_{M_1}^*/ia_M^*} a_{\mathrm{disc}}^{M_1'}(\tau) \Big(\sum_{\xi \in \mathcal{C}_{M_1}} r_{M_1'}^{M'}(\xi\tau_{\lambda'}) \hat{I}_{M'}(\xi\tau_{\lambda'}, f')\, d\lambda' \Big).$$

The expression

$$\sum_{\xi \in \mathcal{C}_{M_1}} r_{M_1'}^{M'}(\xi\tau_{\lambda'}) \hat{I}_{M'}(\xi\tau_{\lambda'}, f')$$

can be written as

$$\sum_{\xi \in \mathcal{C}_{M_1}/\mathcal{C}_M} r_{M_1'}^{M'}(\xi\tau_{\lambda'}) \sum_{\zeta \in \mathcal{C}_M} \hat{I}_{M'}(\zeta\xi\tau_{\lambda'}, f').$$

We observe from a variant of Lemma 8.1 that

$$\sum_{\zeta \in \mathcal{C}_M} \hat{I}_{M'}(\zeta\xi\tau_{\lambda'}, f') = I_M^{\mathcal{E}}\big((\xi\tau)_{\lambda}^{\vee}, f\big) = I_M^{\mathcal{E}}(\overset{\vee}{\tau}_{\lambda}, f).$$

Therefore, by Corollary 11.2, the expression reduces to

$$\ell^{\dim(A_{M_1}/A_M)} r_{M_1}^M(\overset{\vee}{\tau}_{\lambda}) I_M^{\mathcal{E}}(\overset{\vee}{\tau}_{\lambda}, f).$$

Substituting the expansion from Lemma 12.1, we obtain

$$\ell^{\dim(A_{M_1}/A_M)} \sum_{L \in \mathcal{L}(M)} \sum_{\pi_1 \in \Pi^+(M_1(\mathbf{A})^1, t)} \int_{\varepsilon_M + ia_M^*/ia_L^*} r_{M_1}^M(\overset{\vee}{\tau}_{\lambda})$$
$$\times\, r_M^L(\overset{\vee}{\tau}_{\lambda+\mu}, \pi_{1,\lambda+\mu}) I_L^{\mathcal{E}}(\pi_{1,\lambda+\mu}, f)\, d\mu.$$

(We should actually sum over $\pi \in \Pi^+(M(\mathbf{A})^1)$, but the general position of λ means that this sum reduces to one over $\Pi^+(M_1(\mathbf{A})^1)$. Since $\|\mathrm{Im}(\nu_\tau)\| = t$, the summand vanishes unless π_1 actually belongs to $\Pi^+(M_1(\mathbf{A})^1, t)$.) We insert this expression back into the formula for $I_t^{\mathcal{E}}(f)$. Since

$$d\lambda' = \ell^{-\dim(A_{M_1}/A_M)}\, d\lambda,$$

we may write $I_t^{\mathcal{E}}(f)$ as the sum over $\{M_1, L : M_1 \subset L\}$ and $\pi_1 \in \Pi^+(M_1(\mathbf{A})^1, t)$ of the product of 12.4 with

(12.5)

$$\sum_{\{M : M_1 \subset M \subset L\}} \int_{\varepsilon_M + i\mathfrak{a}_{M_1}^*/i\mathfrak{a}_L^*} \sum_{\tau \in \Pi_{\mathrm{disc}}(M_1', t)} a_{\mathrm{disc}}^{M_1'}(\tau) r_{M_1}^M(\check{\tau}_\lambda) r_M^L(\check{\tau}_\lambda, \pi_{1,\lambda}) I_L^{\mathcal{E}}(\pi_{1,\lambda}, f) d\lambda.$$

The summand in (12.5) corresponding to a given τ will vanish unless $\delta(\tau, \pi_1) \neq 0$. Among all such τ fix one, say σ. Then for any other such τ, we can write

$$r_{M_1}^M(\check{\tau}_\lambda) = \sum_{\{L : M_1 \subset L_1 \subset M\}} r_{M_1}^{L_1}(\check{\sigma}_\lambda) \tilde{r}_{L_1}^M(\check{\sigma}_\lambda, \check{\tau}_\lambda).$$

Substitute this expression into (12.5), and take the sum over L_1 outside the integral. By a variant of Lemma 11.3, the function $\tilde{r}_{L_1}^M(\check{\sigma}_\lambda, \check{\tau}_\lambda)$ is slowly increasing and regular for λ in a cylinder about the imaginary space $i\mathfrak{a}_{M_1}^*$. We may therefore deform the contour of integration in (12.5) from $\varepsilon_M + i\mathfrak{a}_{M_1}^*/i\mathfrak{a}_L^*$ to $\varepsilon_{L_1} + i\mathfrak{a}_{M_1}^*/i\mathfrak{a}_L^*$, where ε_{L_1} is a small regular point in $\mathfrak{a}_{L_1}^*$ which depends only on L_1. This leaves us free to bring the sum over M in (12.5) inside the integral and the sum over L_1 and τ. But by (11.5),

$$\sum_{\{M : L_1 \subset M \subset L\}} \tilde{r}_{L_1}^M(\check{\sigma}_\lambda, \check{\tau}_\lambda) r_M^L(\check{\tau}_\lambda, \pi_{1,\lambda}) = \delta(\tau, \pi_1) \tilde{r}_{L_1}^L(\check{\sigma}_\lambda, \pi_{1,\lambda}).$$

Moreover,

$$\sum_{\tau \in \Pi_{\mathrm{disc}}(M_1', t)} a_{\mathrm{disc}}^{M_1'}(\tau) \delta(\tau, \pi_1) = \ell^{\dim(A_{M_1})} a_{\mathrm{disc}}^{M_1, \mathcal{E}}(\pi_1),$$

since π_1 belongs to $\Pi^+(M_1(\mathbf{A})^1, t)$. Therefore (12.5) equals

(12.6)

$$\ell^{\dim(A_{M_1})} \sum_{\{L_1 : M_1 \subset L_1 \subset L\}} \int_{\varepsilon_{L_1} + i\mathfrak{a}_{M_1}^*/i\mathfrak{a}_L^*} a_{\mathrm{disc}}^{M_1, \mathcal{E}}(\pi_1) r_{M_1}^{L_1}(\check{\sigma}_\lambda) \tilde{r}_{L_1}^L(\check{\sigma}_\lambda, \pi_{1,\lambda}) I_L^{\mathcal{E}}(\pi_{1,\lambda}, f) d\lambda.$$

The expression (12.6) clearly simplifies if $M_1 = G$. The contribution of all such terms to the final formula for $I_t^{\mathcal{E}}(f)$ is just

$$\sum_{\pi\in\Pi^{\mathcal{E}}_{\mathrm{disc}}(G,t)} a^{\mathcal{E}}_{\mathrm{disc}} I^{\mathcal{E}}_G(\pi,f). \tag{12.7}$$

If $M_1 \neq G$, we make use of our induction assumption that $a^{M_1,\mathcal{E}}_{\mathrm{disc}}(\pi_1)$ equals $a^{M_1}_{\mathrm{disc}}(\pi_1)$. In particular, we may assume that $\pi_1 \in \Pi_{\mathrm{disc}}(M_1,t)$. But Lemma 11.3 then tells us that $\tilde{r}^L_{L_1}(\check{\sigma}_\lambda, \pi_{1,\lambda})$ is analytic for λ in a cylinder about the imaginary space $i\mathfrak{a}^*_{M_1}$. Moreover, we saw in §10 that

$$I^{\mathcal{E}}_L(\pi_{1,\lambda},f) = I_L(\pi_{1,\lambda},f),$$

and it is known (Lemma 3.4 of [1(g)]) that the function on the right is also analytic for λ near $i\mathfrak{a}^*_{M_1}$. We may therefore deform the contour of integration in (12.6) from $\varepsilon_{L_1} + i\mathfrak{a}^*_{M_1}/i\mathfrak{a}^*_L$ to $i\mathfrak{a}^*_{M_1}/i\mathfrak{a}^*_L$. We can then take the sum over L_1 in (12.6) inside the integral, and we obtain

$$\sum_{\{L_1:M_1\subset L_1\subset L\}} r^{L_1}_{M_1}(\check{\sigma}_\lambda)\tilde{r}^L_{L_1}(\check{\sigma}_\lambda,\pi_{1,\lambda}) = r^L_{M_1}(\pi_{1,\lambda}),$$

from Lemma 11.3. Thus, if $M_1 \neq G$, the expression (12.6) equals

$$\ell^{\dim(A_{M_1})}\int_{i\mathfrak{a}^*_{M_1}/i\mathfrak{a}^*_L} a^{M_1,\mathcal{E}}_{\mathrm{disc}}(\pi_1) r^L_{M_1}(\pi_{1,\lambda}) I^{\mathcal{E}}_L(\pi_{1,\lambda},f)\,d\lambda.$$

Putting these formulas together, we see finally that $I^{\mathcal{E}}_t(f)$ equals the sum of (12.7) and

$$\sum_L \sum_{\{M_1\subset L:M_1\neq G\}} |W^{M_1}_0|\,|W^G_0|^{-1} \sum_{\pi_1\in\Pi_{\mathrm{disc}}(M_1,t)} \int_{i\mathfrak{a}^*_{M_1}/i\mathfrak{a}^*_L} a^{M_1,\mathcal{E}}_{\mathrm{disc}}(\pi_1) r^L_{M_1}(\pi_{1,\lambda}) \times I^{\mathcal{E}}_L(\pi_{1,\lambda},f)\,d\lambda.$$

By definition this is just

$$\sum_{L\in\mathcal{L}} |W^L_0|\,|W^G_0|^{-1}\int_{\Pi^{\mathcal{E}}(L,t)} a^{L,\mathcal{E}}(\pi) I^{\mathcal{E}}_L(\pi,f)\,d\pi,$$

the required formula. ∎

LEMMA 12.3: . *Suppose that $t \geq 0$ and $f \in \mathcal{H}(G(\mathbf{A}))$. Then*

$$I^{\mathcal{E}}_t(f) - I_t(f) = \sum_{\pi\in\Pi+(G(\mathbf{A})^1,t)} \big(a^{\mathcal{E}}_{\mathrm{disc}}(\pi) - a_{\mathrm{disc}}(\pi)\big)\, tr\,\pi(f^1),$$

where f^1 is the restriction of f to $G(\mathbf{A})^1$.

Proof. We use the formulas (9.1) and $(9.1)^{\mathcal{E}}$ to expand the difference between $I_t^{\mathcal{E}}(f)$ and $I_t(f)$. Suppose that $M_1 \subset M \subset G$, with $M_1 \neq G$. Then the induction hypothesis of §9 implies that the set $\Pi_{M_1}^{\mathcal{E}}(M,t)$ is the same as $\Pi_{M_1}(M,t)$, and that

$$a^{M,\mathcal{E}}(\pi) = a^M(\pi), \qquad \pi \in \Pi_{M_1}(M,t).$$

In particular, these numbers are both zero unless π is unitary. But we saw in §10 that for unitary π, $I_M^{\mathcal{E}}(\pi, f)$ was equal to $I_M(\pi, f)$. Therefore, the only contribution to the expansion of $I_t^{\mathcal{E}}(f) - I_t(f)$ comes from $M_1 = M = G$. However, if π is any representation in $\Pi^+(G(\mathbf{A})^1)$, we have

$$I_G^{\mathcal{E}}(\pi, f) = I_G(\pi, f) = \operatorname{tr}\pi(f^1).$$

(The first formula follows, for example, from Theorem 10.2(d).) The corollary follows. ∎

Chapter II has consisted so far of two parallel discussions. Paragraphs 3–7 have dealt with the geometric sides of the two trace formulas, while Paragraphs 8–12 have been concerned with the spectral sides. We should be aware of the similarities between results in these two passages. For example, besides the obvious duality of Theorems A and B, there is the parallelism of Propositions 5.1 and 12.2 and also of Lemmas 5.2 and 12.3. In the next paragraph we shall begin a study of the geometric sides which has no analogue for the spectral sides. This will eventually allow us to exploit the two different formulas for

$$I^{\mathcal{E}}(f) - I(f)$$

given by Lemmas 5.2 and 12.3.

13. The map ε_M

It is known that the trace formula simplifies greatly if the orbital integrals of f are supported on the elliptic sets at two places. We shall exploit a similar idea, but with a progressively less stringent restriction on f. For each $M \in \mathcal{L}$, we define $\mathcal{H}(G(\mathbf{A}), M)$ to be the subspace of $\mathcal{H}(G(\mathbf{A}))$ spanned by functions

$$f = \prod_v f_v, \qquad f_v \in \mathcal{H}(F_v)),$$

which have the following property. For two unramified finite places v_1 and v_2,

$$f_{v_i,L} = 0, \qquad L \in \mathcal{L},\ i = 1,2,$$

unless L contains a conjugate of M. If S is a finite set of places which contains S_{ram} (and at least two other places), we define $\mathcal{H}(G(F_S), M)$ the same way. It is a subspace of $\mathcal{H}(G(\mathbf{A}), M)$.

At this point, we fix a Levi subset $M \in \mathcal{L}$ such that $M \neq G$. We are already carrying the induction hypotheses that Theorems A and B hold if G is replaced by any proper Levi subset. We shall now take on the additional induction assumption that

$$I_L^{\mathcal{E}}(\gamma, f) = I_L(\gamma, f), \qquad \gamma \in L(F_S),\ f \in \mathcal{H}(G(F_S)),$$

for any $S \supset S_{\mathrm{ram}}$ and any $L \in \mathcal{L}$, with $M \subsetneq L$. In §17 we shall show that this formula also applies to M, thereby completing the proof of Theorem A(i). Until then, M will be fixed, and the last induction hypothesis will remain in force. Notice that the induction begins with M maximal (and proper), where the required formula is just (3.9).

LEMMA 13.1: . *For $f \in \mathcal{H}(G(\mathbf{A}), M)$, the distribution*

$$I^{\mathcal{E}}(f) - I(f)$$

equals the sum of

$$|W(\mathfrak{a}_M)|^{-1} \sum_{\gamma \in (M(F))_{M,S}} a^M(S,\gamma)(I_M^{\mathcal{E}}(\gamma, f) - I_M(\gamma, f))$$

and

$$\sum_{\xi \in A_G(F)} \sum_{u \in (\mathcal{U}_G(F))_{G,S}} (a^{\mathcal{E}}(S,u) - a(S,u)) I_G(\xi u, f).$$

(As in §5, S is a finite set of valuations of F that is suitably large in a sense that depends only on supp(f) and $V(f)$.)

Proof. Suppose that S is the disjoint union of S_0, $S_1 = \{v_1\}$ and $S_2 = \{v_2\}$, where S_0 contains S_{ram} and v_1 and v_2 are arbitrary unramified valuations. We can assume that

$$f = \prod_{i=0}^{2} f_i, \qquad f_i \in \mathcal{H}(G(F_{S_i})),$$

where

$$f_{i,L} = 0, \qquad L \in \mathcal{L},\ i = 1, 2,$$

unless L contains a conjugate of M. Applying the formula (5.6) twice, we see that

$$I_L^{\mathcal{E}}(\gamma, f) - I_L(\gamma, f) = \sum_{i=0}^{2} (I_L^{\mathcal{E}}(\gamma, f_i) - I_L(\gamma, f_i)) \prod_{j \neq i} \hat{I}_M^M(\gamma, f_{j,L}),$$

for any $L \in \mathcal{L}$ and $\gamma \in L(F)$. This expression vanishes unless L contains a conjugate of M. On the other hand, it is known (formula (2.4*) of [1(g)]) that

$$I_{wLw^{-1}}(w\gamma w^{-1}, f) = I_L(\gamma, f)$$

for any $w \in W_0$. A similar assertion holds for $I_L^{\mathcal{E}}(\gamma, f)$. It follows from the latest induction assumption that $I_L^{\mathcal{E}}(\gamma, f)$ equals $I_L(\gamma, f)$ unless L is actually conjugate to M. But the number of L which are conjugate to M equals

$$|W_0^M|\,|W_0^G|^{-1}\,|W(\mathfrak{a}_M)|^{-1}.$$

Our lemma therefore follows from Lemma 5.2. ∎

We would like to be able to assert the equality of $I^{\mathcal{E}}(f)$ and $I(f)$. At the moment, however, this is far from clear. In order to go further, we must turn to a technique introduced by Langlands in seminars at the Institute for Advanced Study. (See "Cancellation of singularities" and "Division algebras," Lecture Notes, I.A.S.). We have defined the spaces $\mathcal{H}(G(F_S))^0$ in §7. We shall show that each function

$$\gamma \to I_M^{\mathcal{E}}(\gamma, f) - I_M(\gamma, f), \qquad f \in \mathcal{H}(G(F_S))^0,\ S \supset S_{\mathrm{ram}},$$

is the orbital integral of a suitable function on $M(F_S)$. This will allow us to apply the trace formula for M to the corresponding expansion in Lemma 13.1.

It is best to treat the general situation, in which S is any finite set of valuations with the closure property. Our latest assumption hypothesis means that Theorem 6.1 applies to any $L \in \mathcal{L}(M)$ with $L \neq M$. Therefore, the constants $\varepsilon_L(S)$ are defined, and

$$I_L^{\mathcal{E}}(\gamma, f) = \sum_{L_1 \in \mathcal{L}(L)} \hat{I}_L^{L_1}(\gamma, \varepsilon_{L_1}(S) f_{L_1}), \qquad \gamma \in L(F_S),\ f \in \mathcal{H}(G(F_S)).$$

Proposition 13.2: . *Let S be any finite set of valuations with the closure property. Then there are unique maps*

$$\varepsilon_L : \mathcal{H}(G(F_S))^0 \to \mathcal{I}_{ac}(L(F_S)), \qquad L \in \mathcal{L}(M),$$

such that

(13.1)

$$I_M^{\mathcal{E}}(\gamma, f) = \sum_{L \in \mathcal{L}(M)} \hat{I}_M^L(\gamma, \varepsilon_L(f)), \qquad \gamma \in M(F_S),\ f \in \mathcal{H}(G(F_S))^0.$$

The maps have the descent property

$$\varepsilon_M(f)_{M_1} = \sum_L d_{M_1}^G(M, L) \hat{\varepsilon}_{M_1}^L(f_L), \qquad M_1 \subset M, \tag{13.2}$$

and the splitting property

$$\varepsilon_M(f) = \sum_{L_1, L_2} d_M^G(L_1, L_2) \hat{\varepsilon}_M^{L_1}(f_{1,L_1}) \hat{\varepsilon}_M^{L_2}(f_{2,L_2}) \tag{13.3}$$

for $f = f_1 f_2$ as in (3.4).

Remarks. 1. Since a function in $\mathcal{I}_{ac}(M(F_S))$ is uniquely determined by its orbital integrals, the uniqueness of the map follows inductively from (13.1). Notice that (13.1) also implies that the map ε_M is supported on characters. Therefore the notation $\hat{\varepsilon}_M(f_G) = \varepsilon_M(f)$, which appears in (13.2) and (13.3), makes sense.

2. The proposition is to be regarded as a weaker version of Theorem A(i). For suppose that

$$I_M^{\mathcal{E}}(\gamma, f) = I_M(\gamma, f), \qquad \gamma \in M(F_S),\ f \in \mathcal{H}(G(F_S)),$$

whenever $S \supset S_{\text{ram}}$. Then Theorem (6.1) holds for M, and

$$I_M^{\mathcal{E}}(\gamma, f) = \sum_{L \in \mathcal{L}(M)} \hat{I}_M^L(\gamma, \varepsilon_L(S) f_L)$$

for any S. Proposition 13.2 follows inductively from this, with

$$\varepsilon_M(f) = \varepsilon_M(S) f_M.$$

3. The induction hypothesis at the beginning of this paragraph allows to apply the last remark to any $L \supsetneq M$. We may therefore assume that

$$\varepsilon_L(f) = \varepsilon_L(S)f_L, \qquad L \supsetneq M. \tag{13.4}$$

The defining formula (13.1) then takes the form

$$\hat{I}_M^M(\gamma, \varepsilon_M(f)) = I_M^{\mathcal{E}}(\gamma, f) - \sum_{L \supsetneq M} \hat{I}_M^L(\gamma, \varepsilon_L(S)f_L). \tag{13.1*}$$

If $M_1 \subsetneq M$, the formula (13.2) can be written

$$\varepsilon_M(f)_{M_1} = \varepsilon_{M,M_1}(S)f_{M_1}, \tag{13.2*}$$

where

$$\varepsilon_{M,M_1}(S) = \sum_{L \in \mathcal{L}(M_1)} d_{M_1}^G(M, L)\varepsilon_{M_1}^L(S).$$

Formula (13.3) becomes

$$\varepsilon_M(f) = \varepsilon_M(f_1)f_{2,M} + f_{1,M}\varepsilon_M(f_2) + \tilde{d}_M^G(S_1, S_2)f_M, \tag{13.3*}$$

with

$$\tilde{d}_M^G(S_1, S_2) = \sum_{\substack{L_1, L_2 \in \mathcal{L}(M) \\ L_1, L_2 \neq M}} d_M^G(L_1, L_2)\varepsilon_M^{L_1}(S_1)\varepsilon_M^{L_2}(S_2).$$

4. Suppose that S either contains S_{ram} or consists of one unramified valuation. Then the constant $\varepsilon_L(S)$ vanishes if $M \subsetneq L \subsetneq G$. Formula (13.1*) becomes

$$I_M^{\mathcal{E}}(\gamma, f) - I_M(\gamma, f) = \hat{I}_M^M(\gamma, \varepsilon_M(f)), \qquad f \in \mathcal{H}(G(F_S)). \tag{13.1**}$$

This is the form in which $\varepsilon_M(f)$ will be applied to Lemma 13.1. Also, $\varepsilon_{M_1}^L(S)$ vanishes if $M_1 \subsetneq L \subsetneq G$, so that (13.2*) simplifies to

$$\varepsilon_M(f)_{M_1} = 0, \qquad M_1 \subsetneq M. \tag{13.2**}$$

5. Suppose that S is a disjoint union of S_0 and $S_1 = \{v\}$, where S_0 contains S_{ram} and v is unramified. If $f = f_0 f_1$, the formula (13.3*) yields

$$\varepsilon_M(f) = \varepsilon_M(f_0)f_{1,M} + f_{0,M}\varepsilon_M(f_1). \tag{13.3**}$$

Suppose that f_1 is equal to the characteristic function of $K_v \rtimes \theta$. Then Lemma 4.3 combined with (13.1**) implies that

$$\varepsilon_M(f_1) = 0.$$

Consequently, ε_M will extend to a map from $\mathcal{H}(G(\mathbf{A}))^0$ to $\mathcal{I}_{ac}(M(\mathbf{A}))$.

Proof. We now begin the proof of Proposition 13.2. It will consume most of the next paragraph as well as what remains of this one. Fix S and f. As in §6, set

$$\varepsilon_M(\gamma, f) = I_M^{\mathcal{E}}(\gamma, f) - \sum_{L \supsetneq M} \hat{I}_M^L(\gamma, \varepsilon_L(S) f_L), \qquad \gamma \in M(F_S).$$

This is equal to the right hand side of (13.1*). To prove the existence of the map ε_M, we must show that $\varepsilon_M(\gamma, f)$ is an orbital integral in γ of a function in $\mathcal{H}_{ac}(M(F_S))$. We must also check the properties (13.2) and (13.3). Let us comment on these first.

As we remarked after its proof, Lemma 6.2 is valid under the induction hypothesis we took on at the beginning of this paragraph. The required properties (13.2) and (13.3) will then follow immediately from (6.2*) and (6.3*), once we have established the existence of the map $\varepsilon_M(f)$. Notice that (6.3*) provides a formula for $\varepsilon_M(\gamma, f)$ in terms of functions defined for single valuations $v \in S$. It therefore reduces the proof of the existence of $\varepsilon_M(f)$ to the case that S consists of one valuation. We shall deal with this in §14, treating the real and p-adic cases separately. We conclude this section by discussing some preliminary properties of the functions $\varepsilon_M(\gamma, f)$.

Suppose that γ is a general element in $M(F_S)$. Let us apply the formulas (3.1) and $(3.1)^{\mathcal{E}}$ to $\varepsilon_M(\gamma, f)$. We see that $\varepsilon_M(\gamma, f)$ equals the limit, as a approaches 1 through regular values in $A_M(F_S)$, of

$$\sum_{L_1 \supset M} r_M^{L_1}(\gamma, a)\big(I_{L_1}^{\mathcal{E}}(a\gamma, f) - \sum_{\{L \in \mathcal{L}(L_1): L \neq M\}} \hat{I}_{L_1}^L(a\gamma, \varepsilon_L(f))\big).$$

If $L_1 \neq M$, the sum on the right can be taken over all $L \in \mathcal{L}(L_1)$, and the expression in brackets vanishes. However, when $L_1 = M$, the expression in the brackets is just equal to $\varepsilon_M(a\gamma, f)$. We have shown that

$$\varepsilon_M(\gamma, f) = \lim_{a \to 1} \varepsilon_M(a\gamma, f).$$

This formula is consistent with our hope that $\varepsilon_M(\gamma, f)$ be the orbital integral of a function in $M(F_S)$. In fact, it tells us that we need only consider points γ such that $M_\gamma = G_\gamma$. But properties (3.2) and $(3.2)^{\mathcal{E}}$ tell us that we need only consider points $\gamma \in M(F_S)$ which are G-regular. Thus, we have only to show that

$$\gamma \to \varepsilon_M(\gamma, f), \qquad \gamma \in M(F_S) \cap G_{\mathrm{reg}},$$

equals the orbital integral in γ of a function in $\mathcal{H}_{ac}(M(F_S))$. As agreed above, we may assume that $S = \{v\}$.

We have discussed the distributions ${}^cI_M(\gamma)$ and ${}^cI^{\mathcal{E}}_M(\gamma)$ in §10. If $\gamma \in M(F_S) \cap G_{\mathrm{reg}}$, define

$$ {}^c\varepsilon_M(\gamma, f) = {}^cI^{\mathcal{E}}_M(\gamma, f) - \sum_{\{L \in \mathcal{L}(M): L \neq G\}} \varepsilon^L_M(S)\, {}^cI_L(\gamma, f). $$

In some respects, this function is easier to handle than $\varepsilon_M(\gamma, f)$. For Lemma 4.4 of [1(g)] implies that ${}^c\varepsilon_M(\gamma, f)$ has bounded support as a function of γ in the space of $M^0(F_S)$-orbits in $M(F_S)$. Happily, there is a simple formula relating the two functions.

LEMMA 13.3: . *Suppose that γ belongs to $M(F_S) \cap G_{\mathrm{reg}}$. Then*

$$ \varepsilon_M(\gamma, f) - {}^c\varepsilon_M(\gamma, f) = \hat{I}^M_M\big(\gamma, \theta^{\mathcal{E}}_M(f) - \theta_M(f)\big) = \hat{I}^M_M\big(\gamma, {}^c\theta_M(f) - {}^c\theta^{\mathcal{E}}_M(f)\big). $$

Proof. The function $\varepsilon_M(\gamma, f)$ was defined to be

$$ I^{\mathcal{E}}_M(\gamma, f) - \sum_{L \supsetneq M} \hat{I}^L_M\big(\gamma, \varepsilon_L(S) f_L\big). $$

However, applying the descent properties (3.3) and (6.2) to the sum on the right, we obtain

$$ \varepsilon_M(\gamma, f) = I^{\mathcal{E}}_M(\gamma, f) - \sum_{\{L_1 \in \mathcal{L}(M): L_1 \neq G\}} \varepsilon^{L_1}_M(S) I_{L_1}(\gamma, f). $$

It follows from (10.2) and $(10.2)^{\mathcal{E}}$ that

$$ \begin{aligned} &\varepsilon_M(\gamma, f) - {}^c\varepsilon_M(\gamma, f) \\ &= \big(I^{\mathcal{E}}_M(\gamma, f) - {}^cI^{\mathcal{E}}_M(\gamma, f)\big) - \sum_{\{L_1 \in \mathcal{L}(M): L_1 \neq G\}} \varepsilon^{L_1}_M(S)\big(I_{L_1}(\gamma, f) - {}^cI_{L_1}(\gamma, f)\big) \\ &= \sum_{L \neq G} \Big\{ {}^c\hat{I}^{L,\mathcal{E}}_M\big(\gamma, \theta^{\mathcal{E}}_L(f)\big) - \sum_{L_1 \in \mathcal{L}^L(M)} \varepsilon^{L_1}_M(S)\, {}^c\hat{I}^L_{L_1}\big(\gamma, \theta_L(f)\big) \Big\}. \end{aligned} $$

Consider a summand corresponding to $L \supsetneq M$. Given the induction hypothesis at the beginning of this paragraph, we can apply Theorem 10.2 and Corollary 10.3. We obtain

$$ {}^c\hat{I}^{L,\mathcal{E}}_M\big(\gamma, \theta^{\mathcal{E}}_L(f)\big) = {}^c\hat{I}^{L,\mathcal{E}}_M\big(\gamma, \theta_L(f)\big) = \sum_{L_1 \in \mathcal{L}^L(M)} \varepsilon^{L_1}_M(S)\, {}^c\hat{I}^L_{L_1}\big(\gamma, \theta_L(f)\big), $$

and so the summand vanishes. The summand corresponding to $L = M$ is just

$$ \hat{I}^M_M\big(\gamma, \theta^{\mathcal{E}}_M(f) - \theta_M(f)\big). $$

The first identity of the theorem follows. The second identity is an immediate consequence of the formula

$$\theta_M^{\mathcal{E}}(f) - \theta_M(f) = \mathscr{C}\theta_M(f) - \mathscr{C}\theta_M^{\mathcal{E}}(f),$$

which follows from (10.1), $(10.1)^{\mathcal{E}}$, Theorem 10.2, and our induction hypothesis. ∎

14. Cancellation of singularities

We continue with the proof of Proposition 13.2. It remains for us to establish for any valuation v of F, and any $f \in \mathcal{H}(G(F_v))^0$, that the function

$$\gamma \to \varepsilon_M(\gamma, f), \qquad \gamma \in M(F_v) \cap G_{\mathrm{reg}},$$

is the orbital integral of a function in $\mathcal{H}_{ac}(M(F_v))$.

Suppose first that v is nonArchimedean. Fix a function $f \in \mathcal{H}(G(F_v))^0$, and let σ be a semisimple element in $M(F_v)$. We shall show that

$$(14.1) \qquad \varepsilon_M(\gamma, f) \overset{(M,\sigma)}{\sim} 0, \qquad \gamma \in \sigma M_\sigma(F_v) \cap G_{\mathrm{reg}}.$$

Assume first that $\mathfrak{a}_{M_\sigma} \neq \mathfrak{a}_M$. Then $\sigma M_\sigma(F_v)$ is contained in a proper Levi subset M_1 of M. We have already seen in §13 that if γ belongs to $M_1(F_v) \cap G_{\mathrm{reg}}$, then

$$\varepsilon_M(\gamma, f) = \varepsilon_{M,M_1}(v) \hat{I}^{M_1}_{M_1}(\gamma, f_{M_1}).$$

In particular, the function on the right is an orbital integral on $M(F_v)$, and (14.1) holds in this case.

Next assume that $\mathfrak{a}_{M_\sigma} = \mathfrak{a}_M$. It follows from (3.6) and $(3.6)^{\mathcal{E}}$ that the function

$$\varepsilon_M(\gamma, f) = I^{\mathcal{E}}_M(\gamma, f) - \sum_{L_1 \supsetneq M} \hat{I}^{L_1}_M(\gamma, \varepsilon_{L_1}(f))$$

is (M, σ)-equivalent to

$$\sum_{L \in \mathcal{L}(M)} \sum_{\delta \in \sigma(\mathcal{U}_{L_\sigma}(F_v))} g^L_M(\gamma, \delta)\Big(I^{\mathcal{E}}_L(\delta, f) - \sum_{\substack{L_1 \in \mathcal{L}(L) \\ L_1 \neq M}} \hat{I}^{L_1}_L(\delta, \varepsilon_{L_1}(f))\Big).$$

We can assume that $L \supsetneq M$ in this sum, for if $L = M$ the functions $g^M_M(\gamma, \delta)$ are (M, σ)-equivalent to 0. However, if $L \supsetneq M$, we can sum L_1 over all elements in $\mathcal{L}(L)$ and the expression in the brackets vanishes. It follows that (14.1) holds in general.

By Lemma 13.3, $\varepsilon_M(\gamma, f)$ equals the sum of ${}^c\varepsilon_M(\gamma, f)$ and the invariant orbital integral (in γ) of ${}^c\theta_M(f) - {}^c\theta^{\mathcal{E}}_M(f)$. The functions ${}^c\theta_M(f)$ and ${}^c\theta^{\mathcal{E}}_M(f)$ both belong to $\widetilde{\mathcal{I}}_{ac}(M(F_v))$. But v is nonArchimedean, so the spaces $\widetilde{\mathcal{I}}_{ac}(M(F_v))$ and $\mathcal{I}_{ac}(M(F_v))$ coincide. (See [1(g), §4].) In particular,

$$\hat{I}^M_M(\gamma, {}^c\theta_M(f) - {}^c\theta^{\mathcal{E}}_M(f)) \overset{(M,\sigma)}{\sim} 0, \qquad \gamma \in \sigma M_\sigma(F_v).$$

It follows from (14.1) that

$${}^c\varepsilon_M(\gamma, f) \overset{(M,\sigma)}{\sim} 0, \qquad \gamma \in \sigma M_\sigma(F_v) \cap G_{\mathrm{reg}}.$$

In other words, ${}^c\varepsilon_M(\gamma, f)$ equals the orbital integral of a function in $\mathcal{H}(M(F_v))$ for any regular element $\gamma \in \sigma M_\sigma(F_v)$ which is close to σ. But by Lemma 4.4 of [1(g)], ${}^c\varepsilon_M(\gamma, f)$ has bounded support as a function of γ in the space of $M^0(F_v)$-orbits in $M(F_v)$. Appealing to a partition of unity argument, we find that only finitely many σ need intervene, and that ${}^c\varepsilon_M(\gamma, f)$ is everywhere an orbital integral. In other words, there is a function ${}^c\varepsilon_M(f)$ in $\mathcal{I}(M(F_v))$ such that

$$ {}^c\varepsilon_M(\gamma, f) = \hat{I}_M^M(\gamma, {}^c\varepsilon_M(f)). $$

Therefore

$$ \varepsilon_M(f) = {}^c\varepsilon_M(f) + {}^c\theta_M(f) - {}^c\theta_M^{\mathcal{E}}(f) $$

is the required function.

Now, suppose that v is Archimedean. We adopt the notation of the last part of §4. In particular, we shall regard $G^0(F_v)$ as a real Lie group. Let f be a fixed function in the space $\mathcal{H}(G(F_v))^0$, which in this case equals $\mathcal{H}(G(F_v))$. The main step in proving the existence of $\varepsilon_M(f)$ is to show that $\varepsilon_M(\gamma, f)$ behaves like the orbital integral of a Schwartz function on $M(F_v)$. Let $T = T_0 \rtimes \theta$ be an arbitrary "maximal torus" in G which is defined over F_v. This means that T is the centralizer in G of an element γ in $G_{\text{reg}}(F_v)$. We are going to prove that the restriction of the function $\varepsilon_M(\gamma, f)$ to $T(F_v)$ satisfies two conditions. We shall show that any derivative of $\varepsilon_M(\gamma, f)$ is locally bounded on $T(F_v)$, and that the function has appropriate behaviour across the singular hyperspaces of $T(F_v)$.

Before establishing the first condition, we shall examine the differential equation satisfied by $\varepsilon_M(\gamma, f)$. Suppose that z is an element in $\mathcal{Z}_v = \mathcal{Z}(G(F_v))$. By (3.5) and $(3.5)^{\mathcal{E}}$, the function

$$ \varepsilon_M(\gamma, zf) = I_M^{\mathcal{E}}(\gamma, zf) - \sum_{L_1 \supsetneq M} \hat{I}_M^{L_1}(\gamma, (zf)_{L_1})\varepsilon_{L_1}(v) $$

is equal to

$$ \sum_{L \in \mathcal{L}(M)} \partial_M^L(\gamma, z_L)\big(I_L^{\mathcal{E}}(\gamma, f) - \sum_{\substack{L_1 \in \mathcal{L}(L) \\ L_1 \neq M}} \hat{I}_L^{L_1}(\gamma, f_{L_1})\varepsilon_{L_1}(v)\big). $$

If $L \neq M$, we can sum L_1 over all elements in $\mathcal{L}(L)$, and the expression in the brackets vanishes. If $L = M$, the expression in the brackets is just $\varepsilon_M(\gamma, f)$. Moreover, by Lemma 12.4 of [1(e)],

$$ \partial_M^M(\gamma, z_M) = \partial(h_T(z)), $$

where $\partial(h_T(z))$ is the invariant differential operator on $T(F_v)$ obtained from z by the Harish-Chandra map. It follows that

$$\varepsilon_M(\gamma, zf) = \partial(h_T(z))\varepsilon_M(\gamma, f), \qquad \gamma \in T_{\mathrm{reg}}(F_v). \tag{14.2}$$

This differential equation can be combined with a technique of Harish-Chandra [20(a)] to establish that the derivatives of $\varepsilon_M(\gamma, f)$ are locally bounded. The technique is a fairly standard one, and it has been described clearly in this context in a lecture of Langlands. We shall just sketch the argument.

Let Ω be a compact subset of $T(F_v)$, and set $\Omega_{\mathrm{reg}} = \Omega \cap T_{\mathrm{reg}}(F_v)$. If $\partial(u)$ is an invariant differential operator on $T(F_v)$, there are constants $c(f)$ and q such that

$$|\partial(u) I_M(\gamma, f)| \leq c(f)|D^G(\gamma)|^{-q}, \qquad \gamma \in \Omega_{\mathrm{reg}}.$$

(See [1(g)], formula (2.8).) Both constants depend on $\partial(u)$, but q is independent of f. A similar assertion holds for $I_M^{\mathcal{E}}(\gamma, f)$, and therefore also for $\varepsilon_M(\gamma, f)$. Suppose that $\partial(u_1), \ldots, \partial(u_n)$ are generators over $\partial(h_T(\mathcal{Z}_v))$ of the module of differential operators on $T(F_v)$ of constant coefficients. Then any $\partial(u)$ can be written

$$\partial(u) = \partial(u_1)h_T(z_1) + \ldots + \partial(u_n)h_T(z_n), \qquad z_i \in \mathcal{Z}_v.$$

Applying the differential equation (14.2), we see that

$$\partial(u)\varepsilon_M(\gamma, f) = \sum_{i=1}^{n} \partial(u_i)\varepsilon_M(\gamma, z_i f).$$

It follows that

$$|\partial(u)\varepsilon_M(\gamma, f)| \leq c(f)|D^G(\gamma)|^{-q}, \qquad \gamma \in T_{\mathrm{reg}}(F_v),$$

where q may now be chosen to be independent of $\partial(u)$. This inequality will then lead to the property we want, namely that $\partial(u)\varepsilon_M(\gamma, f)$ is bounded on Ω_{reg}. The result is an immediate consequence of the following elementary lemma. (See the notes of Langlands' lecture, "Cancellation of singularities at the real placcs," I.A.S., p. 21–22.)

LEMMA 14.1.: *Let $\lambda_1, \ldots, \lambda_k$ be a finite set of linear forms on $\mathbf{R}^m$, and let ϕ be a smooth function on the set*

$$B' = \left\{ \xi \in \mathbf{R}^m \ : \ \|\xi\| \leq 1, \ \prod_{i=1}^{n} \lambda_i(\xi) \neq 0 \right\}.$$

Suppose for any differential operator Δ of constant coefficients on $\mathbf{R}^m$ that

$$|\Delta\phi(\xi)| \leq c_\Delta \left|\prod_{i=1}^{k} \lambda_i(\xi)\right|^{-q}, \qquad \xi \in B',$$

where q is independent of Δ. Then $\Delta\phi$ is bounded on B'. ∎

The second condition concerns the behaviour of $\varepsilon_M(\gamma, f)$ across the singular hyperplanes. Recall that a semisimple element $\sigma \in T(F_v)$ is called semiregular if the derived group of G_σ is three dimensional. The condition may be summarized as the requirement that $\varepsilon_M(\gamma, f)$ be (M,σ)-equivalent to 0, for any such σ and for G-regular points γ in $T(F_v)$ near σ. If $M_\sigma = G_\sigma$, this fact follows from (3.2) and $(3.2)^{\mathcal{E}}$.

Suppose then that σ is a semiregular point in $T(F_v)$ with $M_\sigma \neq G_\sigma$. This means that σ lies on a hyperspace in $T(F_v)$ defined by a real root β relative to the action of $T(F_v)$ on the Lie algebra of $G^0(F_v)$. The co-root $\beta^\vee$ belongs to the Lie algebra of $T_0(F_v)$. Set

$$\gamma_r = \sigma \exp(r\beta^\vee), \qquad r \in \mathbf{R}.$$

It is enough to show for any invariant differential operator $\partial(u)$ that the function $\partial(u)\varepsilon_M(\gamma_r, f)$ is smooth at $r = 0$. Associated to T_0 and β, we have a Cayley transform

$$C : T_0 \longrightarrow T_{01}.$$

This is an inner automorphism on G_σ which maps T_0 to a torus T_{01} in G_σ which is F_v-anisotropic modulo the center of G_σ. Let $M_1 \in \mathcal{L}(M)$ be the Levi subset such that

$$A_{M_1} = A_M \cap T_{01}.$$

Then $C\beta$ is a noncompact imaginary root of $M_1^0(F_v)$, and

$$\delta_s = \sigma \exp(sC\beta^\vee), \qquad s \in \mathbf{R},$$

is a G-regular point in

$$T_1(F_v) = \sigma T_{01}(F_v),$$

for s small and nonzero. If w_β is the reflection about $\beta^\vee$,

$$\partial(u_1) = \partial(Cw_\beta u - Cu)$$

is an invariant differential operator on $T_1(F_v)$. By formula (2.7) of [1(g)],

$$\lim_{r \to 0+} \left(\partial(u) I_M^\beta(\gamma_r, f) - \partial(u) I_M^\beta(\gamma_{-r}, f)\right) = n_\beta \lim_{s \to 0} \partial(u_1) I_{M_1}(\delta_s, f),$$

where

$$I_M^\beta(\gamma, f) = I_M(\gamma, f) + \|\beta_M^\vee\| \log(|\gamma^\beta - \gamma^{-\beta}|) I_{M_1}(\gamma, f), \qquad \gamma \in T_{\text{reg}}(F_v).$$

Here n_β is the cosine of the angle between $\beta^\vee$ and $\mathfrak{a}_M$, and $\|\beta_M^\vee\|$ denotes the norm of the projection of $\beta^\vee$ onto $\mathfrak{a}_M$. We shall apply the same formula to G'. The objects β, γ_t, u, etc. can all be mapped to corresponding objects β', γ_t', u', etc. for G', and it is easily seen that $n_{\beta'} = n_\beta$. It follows without difficulty that

$$\lim_{r \to 0+} (\partial(u) I_M^{\mathcal{E},\beta}(\gamma_r, f) - \partial(u) I_M^{\mathcal{E},\beta}(\gamma_{-r}, f)) = n_\beta \lim_{s \to 0} \partial(u_1) I_{M_1}^{\mathcal{E},\beta}(\delta_s, f),$$

where

$$I_M^{\mathcal{E},\beta}(\gamma, f) = I_M^{\mathcal{E}}(\gamma, f) + \|\beta_M^\vee\| \log(|\gamma^\beta - \gamma^{-\beta}|) I_{M_1}^{\mathcal{E}}(\gamma, f), \qquad \gamma \in T_{\text{reg}}(F_v).$$

Now

$$\text{(14.3)} \qquad \lim_{r \to 0+} (\partial(u) \varepsilon_M(\gamma_r, f) - \partial(u) \varepsilon_M(\gamma_{-r}, f))$$

equals the jump at $r = 0$ of

$$\partial(u) I_M^{\mathcal{E}}(\gamma_r, f) - \sum_{L_1 \supsetneq M} \partial(u) \hat{I}_M^{L_1}(\gamma_r, \varepsilon_{L_1}(f)).$$

If L_1 does not contain M_1, σ will be regular in L_1, and the summand will be smooth at $r = 0$. We may therefore take the sum over $L_1 \in \mathcal{L}(M_1)$. Moreover, we have

$$\|\beta_M^\vee\| \log(|\gamma^\beta - \gamma^{-\beta}|) \left(I_{M_1}^{\mathcal{E}}(\gamma, f) - \sum_{L_1 \in \mathcal{L}(M_1)} \hat{I}_{M_1}^{L_1}(\gamma, \varepsilon_{L_1}(f)) \right) = 0,$$

from our induction hypothesis. Therefore, (14.3) equals the jump at $r = 0$ of the expression

$$\partial(u) I_M^{\mathcal{E},\beta}(\gamma_r, f) - \sum_{L_1 \supset M_1} \partial(u) \hat{I}_M^{L_1,\beta}(\gamma_r, \varepsilon_{L_1}(f)).$$

Applying the two formulas above, we can therefore write (14.3) as

$$n_\beta \lim_{s \to 0} \partial(u_1) \left(I_{M_1}^{\mathcal{E}}(\delta_s, f) - \sum_{L_1 \in \mathcal{L}(M_1)} \hat{I}_{M_1}^{L_1}(\delta_s, \varepsilon_{L_1}(f)) \right).$$

But $M_1 \supsetneq M$, so the term in the brackets vanishes by our induction hypothesis. Thus, the function

$$\partial(u) \varepsilon_M(\gamma_r, f)$$

is smooth at $r = 0$. This is equivalent to

$$\varepsilon_M(\gamma, f) \overset{(M,\sigma)}{\sim} 0, \qquad \gamma \in T(F_v) \cap G_{\mathrm{reg}},$$

the required second condition.

We have verified the two conditions. These are two of the three conditions of Shelstad [38(a), Theorem 4.7] that are necessary and sufficient for $\varepsilon_M(\gamma, f)$ to be the orbital integral of a Schwartz function on $M(F_v)$. To avoid introducing extraneous questions in invariant harmonic analysis, we shall not work directly with the Schwartz space. However, the inductive arguments in [38(a)] do suggest how we should proceed.

Let us first recapitulate how $\varepsilon_M(\gamma, f)$ behaves under descent. The descent properties (3.3), $(3.3)^{\mathcal{E}}$, (6.2*) and (13.2*) are purely local. If $S = \{v\}$, they are valid if M_1 belongs to $\mathcal{L}_v$ rather than just $\mathcal{L}$, as stated. (Recall that $\mathcal{L}_v$ consists of Levi subsets over F_v.) Suppose that M_1 is an element in $\mathcal{L}_v$ which is properly contained in M. The induction hypothesis taken on at the beginning of §13 allows us to apply (6.2*). (See the remark following the proof of Lemma 6.2.) Moreover, we are assuming that

$$\hat{\varepsilon}^L_{M_1}(\gamma, f_L) = \hat{I}^{M_1}_{M_1}(\gamma, \hat{\varepsilon}^L_{M_1}(f_L)) = \varepsilon^L_{M_1}(v)\hat{I}^{M_1}_{M_1}(\gamma, f_{M_1}),$$

for any $L \subsetneq G$. It follows that

$$\varepsilon_M(\gamma, f) = \varepsilon_{M,M_1}(v)\hat{I}^{M_1}_{M_1}(\gamma, f_{M_1}), \qquad \gamma \in M_1(F_v) \cap G_{\mathrm{reg}},$$

where

$$\varepsilon_{M,M_1}(v) = \sum_{L \in \mathcal{L}(M_1)} d^G_{M_1}(M, L)\varepsilon^L_{M_1}(v).$$

We claim that the constant $\varepsilon_{M,M_1}(v)$ is independent of M_1. It is enough to show that

$$\varepsilon_{M,M_1'}(v) = \varepsilon_{M,M_1}(v)$$

for any $M_1' \in \mathcal{L}_v$ which is contained in M_1. Now

$$\varepsilon_{M,M_1}(v) = \sum_{L' \in \mathcal{L}(M_1')} \varepsilon^{L'}_{M_1'}(v) \sum_{\{L \in \mathcal{L}(M): L \supset L'\}} d^L_{M_1'}(M_1, L')d^G_{M_1}(M, L),$$

since we can assume that $\varepsilon^L_{M_1}(v)$, $L \subsetneq G$, satisfies the descent property (6.2). It follows from [1(g), (7.1)] that this equals

$$\sum_{L' \in \mathcal{L}(M_1')} \varepsilon^{L'}_{M_1'}(v)d^G_{M_1'}(M, L') = \varepsilon_{M,M_1'}(v),$$

as required. Thus

$$\varepsilon_M(v) = \varepsilon_{M,M_1}(v)$$

is independent of $M_1 \subsetneq M$. We have shown that

$$\varepsilon_M(\gamma, f) = \varepsilon_{M,M_1}(v)\hat{I}_{M_1}^{M_1}(\gamma, f_{M_1}) = \varepsilon_M(v)\hat{I}_M^M(\gamma, f_M),$$

for any $\gamma \in M_1(F_v) \cap G_{\text{reg}}$, as above.

If M/F_v is not minimal, we have just seen how to define the number $\varepsilon_M(v)$, even though we cannot yet apply Theorem 6.1 to M. Set

$$\varepsilon'_M(v) = \begin{cases} \varepsilon_M(v), & \text{if } M/F_v \text{ is not minimal,} \\ 0, & \text{if } M/F_v \text{ is minimal,} \end{cases}$$

and define

$$\varepsilon'_M(\gamma, f) = \varepsilon_M(\gamma, f) - \varepsilon'_M(v)\hat{I}_M^M(\gamma, f_M).$$

Since $\hat{I}_M^M(\gamma, f_M)$ is just the orbital integral of a function in $\mathcal{H}(M(F_v))$, $\varepsilon'_M(\gamma, f)$ satisfies the same two conditions established for $\varepsilon_M(\gamma, f)$. Observe also that if γ belongs to $M_1(F_v) \cap G_{\text{reg}}$, for $M_1 \subsetneq M$ as above, then

$$\varepsilon'_M(\gamma, f) = 0.$$

Now suppose that T is an elliptic "maximal torus" of M over F_v. That is, T is not contained in any proper Levi subset M_1 of M. Then $\varepsilon'_M(\gamma, f)$, suitably normalized, extends to a smooth function of $\gamma \in T(F_v)$. More precisely, there is a locally constant function

$$c : T_{\text{reg}}(F_v) \rightarrow \{z \in \mathbf{C} : |z| = 1\}$$

such that the function

$$\tilde{\varepsilon}'_M(\gamma, f) = c(\gamma)\varepsilon'_M(\gamma, f), \qquad \gamma \in T_{\text{reg}}(F_v),$$

extends to a smooth function on $T(F_v)$ which is skew-symmetric under the Weyl group

$$W^M(T) = \text{Norm}_T(M^0) \,/\, \text{Cent}_T(M^0).$$

This follows in a standard fashion from the two conditions and the vanishing property above.

Observe that for any point $X \in \mathfrak{a}_M$, the set

$$T(F_v)^X = M(F_v)^X \cap T(F_v) = \{x \in M(F_v) : H_M(x) = X\} \cap T(F_v)$$

is compact. Let us write $\Pi^+_{\text{disc}}(M(F_v))$ for the set of representations in $\Pi^+_{\text{temp}}(M(F_v))$ which are not of the form

$$\pi_1^M, \qquad \pi_1 \in \Pi^+_{\text{temp}}(M_1(F_v)),\ M_1 \in \mathcal{L}_v,\ M_1 \subsetneq M.$$

The function $c(\gamma)$ above has the additional property that the set of functions

$$\Phi_\pi(\gamma) = c(\gamma)|D^M(\gamma)|^{1/2}\Theta_\pi(\gamma), \qquad \gamma \in T(F_v),$$

in which π ranges over a set of representatives of orbits of $i\mathfrak{a}_M^*$ in $\Pi_{\text{disc}}^+(M(F_v))$, forms an orthogonal basis of a Hilbert space of functions on $T(F_v)$ which includes $\widetilde{\varepsilon}'_M(\cdot, f)$. Indeed, each Φ_π is just a fixed multiple of the skew-symmetrization of a 1-dimensional character on $T(F_v)$. This is a well known result of Harish-Chandra if $G = G^0$; if $G \neq G^0$, it follows from [11(a), Theorem 8.1] and the corresponding fact for $G = G^0$. (In fact, the existence of the elliptic torus T means that M^0 is a product of several copies of $GL(2)$ and $GL(1)$, so the property actually follows from local Archimedean base change for $GL(2)$.) Define

$$(14.4) \qquad \varepsilon'_M(f, \pi, X) = |W^M(T)|^{-1} \int_{T(F_v)^X} \widetilde{\varepsilon}'_M(\gamma, f)\overline{\Phi_\pi(\gamma)}\, d\gamma,$$

if π is any representation in $\Pi_{\text{disc}}^+(M(F_v))$. If π belongs to the complement of $\Pi_{\text{disc}}^+(M(F_v))$ in $\Pi_{\text{temp}}^+(M(F_v))$, we shall simply set $\varepsilon'_M(f, \pi, X) = 0$.

We claim that $\varepsilon'_M(f, \pi, X)$ is a Schwartz function of X. Since the function is smooth, it suffices to show that for any invariant differential operator Δ on $\mathfrak{a}_M$, the function

$$\Delta\varepsilon'_M(f, \pi, X), \qquad X \in \mathfrak{a}_M,$$

is rapidly decreasing. Observe first that $\varepsilon'_M(f, \pi, X)$ equals an integral, over the set of $M^0(F_v)$-orbits in $M(F_v)^X$, of the product of $\varepsilon'_M(\gamma, f)$ with $|D^M(\gamma)|^{1/2}\Theta_\pi(\gamma)$. But Lemma 13.3 tells us that we can write $\varepsilon'_M(\gamma, f)$ as the sum of

$${}^c\varepsilon_M(\gamma, f) - \varepsilon'_M(v)\hat{I}_M^M(\gamma, f_M)$$

and

$$\hat{I}_M^M(\gamma, {}^c\theta_M(f) - {}^c\theta_M^{\mathcal{E}}(f)),$$

for any G-regular element γ in $M(F_v)^X$. The first function has bounded support (on the $M^0(F_v)$-orbits in $M(F_v)^X$), and vanishes if X lies outside a compact set. The integral of the product of the second function with $|D^M(\gamma)|^{1/2}\Theta_\pi(\gamma)$ equals

$${}^c\theta_M(f, \pi, X) - {}^c\theta_M^{\mathcal{E}}(f, \pi, X).$$

It follows that

$$\varepsilon'_M(f, \pi, X) = {}^c\theta_M(f, \pi, X) - {}^c\theta_M^{\mathcal{E}}(f, \pi, X),$$

for any point $X \in \mathfrak{a}_M$ outside a fixed compact set. Now Corollary 5.3 of [1(g)] tells us that $\Delta{}^c\theta_M(f, \pi, X)$ is a rapidly decreasing function of $X \in \mathfrak{a}_M$. A similar assertion applies to $\Delta{}^c\theta_M^{\mathcal{E}}(f, \pi, X)$. Consequently, the function

$\Delta\varepsilon'_M(f,\pi,X)$ is rapidly decreasing, and $\varepsilon'_M(f,\pi,X)$ is indeed a Schwartz function of X.

We would like to show that as a function of (π, X), $\varepsilon'_M(f,\pi,X)$ belongs to the space $\mathcal{I}_{ac}(M(F_v))$. There are two further properties to establish. We must show that the function is $(K_v \cap M^0(F_v))$-finite. That is, we need to find a finite subset Γ_M of $\Pi(K_v \cap M^0(F_v))$ such that $\varepsilon'_M(f,\pi,X)$ vanishes unless the restriction of π to $(K_v \cap M^0(F_v))$ contains a representation in Γ_M. We must also show that for each $X \in \mathfrak{a}_M$, the function $\varepsilon'_M(f,\pi,X)$ belongs to the Paley-Wiener space in the natural coordinates (taken modulo $i\mathfrak{a}^*_M$) on $\Pi^+_{\text{temp}}(M(F_v))$. This second property poses no problem. For the Paley-Wiener requirement is trivial unless π is properly induced, in which case $\varepsilon'_M(f,\pi,X) = 0$. To establish the first property, we shall use the differential equation (14.2). Set

$$\varepsilon'_M(f,\pi) = \int_{\mathfrak{a}_M} \varepsilon'_M(f,\pi,X)\,dX, \qquad \pi \in \Pi^+_{\text{temp}}(M(F_v)).$$

It follows from the differential equations, the definition (14.4), and the fact that $\varepsilon'_M(f,\pi,X)$ is a Schwartz function of X, that

$$\varepsilon'_M(zf,\pi) = \pi(z_M)\varepsilon'_M(f,\pi), \qquad \pi \in \Pi^+_{\text{temp}}(M(F_v)),$$

where $\pi(z_M)$ denotes the infinitesimal character of π evaluated at z_M. Thus, as a function of f, $\varepsilon'_M(f,\pi)$ is an invariant eigendistribution of $\mathcal{Z}_v$. By first taking π to be in general position, one sees easily that

$$\varepsilon_M(f,\pi) = c(\pi)tr(\mathcal{I}_P(\pi,f)),$$

where $P \in \mathcal{P}(M)$ and $c(\pi)$ is a smooth function on $\Pi^+_{\text{temp}}(M(F_v))$. Since f is K_v-finite, there is a finite subset Γ of $\Pi(K_v)$ such that $tr(\mathcal{I}_P(\pi,f))$ vanishes unless the K_v-spectrum of $\mathcal{I}_P(\pi)$ meets Γ. The first property then holds if we take Γ_M to be the set of irreducible constituents of restrictions of elements in Γ to $K_v \cap M^0(F_v)$. This proves that the function

$$\varepsilon'_M(f) : (\pi, X) \to \varepsilon'_M(f,\pi,X)$$

belongs to $\mathcal{I}_{ac}(M(F_v))$. In particular, the orbital integral $\hat{I}^M_M(\gamma, \varepsilon'_M(f))$ is defined, for any $\gamma \in M(F_v)$. Applying Fourier inversion on $T(F_v)^X$ to (14.4), one sees without difficulty that

$$\hat{I}^M_M(\gamma, \varepsilon'_M(f)) = \varepsilon'_M(\gamma, f).$$

We are almost done. Define

$$\varepsilon_M(f) = \varepsilon'_M(f) + \varepsilon'_M(v) f_M.$$

Since f_M belongs to $\mathcal{I}(M(F_v))$, the function $\varepsilon_M(f)$ belongs to $\mathcal{I}_{ac}(M(F_v))$. The formula

$$\hat{I}_M^M(\gamma, \varepsilon_M(f)) = \varepsilon_M(\gamma, f)$$

follows from the definitions and the analogous formula for $\varepsilon'_M(f)$. We have thus defined the required map ε_M when S consists of one Archimedean valuation. This was our final step, so the proof of Proposition 13.2 is at last complete. ∎

COROLLARY 14.2: *If S is any finite set with the closure property,*

$$\varepsilon_M(f) = {}^c\varepsilon_M(f) + {}^c\theta_M(f) - {}^c\theta_M^{\mathcal{E}}(f), \qquad f \in \mathcal{H}(G(F_S))^0,$$

where ${}^c\varepsilon_M(f)$ is a function in $\widetilde{\mathcal{I}}(M(F_S))$. In particular, for any $\pi \in \Pi^+(M(F_S))$, $\varepsilon_M(f, \pi, X)$ is a Schwartz function of $X \in \mathfrak{a}_{M,S}$.

Proof. By Lemma 13.3,

$${}^c\varepsilon_M(f) = \varepsilon_M(f) - {}^c\theta_M(f) + {}^c\theta_M^{\mathcal{E}}(f)$$

is a function in $\widetilde{\mathcal{I}}_{ac}(M(F_S))$ whose orbital integral at any $\gamma \in M(F_S) \cap G_{\text{reg}}$ equals ${}^c\varepsilon_M(\gamma, f)$. But it follows inductively from [1(g), Lemma 4.4] that ${}^c\varepsilon(\gamma, f)$ vanishes if $X = H_M(\gamma)$ lies outside a compact set. Since ${}^c\varepsilon_M(f, \pi, X)$ equals the integral of the normalized character of π against ${}^c\varepsilon_M(\gamma, f)$, this function also vanishes if X is large. Therefore, ${}^c\varepsilon_M(f)$ belongs to $\widetilde{\mathcal{I}}(M(F_S))$. The second assertion of the corollary follows from [1(g), Corollary 5.3], as we saw above in the special case that S consists of one Archimedean prime. ∎

As we noted in §13, ε_M extends to a map from $\mathcal{H}(G(\mathbf{A}))^0$ to $\mathcal{I}_{ac}(M(\mathbf{A}))$. In [1(f), §11] we introduced a space of *moderate* functions, which lies between $\mathcal{I}(M(\mathbf{A}))$ and $\mathcal{I}_{ac}(M(\mathbf{A}))$. (See also the appendix to [1(h)].) There is no need to repeat the definition here. Let us say only that for a function $\phi \in \mathcal{I}_{ac}(M(\mathbf{A}))$ to be moderate it must satisfy a weak growth condition and an equally weak support condition.

COROLLARY 14.3: *For each $f \in \mathcal{H}(G(\mathbf{A}))^0$, $\varepsilon_M(f)$ is a moderate function.*

Proof. We can assume that f is of the form $\prod_v f_v$. By (13.3*),

$$\varepsilon_M(f) = \sum_v \varepsilon_{M,v}(f) + d_0 f_M,$$

where d_0 is a constant and

$$\varepsilon_{M,v}(f) = \varepsilon_M(f_v) \prod_{w \neq v} f_{w,M}.$$

Almost all the functions $\varepsilon_{M,v}(f)$ vanish. The function f_M belongs to $\mathcal{I}(M(\mathbf{A}))$, and is certainly moderate. It is therefore enough to fix a valuation v and prove that for a fixed function $f \in \mathcal{H}(G(F_v))^0$, $\varepsilon_M(f)$ is a moderate function in $\mathcal{I}_{ac}(M(F_v))$.

There are two conditions to check. They must be established for any function

$$X_1 \to \int_{i\mathfrak{a}^*_{M_1,v}/i\mathfrak{a}^*_{M,v}} \varepsilon_M(f, \pi^M_{1,\Lambda}, X)e^{-\Lambda(X_1)}\, d\Lambda,$$

in which M_1 is is a Levi subset of M over F_v, π_1 belongs to $\Pi^+_{\text{temp}}(M_1(F_v))$, and X_1 is a point in $\mathfrak{a}_{M_1,v}$ whose projection onto $\mathfrak{a}_{M,v}$ equals X. If $M_1 \subsetneq M$, the formula (13.2*) implies that the function is compactly supported. If $M_1 = M$, Corollary 14.2 asserts that the function belongs to the Schwartz space. In each case, the required growth and support conditions hold. ∎

Finally, we shall show that ε_M behaves nicely under multipliers. Let $\mathfrak{h}^1$ be the orthogonal complement of $\mathfrak{a}_G$ in the space $\mathfrak{h}$ defined in §9. Multipliers are attached to elements α in $\mathcal{E}(\mathfrak{h}^1)^W$, the convolution algebra of compactly supported, W-invariant distributions on $\mathfrak{h}^1$. Recall that there is an action $f \to f_\alpha$ of the algebra $\mathcal{E}(\mathfrak{h}^1)^W$ on $\mathcal{H}(G(\mathbf{A}))$ such that

$$\mathcal{I}_P(\pi, f_\alpha) = \hat{\alpha}(\nu_\pi)\mathcal{I}_P(\pi, f), \qquad \pi \in \Pi^+(M(\mathbf{A})).$$

There is also a compatible action $\phi \to \phi_\alpha$ of $\mathcal{E}(\mathfrak{h}^1)^W$ on $\mathcal{I}_{ac}(M(\mathbf{A}))$ which for any $\pi \in \Pi^+_{\text{temp}}(M(\mathbf{A}))$ and $X \in \mathfrak{a}_M$ is given by

$$\phi_\alpha(\pi, X) = \int_{\mathfrak{a}^G_M} \phi(\pi, Y) \int_{i\mathfrak{a}^*_M/i\mathfrak{a}^*_G} \hat{\alpha}(\nu_\pi + \mu)e^{-\mu(X-Y)}\, d\mu\, dY.$$

The reader can check that there is a natural map $\alpha \to \alpha'$ from $\mathcal{E}(\mathfrak{h}^1)^W$ to $\mathcal{E}(\mathfrak{h}^1 \cap \mathfrak{h}')^{W'}$ such that

$$(f_\alpha)' = f'_{\alpha'}, \qquad f \in \mathcal{H}(G(\mathbf{A})).$$

COROLLARY 14.4.: $\varepsilon_M(f_\alpha) = \varepsilon_M(f)_\alpha, \quad f \in \mathcal{H}(G(\mathbf{A}))^0,\ \alpha \in \mathcal{E}(\mathfrak{h}^1)^W.$

Proof. Let us fix a function

$$f_0 = \prod_{v \text{ finite}} f_v, \qquad f_v \in \mathcal{H}(G(F_v)),$$

with the property that for any $f_\infty \in \mathcal{H}(G(F_\infty))$, the function $f = f_\infty f_0$ belongs to $\mathcal{H}(G(\mathbf{A}))^0$. We shall vary f_∞. Suppose that $\pi = \pi_\infty \otimes \pi_0$ is a representation in $\Pi^+_{\text{temp}}(M(\mathbf{A}))$. Using the differential equations (14.2), we can argue as above to show that if

$$\varepsilon_M(f,\pi) = \int_{\mathfrak{a}_M} \varepsilon_M(f,\pi,X)\,dX,$$

then

$$\varepsilon_M(f,\pi) = c(\pi, f_0)\,tr(\mathcal{I}_P(\pi_\infty, f_\infty)),$$

where $P \in \mathcal{P}(M)$ and $c(\pi, f_0)$ is a scalar which is independent of f_∞. It follows that for any $\alpha \in \mathcal{E}(\mathfrak{h}^1)^W$,

$$\begin{aligned}\varepsilon_M(f_\alpha,\pi) &= c(\pi, f_0)\,tr\big(\mathcal{I}_P(\pi_\infty,(f_\infty)_\alpha)\big)\\ &= \hat{\alpha}(\nu_\pi)c(\pi, f_0)\,tr(\mathcal{I}_P(\pi_\infty, f_\infty))\\ &= \hat{\alpha}(\nu_\pi)\varepsilon_M(f,\pi).\end{aligned}$$

We obtain

$$\begin{aligned}\varepsilon_M(f_\alpha,\pi,X) &= \int_{i\mathfrak{a}_M^*} \varepsilon_M(f_\alpha,\pi_\lambda)e^{-\lambda(X)}\,d\lambda\\ &= \int_{\mathfrak{a}_M^G} \varepsilon_M(f_\alpha,\pi,Y)\int_{i\mathfrak{a}_M^*/i\mathfrak{a}_G^*} \hat{\alpha}(\nu_\pi+\mu)e^{-\mu(X-Y)}\,d\mu\,dY,\end{aligned}$$

as required. ∎

15. Separation by infinitesimal character

We can now apply the map ε_M to the formula for $I^{\mathcal{E}}(f) - I(f)$ in Lemma 13.1. Let us define $\mathcal{H}(G(\mathbf{A}), M)^0$ to be the space of functions f in

$$\mathcal{H}(G(\mathbf{A}), M) \cap \mathcal{H}(G(\mathbf{A}))^0$$

which satisfy one additional condition. We ask that f vanish at any element in $G(\mathbf{A})$ whose component at each finite place v belongs to $A_G(F_v)$. This last condition is of course vacuous unless $\ell = 1$. Combined with the earlier definition of $\mathcal{H}(G(\mathbf{A}))^0$, it is designed to ensure that the orbital integrals of f vanish at any element

$$\gamma = \xi u, \qquad \xi \in A_G(F),\ u \in \mathcal{U}_G(F).$$

Notice that f may be modified at the Archimedean places, and the function will still remain in $\mathcal{H}(G(\mathbf{A}), M)^0$.

LEMMA 15.1.: *Suppose that f belongs to $\mathcal{H}(G(\mathbf{A}), M)^0$. Then*

$$I^{\mathcal{E}}(f) - I(f) = |W(\mathfrak{a}_M)|^{-1} \hat{I}^M(\varepsilon_M(f)),$$

where I^M is of course the analogue for M of I.

Proof. Consider the formula for $I^{\mathcal{E}}(f) - I(f)$ provided by Lemma 13.1. The conditions on f imply that the second term in the formula vanishes. By formula (13.1**), the first term equals

$$|W(\mathfrak{a}_M)|^{-1} \sum_{\gamma \in (M(F))_{M,S}} a^M(S,\gamma) \hat{I}^M_M(\gamma, \varepsilon_M(f)),$$

where S is a large finite set of valuations. Now $\varepsilon_M(f)$ is a function in $\mathcal{I}_{ac}(M(\mathbf{A}))$ which is cuspidal at two places. In other words, $\varepsilon_M(f)$ is a finite sum of functions $\prod_v \phi_v$ in $\mathcal{I}_{ac}(M(\mathbf{A}))$ such that for two unramified places v_1 and v_2, and any $M_1 \subsetneq M$,

$$\phi_{v_i, M_1} = 0, \qquad i = 1, 2.$$

This follows from (13.2*), (13.3*) and the fact that f belongs to $\mathcal{H}(G(\mathbf{A}), M)$. Applying (3.4) to M, we find that

$$\hat{I}^M_{M_1}(\gamma, \varepsilon_M(f)) = 0, \qquad \gamma \in M_1(F),$$

for any such M_1. It follows from (5.1), applied to M, that

$$\sum_{\gamma\in(M(F))_{M,S}} a^M(S,\gamma)\hat{I}_M^M(\gamma,\varepsilon_M(f))$$

$$= \sum_{M_1\in\mathcal{L}^M} |W_0^{M_1}||W_0^M|^{-1} \sum_{\gamma\in(M_1(F))_{M_1,S}} a^{M_1}(S,\gamma)\hat{I}_{M_1}^M(\gamma,\varepsilon_M(f))$$

$$= \hat{I}^M(\varepsilon_M(f)).$$

The lemma follows. ∎

We fix the function f in $\mathcal{H}(G(\mathbf{A}),M)^0$. Combined with the expansions $I^{\mathcal{E}} = \sum_t I_t^{\mathcal{E}}$ and $I = \sum_t I_t$, Lemma 15.1 yields the formula

$$\sum_{t\geq 0}\Big(I_t^{\mathcal{E}}(f) - I_t(f) - |W(\mathfrak{a}_M)|^{-1}\hat{I}_t^M(\varepsilon_M(f))\Big) = 0. \tag{15.1}$$

We are going to apply the spectral expansions of the distributions on the left. We will then try to deduce what remains of Theorem B. As we remarked in §9, however, our control over the convergence is very weak. In this section we shall simply isolate the terms in (15.1) according to their Archimedean infinitesimal character.

We shall use an argument based on multipliers. Associated to the real Lie group $G^0(F_\infty)^1$, we have the real vector space $\mathfrak{h}^1$, defined in §14. It is convenient to work with a subset of the complex dual space $\mathfrak{h}_{\mathbf{C}}^*/\mathfrak{a}_{G,\mathbf{C}}^*$ of $\mathfrak{h}^1$ which contains the infinitesimal character of any unitary representation of $G^0(F_\infty)^1$. Let $\mathfrak{h}_u^*$ denote the set of points ν in $\mathfrak{h}_{\mathbf{C}}^*/i\mathfrak{a}_G^*$ such that $\overline{\nu} = -s\nu$ for some element $s\in W$ of order 2. Here $\overline{\nu}$ denotes the conjugation on $\mathfrak{h}_{\mathbf{C}}^*$ relative to $\mathfrak{h}^*$. The Archimedean infinitesimal character ν_π associated to any $\pi\in\Pi_{\mathrm{unit}}^+(G(\mathbf{A})^1)$ belongs to the subset

$$(\mathfrak{h}')_u^* = \mathfrak{h}_u^* \cap (\mathfrak{h}'\cap\mathfrak{h}^1)_{\mathbf{C}}^*$$

of $\mathfrak{h}_u^*$. It is clear that for any nonnegative numbers r and T, the set

$$\mathfrak{h}_u^*(r,T) = \{\nu\in\mathfrak{h}_u^* : \|\mathrm{Re}\,(\nu)\|\leq r,\ \|\mathrm{Im}\,\nu\|\geq T\}$$

is invariant under W. (An element $\nu\in\mathfrak{h}_u^*$ is just a coset of $i\mathfrak{a}_G^*$ in $\mathfrak{h}_{\mathbf{C}}^*$, but $\|\nu\|$ is understood to be the minimum value of the norm on the coset.) The multipliers enter through an estimate from [1(h)]. The result pertains to functions $\phi\in\mathcal{I}_{ac}(G(\mathbf{A}))$ which are moderate in the sense described at the end of the last section. Suppose that ϕ is a given moderate function. Then Corollary 6.3 of [1(h)] provides positive constants C, k, and r such that for any $T > 0$ and any $\alpha \in C_N^\infty(\mathfrak{h}^1)^W$, with $N > 0$, the inequality

$$\sum_{t>T} |\hat{I}_t(\phi_\alpha)| \leq Ce^{kN} \sup_{\nu \in \mathfrak{h}_u^*(r,T)} (|\hat{\alpha}(\nu)|), \tag{15.2}$$

holds.

We return to our original function $f \in \mathcal{H}(G(\mathbf{A}), M)^0$. It follows easily from the definitions that if $\alpha \in \mathcal{E}(\mathfrak{h}^1)^W$, the function f_α also belongs to $\mathcal{H}(G(\mathbf{A}), M)^0$. In particular, f_α satisfies (15.1). Therefore, for any $T \geq 0$, the expression

$$\left| \sum_{t \leq T} \left(I_t^{\mathcal{E}}(f_\alpha) - I_t(f_\alpha) - |W(\mathfrak{a}_M)|^{-1} \hat{I}_t^M(\varepsilon_M(f_\alpha)) \right) \right| \tag{15.3}$$

is bounded by

$$\sum_{t>T} \left(|I_t(f_\alpha)| + |I_t^{\mathcal{E}}(f_\alpha)| + |W(\mathfrak{a}_M)|^{-1} |\hat{I}_t^M(\varepsilon_M(f_\alpha))| \right).$$

We can write

$$I_t(f_\alpha) = \hat{I}_t(f_{G,\alpha}),$$

$$I_t^{\mathcal{E}}(f_\alpha) = \hat{I}_t((f_\alpha)') = \hat{I}_t(f'_{\alpha'}),$$

and

$$\varepsilon_M(f_\alpha) = \varepsilon_M(f)_\alpha,$$

by Corollary 14.4. Consequently, (15.3) is bounded by

$$\sum_{t>T} |\hat{I}_t^G(f_{G,\alpha})| + \sum_{t>T} |\hat{I}_t^{G'}(f'_{\alpha'})| + |W(\mathfrak{a}_M)|^{-1} \sum_{t>T} |\hat{I}_t^M(\varepsilon_M(f)_\alpha)|.$$

The functions $f_G \in \mathcal{I}(G(\mathbf{A}))$ and $f' \in \mathcal{I}(G'(\mathbf{A}))$ are of course both moderate, and by Corollary 14.3, the same is true of the function $\varepsilon_M(f) \in \mathcal{I}_{ac}(M(\mathbf{A}))$. We can therefore apply the estimate (15.2) to these three different functions. Observe that $(\mathfrak{h}')_u^*(r,T)$, the set defined above but with G replaced by G', is actually contained in $\mathfrak{h}_u^*(r,T)$. It follows that there are positive constants C, k, and r such that for any $\alpha \in C_N^\infty(\mathfrak{h}^1)^W$, with $N > 0$, and for any $T > 0$, the expression (15.3) is bounded by

$$Ce^{kN} \sup_{\nu \in \mathfrak{h}_u^*(r,T)} (|\hat{\alpha}(\nu)|). \tag{15.4}$$

Let ν_1 be an arbitrary but fixed point in $\mathfrak{h}_u^*$. Enlarging the constant r in (15.4) if necessary, we may assume that ν_1 belongs to the cylinder

$$\mathfrak{h}_u^*(r) = \mathfrak{h}_u^*(r,0).$$

LEMMA 15.2.: *There is a function $\alpha_1 \in C_c^\infty(\mathfrak{h}^1)^W$ such that $\hat{\alpha}_1$ maps $\mathfrak{h}_u^*(r)$ to the unit interval, and such that the inverse image of 1 under $\hat{\alpha}_1$ is the finite set*

$$W(\nu_1) = \{s\nu_1 : s \in W\}.$$

Proof. Consider the space of functions

$$\nu \to \hat{\alpha}(\nu), \qquad \nu \in \mathfrak{h}_u^*, \ \alpha \in C_c^\infty(\mathfrak{h}^1)^W,$$

on $\mathfrak{h}_u^*$. It is clear from the definition of $\mathfrak{h}_u^*$ that the real and imaginary parts of any such function also belong to the space. We can therefore find a function $\alpha_0 \in C_c^\infty(\mathfrak{h}^1)^W$, with $\hat{\alpha}_0(\nu_1) \neq 0$, such that $\hat{\alpha}_0$ is real valued on $\mathfrak{h}_u^*$. Let $p_0 = 1, p_1, \ldots, p_m$ be a set of generators of the algebra of W-invariant polynomials on $\mathfrak{h}_{\mathbf{C}}^*/\mathfrak{a}_{G,\mathbf{C}}^*$. We can assume that each p_i is real valued on $\mathfrak{h}_u^*$. Since $\hat{\alpha}_0$ is rapidly decreasing at infinity on $\mathfrak{h}_u^*(r)$, the function

$$\beta(\nu) = (p_0(\nu)\hat{\alpha}_0(\nu), \ldots, p_m(\nu)\hat{\alpha}_0(\nu))$$

maps $\mathfrak{h}_u^*(r)$ continuously to a compact rectangle $[a,b]^{m+1}$ in $\mathbf{R}^{m+1}$. Set

$$s = (s_0, s_1, \ldots, s_m) = \beta(\nu_1).$$

Then

$$\beta^{-1}(\{s\}) = W(\nu_1).$$

For each i, let

$$q_i : [a,b] \to [0,1]$$

be a real polynomial such that $q_i^{-1}(\{1\})$ equals $\{s_i\}$. Since $s_0 \neq 0$, we can assume that q_0 has no constant term. Consequently

$$\hat{\alpha}_1(\nu) = \prod_{i=0}^{m} q_i(p_i(\nu)\hat{\alpha}_0(\nu))$$

is the Fourier-Laplace transform of a function $\alpha_1 \in C_c^\infty(\mathfrak{h}^1)^W$. It clearly satisfies the requirements of the lemma. ∎

Fix α_1 as in the last lemma. Then α_1 belongs to $C_{N_1}^\infty(\mathfrak{h}^1)^W$ for some $N_1 > 0$. If r and k are as in (15.4), choose $T > 0$ so that

$$|\hat{\alpha}_1(\nu)| \leq e^{-2kN_1}$$

for all $\nu \in \mathfrak{h}_u^*(r,T)$. This is certainly possible, since $\hat{\alpha}_1$ is rapidly decreasing on $\mathfrak{h}_u^*(r)$. For each positive integer m, define

$$\alpha_m = \underbrace{\alpha_1 * \cdots * \alpha_1}_{m}.$$

Then α_m belongs to $C^\infty_{mN_1}(\mathfrak{h}^1)^W$, and

$$\hat{\alpha}_m(\nu) = \hat{\alpha}_1(\nu)^m.$$

Taking $\alpha = \alpha_m$ above, we see that the expression

$$\sum_{t \leq T} \Big(I_t^{\mathcal{E}}(f_{\alpha_m}) - I_t(f_{\alpha_m}) - |W(\mathfrak{a}_M)|^{-1} \hat{I}_t^M(\varepsilon_M(f)_{\alpha_m}) \Big) \tag{15.5}$$

is bounded in absolute value by

$$Ce^{-kN_1 m}.$$

Consequently, (15.5) approaches 0 as m approaches ∞. This assertion is a signficant improvement over the formula (15.1). For the sum in (15.5) can be taken over a finite set which is independent of m. This will allow us to take the expansions of the terms in (15.5) and study the limit as m approaches ∞.

Apply Lemma 12.3 to the function f_{α_m}. Since

$$tr\, \pi((f_{\alpha_m})^1) = tr\,(\pi(f^1))\hat{\alpha}_1(\nu_\pi)^m,$$

we see that

$$\sum_{t \leq T} (I_t^{\mathcal{E}}(f_{\alpha_m}) - I_t(f_{\alpha_m}))$$

equals

$$\sum_{t \leq T} \sum_{\pi \in \Pi^+(G(\mathbf{A})^1, t)} (a^{\mathcal{E}}_{\text{disc}}(\pi) - a_{\text{disc}}(\pi)) tr\,(\pi(f^1))\hat{\alpha}_1(\nu_\pi)^m. \tag{15.6}$$

Next, we expand $I_t^M(\varepsilon_M(f)_{\alpha_m})$. The function $\varepsilon_M(f)_{\alpha_m}$ is a finite sum of functions which are cuspidal at two places. It follows from Theorem 7.1(a) of [1(h)] that

$$\hat{I}_t^M(\varepsilon_M(f)_{\alpha_m}) = \sum_{\pi \in \Pi_{\text{disc}}(M,t)} a^M_{\text{disc}}(\pi) \hat{I}^M_M(\pi, \varepsilon_M(f)_{\alpha_m}).$$

We require a lemma.

LEMMA 15.3.: *Suppose that $\pi \in \Pi^+_{\text{unit}}(M(\mathbf{A})^1)$. Then there is a Schwartz function*

$$\lambda \to \varepsilon_M(f^1, \pi, \lambda), \qquad \lambda \in i\mathfrak{a}^*_M / i\mathfrak{a}^*_G,$$

on $i\mathfrak{a}^*_M/i\mathfrak{a}^*_G$ such that for any $\alpha \in C_c^\infty(\mathfrak{h}^1)^W$,

$$\hat{I}^M_M(\pi, \varepsilon_M(f)_\alpha) = \int_{i\mathfrak{a}^*_M/i\mathfrak{a}^*_G} \varepsilon_M(f^1, \pi, \lambda)\hat{\alpha}(\nu_\pi + \lambda)\, d\lambda.$$

Proof. As a function in $\mathcal{I}_{ac}(M(\mathbf{A}))$, $\varepsilon_M(f)_\alpha$ is *a priori* defined only on $\Pi^+_{\text{temp}}(M(\mathbf{A})) \times \mathfrak{a}_M$, but it may be naturally extended to a function on $\Pi^+(M(\mathbf{A})) \times \mathfrak{a}_M$. (This is a reflection of the fact that a function in $\mathcal{H}_{ac}(M(\mathbf{A}))$ is compactly supported on any set $M(\mathbf{A})^X$, and can therefore be integrated against a nontempered character.) Identify π with an orbit $\{\pi_\lambda\}$ of $i\mathfrak{a}^*_M$ in $\Pi^+_{\text{unit}}(M(\mathbf{A}))$. By definition, $\hat{I}^M_M(\pi, \varepsilon_M(f)_\alpha)$ is the value of $\varepsilon_M(f)_\alpha$ at $(\pi_\lambda, 0)$. It follows from the formula (6.1) of [1(h)] that

$$\hat{I}^M_M(\pi, \varepsilon_M(f)_\alpha) = \int_{\mathfrak{a}^G_M} \varepsilon_M(f, \pi_\lambda, Y)\alpha_M(\pi_\lambda, -Y)\, dY,$$

for any λ, where

$$\alpha_M(\pi_\lambda, -Y) = \int_{i\mathfrak{a}^*_M / i\mathfrak{a}^*_G} \hat{\alpha}(\nu_\pi + \lambda + \mu)e^{\mu(Y)} d\mu.$$

This last expression is compactly supported as a function of $Y \in \mathfrak{a}^G_M$, so the integral over $\mathfrak{a}^G_M$ above converges. Our lemma will follow from Fourier inversion if we can show that

$$\varepsilon_M(f, \pi_\lambda, X), \qquad X \in \mathfrak{a}_M,$$

is a Schwartz function on $\mathfrak{a}_M$. This is actually a sensitive point. What saves us is the unitarity of π.

We can assume that f is of the form $\prod_v f_v$. By (13.3*),

$$\varepsilon_M(f) = \sum_v \varepsilon_{M,v}(f) + d_0 f_M,$$

where d_0 is a constant and

$$\varepsilon_{M,v}(f) = \varepsilon_M(f_v) \prod_{w \neq v} f_{w,M}.$$

Since

$$f_M(\pi_{\lambda+\mu}), \qquad \mu \in i\mathfrak{a}^*_M,$$

is a Schwartz function of μ,

$$f_M(\pi_\lambda, X) = \int_{i\mathfrak{a}^*_M} f_M(\pi_{\lambda+\mu})e^{-\mu(X)}\, d\mu, \qquad X \in \mathfrak{a}_M,$$

is a Schwartz function of X. This leaves the functions $\varepsilon_{M,v}(f)$. Almost all of them vanish, so we have only to show that for a fixed v, $\varepsilon_{M,v}(f, \pi_\lambda, X)$ is a Schwartz function.

Fix λ, and write

$$\pi_\lambda = \pi_v \otimes \pi^v, \qquad \pi_v \in \Pi^+_{\text{unit}}(M(F_v)),$$

where

$$\pi^v = \bigotimes_{w \neq v} \pi_w, \qquad \pi_w \in \Pi^+_{\mathrm{unit}}(M(F_w)).$$

Then

$$\varepsilon_{M,v}(f, \pi_\lambda, X) = \sum_{\rho_v \in \Sigma^+(M(F_v))} \Delta(\pi_v, \rho_v) \varepsilon_{M,v}(f, \rho_v \otimes \pi^v, X),$$

where, as we recall,

$$tr(\pi_v) = \sum_{\rho_v \in \Sigma^+(M(F_v))} \Delta(\pi_v, \rho_v) tr(\rho_v)$$

is the decomposition of π_v into standard representations. The unitarity of π_v implies that any ρ_v with $\Delta(\pi_v, \rho_v) \neq 0$ is either tempered or induced from a proper parabolic subset. If ρ_v is properly induced, it follows easily from (13.2*) that

$$\varepsilon_{M,v}(f, \rho_v \otimes \pi^v, X) = \varepsilon_{M,M_1}(v) \int_{i\mathfrak{a}^*_M} f_M((\rho_v \otimes \pi^v)_\mu) e^{-\mu(X)} d\mu,$$

for some proper Levi subset $M_1 \subsetneq M$. This is a Schwartz function of $X \in \mathfrak{a}_M$. On the other hand, suppose that ρ_v is tempered. Then Corollary 14.2 insures that $\varepsilon_M(f_v, \rho_v, X_v)$ is a Schwartz function of $X_v \in \mathfrak{a}_{M,v}$. It follows easily that $\varepsilon_{M,v}(f, \rho_v \otimes \pi^v, X)$ is a Schwartz function of $X \in \mathfrak{a}_M$. We have thus established that $\varepsilon_{M,v}(f, \pi_\lambda, X)$ is a Schwartz function. The lemma follows. ∎

Apply the lemma to the formula for $\hat{I}^M_t(\varepsilon_M(f)_{\alpha_m})$ above. We see that

$$\sum_{t \leq T} |W(\mathfrak{a}_M)|^{-1} \hat{I}^M_t(\varepsilon_M(f)_{\alpha_m})$$

equals

(15.7)

$$|W(\mathfrak{a}_M)| \sum_{t \leq T} \sum_{\pi \in \Pi_{\mathrm{disc}}(M,t)} a^M_{\mathrm{disc}}(\pi) \int_{i\mathfrak{a}^*_M / i\mathfrak{a}^*_G} \varepsilon_M(f^1, \pi, \lambda) \hat{\alpha}_1(\nu_\pi + \lambda)^m \, d\lambda.$$

We have shown that (15.5) equals the difference between (15.6) and (15.7). Consequently, this difference approaches 0 as m approaches ∞. In each of the expressions (15.6) and (15.7), the sums over t and π are finite. We first apply the dominated convergence theorem to (15.7). Since $\varepsilon_M(f^1, \pi, \lambda)$ is a Schwartz function on $i\mathfrak{a}^*_M / i\mathfrak{a}^*_G$, and

(15.8)
$$0 \leq \hat{\alpha}_1(\nu_\pi + \lambda) < 1,$$

except possibly at a finite number of λ, we see that the expression (15.7) approaches 0 as m approaches ∞. The same is therefore true of (15.6). We next consider the terms in (15.6). If π does not belong to the set

$$\Pi^+_{\nu_1}(G(\mathbf{A})^1) = \{\pi \in \Pi^+(G(\mathbf{A})^1) : \nu_\pi \in W(\nu_1)\},$$

the inequality (15.8) holds. Consequently, the corresponding term in (15.6) approaches 0 as m approaches ∞. On the other hand, if π belongs to $\Pi^+_{\nu_1}(G(\mathbf{A})^1)$, the term simply equals

$$\left(a^{\mathcal{E}}_{\text{disc}}(\pi) - a_{\text{disc}}(\pi)\right) tr\, \pi(f^1).$$

We can certainly assume that $||\text{Im}(\nu_1)|| \leq T$. This insures that all such terms will be included in (15.6). Letting m approach ∞, we obtain the following important result.

LEMMA 15.4.: *For each $f \in \mathcal{H}(G(\mathbf{A}), M)^0$ and $\nu_1 \in \mathfrak{h}^*_u$, we have*

$$\sum_{\pi \in \Pi^+_{\nu_1}(G(\mathbf{A})^1)} (a^{\mathcal{E}}_{\text{disc}}(\pi) - a_{\text{disc}}(\pi))\, tr\, \pi(f^1) = 0.$$

∎

16. Elimination of restrictions on f

With Lemma 15.4 we have reached a watershed. For certain functions f we will be able to prove the equality of $I^{\mathcal{E}}(f)$ and $I(f)$. However, for this to be effective, we must first extend the formula of Lemma 15.4 to a larger class of functions.

As in the last section, let ν_1 be an arbitrary but fixed point in $\mathfrak{h}_u^*$. Let

$$K_1 = \prod_{v \text{ finite}} K_{1,v}$$

be an open compact subgroup of $G^0(\mathbf{A}_{\text{fin}})$. (We are writing $\mathbf{A}_{\text{fin}}$ for the finite adèles.) We shall write $\Pi^+_{\nu_1,K_1}(G(\mathbf{A})^1)$ for the set of representations $\pi \in \Pi^+(G(\mathbf{A})^1)$ such that $\nu_\pi \in W(\nu_1)$, and such that π contains a K_1-fixed vector. By Lemma 4.2 of [1(h)], there are only finitely many $\pi \in \Pi^+_{\nu_1,K_1}(G(\mathbf{A})^1)$ such that $a_{\text{disc}}(\pi) \neq 0$. Now

$$a^{\mathcal{E}}_{\text{disc}}(\pi) = \ell^{-\dim(A_G)} \sum_{\tau \in \Pi(G'(\mathbf{A})^1)} a^{G'}_{\text{disc}}(\tau)\delta(\tau,\pi).$$

Using Corollary 8.3, one sees easily that there is an open compact subgroup K_1' of $G'(\mathbf{A}_{\text{fin}})$ such that if $\delta(\tau,\pi) \neq 0$ for some $\pi \in \Pi^+_{\nu_1,K_1}(G(\mathbf{A})^1)$, then τ belongs to $\Pi^+_{\nu_1',K_1'}(G(\mathbf{A})^1)$. But there are only finitely many $\tau \in \Pi^+_{\nu_1',K_1'}(G(\mathbf{A})^1)$ with $a^{G'}_{\text{disc}}(\tau) \neq 0$, by Lemma 4.2 of [1(h)] again. Consequently, there are only finitely many $\pi \in \Pi^+_{\nu_1,K_1}(G(\mathbf{A})^1)$ with $a^{\mathcal{E}}_{\text{disc}}(\pi) \neq 0$.

Write $\mathcal{H}(G(\mathbf{A}),M)_{K_1}$ for the subspace of functions in $\mathcal{H}(G(\mathbf{A}),M)$ which are bi-invariant under K_1, and set

$$\mathcal{H}(G(\mathbf{A}),M)^0_{K_1} = \mathcal{H}(G(\mathbf{A}),M)_{K_1} \cap \mathcal{H}(G(\mathbf{A}),M)^0.$$

Then Lemma 15.4 tells us that

$$\sum_{\pi \in \Pi^+_{\nu_1,K_1}(G(\mathbf{A})^1)} (a^{\mathcal{E}}_{\text{disc}}(\pi) - a_{\text{disc}}(\pi)) \operatorname{tr} \pi(f^1) = 0,$$

for any $f \in \mathcal{H}(G(\mathbf{A}),M)^0_{K_1}$. The sum can be taken over a finite set which depends only on (ν_1, K_1). We can write

$$\operatorname{tr} \pi(f^1) = \sum_{\rho \in \Sigma^+_{\nu_1,K_1}(G(\mathbf{A})^1)} \Delta(\pi,\rho) \operatorname{tr} \rho(f^1),$$

where $\Sigma^+_{\nu_1,K_1}(G(\mathbf{A})^1)$ is the set of representations $\rho \in \Sigma^+(G(\mathbf{A})^1)$ such that $\nu_\rho \in W(\nu_1)$ and such that ρ contains a K_1-fixed vector. Then

$$(16.1) \qquad \sum_{\rho \in \Sigma^+_{\nu_1,K_1}(G(\mathbf{A})^1)} A(\rho)\,\mathrm{tr}\,\rho(f^1) = 0, \qquad f \in \mathcal{H}(G(\mathbf{A}), M)^0_{K_1},$$

where

$$(16.2) \qquad A(\rho) = \sum_{\pi \in \Pi^+_{\nu_1,K_1}(G(\mathbf{A})^1)} (a^{\mathcal{E}}_{\mathrm{disc}}(\pi) - a_{\mathrm{disc}}(\pi))\Delta(\pi, \rho).$$

Our goal in this section is to show that (16.1) holds if f belongs to $\mathcal{H}(G(\mathbf{A}), M)_{K_1}$, rather than just $\mathcal{H}(G(\mathbf{A}), M)^0_{K_1}$. In so doing we may assume that $\ell = 1$, since the two spaces are otherwise equal. Then $G = G^0$, and G is the group of units of a central simple algebra. We shall use an approximation argument. Let v be a fixed valuation from the exceptional set S_G, and write $\kappa = K_{1,v}$. Then $\mathcal{H}(G(F_v))^0_\kappa$ is the space of compactly supported functions f_v on $G(F_v)$ which are bi-invariant under κ and such that

$$I_G(\xi u, f_v) = 0$$

for any $\xi \in A_G(F_v)$ and any element $u \neq 1$ in $\mathcal{U}_G(F_v)$. Write $\mathbf{A}^v$ for the ring of adèles which are 0 at v.

LEMMA 16.1.: *Suppose that f^v is a smooth, compactly supported function on $G(\mathbf{A}^v)$ such that (16.1) holds for any function $f = f^v f_v$ with $f_v \in \mathcal{H}(G(F_v))^0_\kappa$. Then (16.1) also holds for $f = f^v f_v$ with $f_v \in \mathcal{H}(G(F_v))_\kappa$.*

Proof. Let C be a fixed finite subset of the lattice

$$\mathfrak{a}_{G,v} = \{H_G(x) : x \in G(F_v)\}.$$

We shall simply write $\mathcal{H}_v$ for the subspace of functions in $\mathcal{H}(G(F_v))_\kappa$ which are supported on

$$\{x \in G(F_v) : H_G(x) \in C\},$$

and we shall write

$$\mathcal{H}^0_v = \mathcal{H}_v \cap \mathcal{H}(G(F_v))^0_\kappa.$$

Let I be the finite set of pairs $i = (\xi, u)$, in which ξ ranges over the elements in $A_G(F_v)/A_G(F_v) \cap \kappa$ such that $H_G(\xi)$ belongs to C, and u ranges over the nontrivial unipotent conjugacy classes in $G(F_v)$. Set $J_i = I_G(\xi u)$. Then $\{J_i : i \in I\}$ is a linearly independent set of linear forms on $\mathcal{H}_v$ whose kernel is $\mathcal{H}^0_v$. Choose elements $\{f^j_v : j \in I\}$ in $\mathcal{H}_v$ such that

$$J_i(f^j_v) = \begin{cases} 1, & i = j \\ 0, & i \neq j. \end{cases}$$

Then

$$f_v \to \overline{f}_v = f_v - \sum_{i \in I} J_i(f_v) f_v^i, \qquad f_v \in \mathcal{H}_v,$$

is a projection of $\mathcal{H}_v$ onto $\mathcal{H}_v^0$. By assumption,

$$\sum_\rho A(\rho) \operatorname{tr} \rho(f^v \overline{f}_v) = 0, \qquad f_v \in \mathcal{H}_v.$$

We therefore obtain

$$(16.3) \qquad \sum_\rho A(\rho) \operatorname{tr} \rho(f^v f_v) = \sum_i \beta^i J_i(f_v), \qquad f_v \in \mathcal{H}_v,$$

where

$$\beta^i = \sum_\rho A(\rho) \operatorname{tr} \rho(f^v f_v^i).$$

We must show that each side of (16.3) vanishes.

There are only finitely many pairs $(L, \mathfrak{o})$, in which L is a group in $\mathcal{L}_v$ and $\mathfrak{o}$ is an orbit of the compact group

$$i\mathfrak{a}_{L,v}^* = i(\mathfrak{a}_L^* / \operatorname{Hom}(\mathfrak{a}_{L,v}, \mathbf{Z}))$$

in $\Pi_{\text{disc}}(L(F_v))_\kappa$. (Here $\Pi_{\text{disc}}(L(F_v))_\kappa$ denotes the set of representations in $\Pi_{\text{temp}}(L(F_v))$ which are square integrable modulo the center and which contain a $\kappa \cap L(F_v)$-fixed vector.) For any such orbit, let $W_\mathfrak{o}$ be the stabilizer of $\mathfrak{o}$ in $W(\mathfrak{a}_L)$. Let $\Sigma_\mathfrak{o}$ be the set of $\rho \in \Sigma_{\nu_1,K_1}^+(G(\mathbf{A})^1)$ which are restrictions to $G(\mathbf{A})^1$ of representations of the form

$$\rho^v \otimes \mathcal{I}_P(\sigma_\mu),$$

with $\rho^v \in \Sigma_{\nu_1,K_1}^+(G(\mathbf{A}^v))$, $P \in \mathcal{P}(L)$, $\sigma \in \mathfrak{o}$, and $\mu \in \mathfrak{a}_L^*$. The point σ_μ in $\mathfrak{o} \times \mathfrak{a}_L^*$ is uniquely determined as a $W_\mathfrak{o}$-orbit, modulo translation by $i\mathfrak{a}_{G,v}^*$ in $\mathfrak{o}$. We shall write $X_\rho = \sigma_\mu$. It is clear that two sets $\Sigma_\mathfrak{o}$ and $\Sigma_{\mathfrak{o}'}$ are either equal or disjoint, depending on whether $\mathfrak{o}$ and $\mathfrak{o}'$ are W_0-conjugate or not. It is also clear that $\Sigma_{\nu_1,K_1}^+(G(\mathbf{A})^1)$ is a union of sets $\Sigma_\mathfrak{o}$. It will therefore be enough to show that for each $\mathfrak{o}$, and $f_v \in \mathcal{H}_v$, the number

$$(16.4) \qquad \sum_{\rho \in \Sigma_\mathfrak{o}} A(\rho) \operatorname{tr} \rho(f^v f_v)$$

vanishes.

We shall fix $(L, \mathfrak{o})$ and the function $f_v \in \mathcal{H}_v$. Let $\mathcal{I}_\mathfrak{o}$ be the space of functions

$$\phi : \mathfrak{o} \to \mathbf{C}$$

which satisfy the following three conditions.

(i) For each $\sigma \in \mathfrak{o}$, $\phi(\sigma_\Lambda)$ is a finite Fourier series in $\Lambda \in i\mathfrak{a}^*_{L,v}$.
(ii) $\phi(w\sigma) = \phi(\sigma)$, $\qquad \sigma \in \mathfrak{o},\ w \in W_{\mathfrak{o}}$.
(iii) $\phi(\sigma_\lambda) = \phi(\sigma)$, $\qquad \lambda \in i\mathfrak{a}^*_{G,v}$.

Notice that the second and third conditions insure that the number $\phi(X_\rho)$ are well defined. For each $\phi \in \mathcal{I}_{\mathfrak{o}}$ there is a function $f_v^\phi \in \mathcal{H}_v$ such that the number

$$\operatorname{tr} \rho(f^v f_v^\phi), \qquad \rho \in \Sigma^+_{\nu_1, K_1}(G(\mathbf{A})^1),$$

is zero unless ρ belongs to $\Sigma_{\mathfrak{o}}$, in which case

$$\operatorname{tr} \rho(f^v f_v^\phi) = \operatorname{tr} \rho(f^v f_v)\phi(X_\rho).$$

The existence of f_v^ϕ follows from the trace Paley-Wiener theorem for $G(F_v)$ ([6]). We replace f_v by f_v^ϕ in (16.3). The left hand side becomes

$$\sum_{\rho \in \Sigma_{\mathfrak{o}}} c_\rho \phi(X_\rho),$$

where

$$c_\rho = A(\rho) \operatorname{tr} \rho(f^v f_v).$$

To evaluate the right hand side, we use the fact that every unipotent class in $G(F_v)$ is induced. For any $u \in (\mathcal{U}_G(F_v))$ there is a W_0-orbit $\mathcal{L}(u)$ in $\mathcal{L}_v$ such that for each $L_1 \in \mathcal{L}(u)$ and $Q \in \mathcal{P}(L_1)$, $u \cap N_Q$ is dense in N_Q. If $i = (\xi, u)$,

$$J_i(f_v^\phi) = \int_{\mathfrak{o}} B_i(\sigma)\phi(\sigma)\, d\sigma,$$

where B_i is a smooth function on $\mathfrak{o}$. More precisely,

$$B_i(\sigma) = \chi_\sigma(\xi)^{-1} \mu_L^u(\sigma) \operatorname{tr}(\mathcal{I}_P(\sigma, f_v)), \qquad P \in \mathcal{P}(L),$$

where $\chi_\sigma(\xi)$ is the central character of σ at ξ, and $\mu_L^u(\sigma)$ vanishes unless L is contained in an element $L_1 \in \mathcal{L}(u)$, in which case $\mu_L^u(\sigma)$ is the Plancherel density associated to the Levi subgroup L of L_1. Notice that since $u \neq 1$, $B_i(\sigma) = 0$ if L equals G. The equation (16.3) becomes

$$\sum_{\rho \in \Sigma_{\mathfrak{o}}} c_\rho \phi(X_\rho) = \int_{\mathfrak{o}} B(\sigma)\phi(\sigma)\, d\sigma, \tag{16.5}$$

where

$$B(\sigma) = \sum_i \beta^i B_i(\sigma),$$

a smooth function on $\mathfrak{o}$ which vanishes if $L = G$.

Our final step is to show that each side of (16.5) is zero. This is almost obvious. We can assume that $L \neq G$, so that

$$\mathfrak{o}^1 = \mathfrak{o}/i\mathfrak{a}^*_{G,v}$$

is a compact torus of positive dimension. On the right hand side of (16.5) we have a distribution on $\mathfrak{o}^1$ which is a smooth function, and on the left we have a finite sum of point distributions on the complexification of $\mathfrak{o}^1$. Since the points $\{X_\rho\}$ are only defined as $W_\mathfrak{o}$-orbits anyway, and B is symmetric under $W_\mathfrak{o}$, we do not even need to assume that ϕ is symmetric under $W_\mathfrak{o}$. It can be any finite Fourier series on $\mathfrak{o}^1$. Its Fourier transform can be any compactly supported function on the dual lattice. Consider the Fourier transform of each side of (16.5) as a distribution on the dual lattice. The left hand side is a finite sum of exponentials, while the right hand side is a rapidly decreasing function. It is clear from this that each side vanishes.

Having shown that each side of (16.5) is zero, we take $\phi = 1$. We obtain

$$\sum_{\rho \in \Sigma_\mathfrak{o}} c_\rho = 0.$$

The expression on the left is just (16.4), so the proof of the lemma is complete. ∎

We apply the lemma to each place in S_G. It follows inductively that (16.1) holds for any function in $\mathcal{H}(G(\mathbf{A}), M)_{K_1}$ which vanishes on $G(F_\infty)A_G(\mathbf{A}_{\text{fin}})$. It is easy to remove this last restriction. For we are free to modify an arbitrary function $f \in \mathcal{H}(G(\mathbf{A}), M)_{K_1}$ outside a finite set S of valuations. Choose any unramified place w outside of S such that $K_{1,w}$ equals K_w, the standard maximal compact subgroup. Let h be a variable function in $\mathcal{H}(G(F_w))_{K_w}$, and evaluate the left hand side of (16.1) on the function

$$f^h(x) = f(x)h(x_w), \qquad x \in G(\mathbf{A}).$$

The expression vanishes if h is zero on $A_G(F_w)$, so as a linear form in h it may be expressed in terms of the Plancherel density. On one hand, the Plancherel density is a continuous function on the unramified representations in $\Pi_{\text{temp}}(G(F_w))$, while on the other hand, the sum in (16.1) may be taken over a finite set. It follows that the linear form vanishes on any h. Therefore, (16.1) holds for any function $f \in \mathcal{H}(G(\mathbf{A}), M)_{K_1}$.

We return to the case that ℓ is arbitrary. It is best to translate (16.1) back into a sum over irreducible representations. Given $f \in \mathcal{H}(G(\mathbf{A}), M)_{K_1}$, we substitute (16.2) back into (16.1). We obtain

$$\sum_{\pi\in\Pi^+_{\nu_1,K_1}(G(\mathbf{A})^1)} (a^{\mathcal{E}}_{\mathrm{disc}}(\pi) - a_{\mathrm{disc}}(\pi))\,\mathrm{tr}\,\pi(f^1) = 0. \tag{16.6}$$

If π belongs to the complement of $\Pi^+_{\nu_1,K_1}(G(\mathbf{A})^1)$ in $\Pi^+_{\nu_1}(G(\mathbf{A})^1)$, $\mathrm{tr}\,\pi(f^1)$ equals 0, so we can certainly take the sum over the larger set. But any function in $\mathcal{H}(G(\mathbf{A}),M)$ belongs to $\mathcal{H}(G(\mathbf{A}),M)_{K_1}$ for some K_1. It follows that

$$\sum_{\pi\in\Pi^+_{\nu_1}(G(\mathbf{A})^1)} (a^{\mathcal{E}}_{\mathrm{disc}}(\pi) - a_{\mathrm{disc}}(\pi))\,\mathrm{tr}\,\pi(f^1) = 0, \qquad f\in\mathcal{H}(G(\mathbf{A}),M). \tag{16.7}$$

Proposition 16.2.: *For any $f\in\mathcal{H}(G(\mathbf{A}),M)$, we have*

$$I^{\mathcal{E}}(f) = I(f).$$

Proof. Let t be any nonnegative real number. Then for any $f\in\mathcal{H}(G(\mathbf{A}),M)$,

$$I^{\mathcal{E}}_t(f) - I_t(f) = \sum_{\{\nu_1:\|\mathrm{Im}\,\nu_1\|=t\}}\ \sum_{\pi\in\Pi^+_{\nu_1}(G(\mathbf{A})^1)} (a^{\mathcal{E}}_{\mathrm{disc}}(\pi) - a_{\mathrm{disc}}(\pi))\,\mathrm{tr}\,\pi(f^1) = 0,$$

by Lemma 12.3 and (16.7). We therefore obtain

$$I^{\mathcal{E}}(f) = \sum_t I^{\mathcal{E}}_t(f) = \sum_t I_t(f) = I(f),$$

as required. ∎

17. Completion of the proofs of Theorems A and B

Having established Proposition 16.2, we shall return to the geometric sides of the trace formulas. We are at last ready to deduce the equality of $I_M^{\mathcal{E}}(\gamma)$ and $I_M(\gamma)$.

Suppose that $f \in \mathcal{H}(G(\mathbf{A}), M)$. Then by Lemma 13.1 and Proposition 16.2, the sum of the expressions

$$(17.1) \qquad |W(\mathfrak{a}_M)|^{-1} \sum_{\gamma \in (M(F))_{M,S}} a^M(S,\gamma)(I_M^{\mathcal{E}}(\gamma, f) - I_M(\gamma, f))$$

and

$$(17.2) \qquad \sum_{\xi \in A_G(F)} \sum_{u \in (\mathcal{U}_G(F))_{G,S}} (a^{\mathcal{E}}(S,u) - a(S,u)) I_G(\xi u, f)$$

vanishes. As usual, $S \supset S_{\mathrm{ram}}$ is a large finite set of valuations depending only on $\mathrm{supp}\,(f)$ and $V(f)$, and the sums in (17.1) and (17.2) can each be taken over finite sets that also depend only on $\mathrm{supp}\,(f)$ and $V(f)$.

We can assume that S is the disjoint union of a given finite set $S_0 \supset S_{\mathrm{ram}}$ with further sets

$$S_i = \{v_i\}, \qquad 1 \leq i \leq k,$$

where v_1 and v_2 are fixed valuations at which G splits, and $\{v_3, \dots, v_k\}$ is a large additional finite set of unramified places. If

$$f = \prod_{i=0}^{k} f_i, \qquad f_i \in \mathcal{H}(G(F_{S_i})),$$

it follows inductively from (5.6) that

$$(17.3) \quad I_M^{\mathcal{E}}(\gamma, f) - I_M(\gamma, f) = \sum_{i=0}^{k} (I_M^{\mathcal{E}}(\gamma, f_i) - I_M(\gamma, f_i)) \prod_{j \neq i} \hat{I}_M^M(\gamma, f_{j,M}),$$

for any element $\gamma \in M(F)$. We shall take γ to be a regular element in $M(F)$ which is elliptic at v_1 and v_2. This means that the torus M_γ / A_M is anisotropic over v_1 and v_2. We shall use v_1 and v_2 to isolate the contributions from γ to (17.1). Indeed, for $i = 1, 2$, we can choose f_i to be supported on a very small neighborhood of γ in $G(F_{S_i})$, and so that

$$\hat{I}_M^M(\gamma, f_{i,M}) = I_G(\gamma, f_i) = 1.$$

Then f_i will be supported on the F_{v_i}-elliptic set in M, and the function f above will belong to $\mathcal{H}(G(F_S), M)$. Apply the splitting formula (17.3) to the terms in (17.1). Shrinking the functions f_1 and f_2 around γ does not increase the support of f or the set $V(f)$. Therefore, the set S may be chosen independently of f_1 and f_2, and the sums in (17.1) and (17.2) may be taken over fixed finite sets. It is thus clear that f_1 and f_2 may be chosen so that (17.2) vanishes and so that the only contributions to (17.1) come from conjugates of γ. But

$$I_M^{\mathcal{E}}(w\gamma w^{-1}, f) - I_M(w\gamma w^{-1}, f) = I_M^{\mathcal{E}}(\gamma, f) - I_M(\gamma, f), \qquad w \in W(\mathfrak{a}_M),$$

so we actually need consider only the summand in (17.1) corresponding to γ. Moreover, γ is semisimple, so if S is large enough (in a sense that depends only on γ), we have

$$a^M(S, \gamma) = \mathrm{vol}(M_\gamma(F)\backslash M_\gamma(\mathbf{A})^1),$$

by Theorem 8.2 of [1(d)]. In particular, this constant is not zero. It follows that

$$(17.4) \qquad \sum_{i=0}^{k} (I_M^{\mathcal{E}}(\gamma, f_i) - I_M(\gamma, f_i)) \prod_{j \neq i} \hat{I}_M^M(\gamma, f_{j,M}) = 0,$$

for γ, f_1 and f_2 as above.

Suppose now that V is any finite set of valuations of F which either contains S_{ram} or consists of one unramified valuation v. We can obviously arrange that V equals one of the sets S_i above, with $i \neq 1, 2$. Choose an element $\gamma \in M(F)$ as in (17.4), and let f_V be an arbitrary function in $\mathcal{H}(G(F_V))$. We suppose first that

$$\hat{I}_M^M(\gamma, f_{V,M}) = I_G(\gamma, f_V) = 0.$$

Then the only contribution to (17.4) will be the summand corresponding to $V = S_i$. For the sets S_j other than S_1, S_2 and S_i, choose f_j to by any function such that

$$\hat{I}_M^M(\gamma, f_{j,M}) \neq 0.$$

The left-hand side of (17.4) becomes a nonzero multiple of

$$I_M^{\mathcal{E}}(\gamma, f_V) - I_M(\gamma, f_V).$$

We conclude that this distribution vanishes for any f_V whose orbital integral vanishes at γ. It follows that there is a constant $\varepsilon_M(\gamma)$ such that

$$(17.5) \qquad I_M^{\mathcal{E}}(\gamma, f_V) - I_M(\gamma, f_V) = \varepsilon_M(\gamma) I_G(\gamma, f_V),$$

for any function $f_V \in \mathcal{H}(G(F_V))$. Let V^+ be the union of V with the valuations v_1 and v_2 above. Write $U_{V^+}(M)$ for the set of elements

$$\prod_{v \in V^+} \gamma_v, \qquad \gamma_v \in M(F_v) \cap G_{\mathrm{reg}},$$

such that for $i = 1, 2$, γ_{v_i} is F_{v_i}-elliptic in $M(F_{v_i})$. Then $U_{V^+}(M)$ is open in

$$M(F_{V^+}) = \prod_{v \in V^+} M(F_v).$$

The set $M(F)$ is dense in $M(F_{V^+})$, so the intersection of $M(F)$ with $U_{V^+}(M)$ is dense in $U_{V^+}(M)$. It follows that we can approximate any G-regular element $\gamma_V \in M(F_V)$ by elements γ which occur in (17.5). Since $I_M^{\mathcal{E}}(\gamma_V, f_V)$, $I_M(\gamma_V, f_V)$ and $I_G(\gamma_V, f_V)$ are smooth on $M(F_V) \cap G_{\mathrm{reg}}$, we see that ε_M extends to a smooth function on this space, and that
(17.6)

$$I_M^{\mathcal{E}}(\gamma_V, f_V) - I_M(\gamma_V, f_V) = \varepsilon_M(\gamma_V) I_G(\gamma_V, f_V), \qquad \gamma_V \in M(F_V) \cap G_{\mathrm{reg}}.$$

We want to show that $\varepsilon_M(\gamma_V) = 0$.

Consider first the case that V consists of one unramified valuation v. Take $f_V = f_v$ to be the characteristic function of $K_v \rtimes \theta$. Then by Lemma 4.3, the left hand side of (17.6) vanishes. On the other hand, if γ_v belongs to $K_v \rtimes \theta$, the orbital integral $I_G(\gamma_v, f_v)$ does not vanish. It follows that $\varepsilon_M(\gamma_v) = 0$ for any such γ_v. Suppose in addition to being unramified, that G splits completely at v. Then by Lemma 4.2, the left hand side of (17.6) vanishes if f_v is any K_v-bi-invariant function in $\mathcal{H}(G(F_v))$. For a given γ_v we can always choose such an f_v so that $I_G(\gamma_v, f_v) \neq 0$. It follows that $\varepsilon_M(\gamma_v) = 0$ in this case for all γ_v.

Now take $V = S_0$ to be any arbitrary finite set which contains S_{ram}, and let

$$S = \bigcup_{i=0}^{k} S_i = S_0 \cup \{v_1, \dots, v_k\}$$

as at the beginning of the argument. Choose $\gamma \in M(F)$ as in (17.4), and let γ_i be the image of γ in $M(F_{S_i})$. We then substitute the formula (17.6) (with V replaced by S_i) into (17.4). Choosing the functions f_i appropriately, we find that

$$\sum_{i=0}^{k} \varepsilon_M(\gamma_i) = 0.$$

We are free to drop any of the terms in this sum corresponding to unramified valuations at which γ is integral. This means that we can take γ to be any

G-regular element in $M(F)$ which is elliptic at v_1 and v_2, and which is integral outside S. Suppose that G splits completely at each of the places $v_1, \dots, v_k$. Then

$$\varepsilon_M(\gamma_i) = 0, \qquad 1 \le i \le k.$$

It follows that

$$\varepsilon_M(\gamma_0) = 0$$

for any such γ. But as long as k is large enough, the set of elements $\gamma \in M(F)$ which are integral outside of S, and which are elliptic at v_1 and v_2, projects onto a dense subset of $M(F_{S_0}) = M(F_V)$. It follows that

$$\varepsilon_M(\gamma_V) = 0, \qquad \gamma_V \in M(F_V) \cap G_{\text{reg}}.$$

We have thus established the formula

$$I_M^{\mathcal{E}}(\gamma, f) = I_M(\gamma, f), \qquad f \in \mathcal{H}(G(F_V)),$$

where V is any finite set of valuations which contains S_{ram}, and γ is a G-regular element in $M(F_V)$. It then follows from Lemma 3.6 that the formula holds for any element $\gamma \in M(F_V)$. So we have finally finished the induction argument begun in §13, where we first fixed M. In other words, the formula holds for any $M \in \mathcal{L}$. This completes the proof of the local assertion (i) of Theorem A.

We agreed that the global assertion (ii) of Theorem A was a consequence of the induction hypothesis of §5 unless $M = G$ and

$$\gamma = \xi u, \qquad \xi \in A_G(F),\ u \in (\mathcal{U}_G(F))_{G,S},$$

for any large finite set S. To deal with this last case, we go back to the discussion at the beginning of this paragraph, with M a minimal element in $\mathcal{L}$. Then $\mathcal{H}(G(\mathbf{A}), M)$ equals $\mathcal{H}(G(\mathbf{A}))$. Since we have established the local assertion of Theorem A, the expression (17.1) vanishes. Therefore so does (17.2). Now G is such that $(\mathcal{U}_G(F))_{G,S}$ equals $(\mathcal{U}_G(F))$, the set of unipotent classes in G defined over F. It follows that

$$\sum_{\xi \in A_G(F)} \sum_{u \in (\mathcal{U}_G(F))} (a^{\mathcal{E}}(S, u) - a(S, u)) I_G(\xi u, f) = 0,$$

for each $f \in \mathcal{H}(G(\mathbf{A}))$. Fix an arbitrary element $u_1 \in (\mathcal{U}_G(F))$, and choose $f \in \mathcal{H}(G(\mathbf{A}))$ such that

$$I_G(\xi u, f) = \begin{cases} 1, & \text{if } (\xi, u) = (1, u_1), \\ 0, & \text{otherwise.} \end{cases}$$

We then see that $a^{\mathcal{E}}(S,u_1)$ equals $a(S,u_1)$. This finishes the remaining case of the global assertion (ii) of Theorem A. The proof of the theorem is therefore complete. ∎

We proved the local assertion (i) of Theorem B in §10. The induction hypothesis of §9 reduces the global assertion (ii) of Theorem B to proving the equality of $a^{\mathcal{E}}_{\text{disc}}(\pi)$ and $a_{\text{disc}}(\pi)$, for $\pi \in \Pi(G(\mathbf{A})^1)$. Any such π belongs to a set $\Pi^+_{\nu_1,K_1}(G(\mathbf{A})^1)$ so we shall fix ν_1 and K_1. Since we have now established Theorem A, we are at liberty to apply (16.6) with any $M \in \mathcal{L}$. Taking M to be minimal, and then noting that $\mathcal{H}(G(\mathbf{A}),M)_{K_1}$ equals $\mathcal{H}(G(\mathbf{A}))_{K_1}$, we obtain

$$\sum_{\nu\in\Pi^+_{\nu_1,K_1}(G(\mathbf{A})^1)} (a^{\mathcal{E}}_{\text{disc}}(\pi) - a_{\text{disc}}(\pi))\,\text{tr}\,\pi(f^1) = 0, \qquad f \in \mathcal{H}(G(\mathbf{A}))_{K_1}.$$

The sum may be taken over a finite set. However, the set of linear forms

$$f \to \text{tr}\,\pi(f^1), \qquad f \in \mathcal{H}(G(\mathbf{A}))_{K_1},$$

parametrized by $\Pi^+_{\nu_1,K_1}(G(\mathbf{A})^1)$, is linearly independent. This follows from the linear independence of Archimedean characters, and the non-Archimedean trace Paley-Wiener theorem ([6], [33(c)]). It follows that

$$a^{\mathcal{E}}_{\text{disc}}(\pi) - a_{\text{disc}}(\pi) = 0, \qquad \pi \in \Pi^+_{\nu_1,K_1}(G(\mathbf{A})^1).$$

This completes the proof of Theorem B. ∎

It is of course the global assertion (ii) of Theorem B which is relevant to the comparison of automorphic representations. It tells us that

$$a^{G,\mathcal{E}}_{\text{disc}}(\pi) = a^{G}_{\text{disc}}(\pi), \qquad \pi \in \Pi^+(G(\mathbf{A})^1).$$

Recall that $I_{\text{disc},t}(f)$ is the linear combination of characters on $G(\mathbf{A})^1$ given explicitly by the expression (9.2). Then

$$\begin{aligned} I_{\text{disc},t}(f) &= \sum_{\pi\in\Pi^+(G(\mathbf{A})^1,t)} a^{G}_{\text{disc}}(\pi) I_G(\pi,f) \\ &= \sum_{\pi\in\Pi^+(G(\mathbf{A})^1,t)} a^{G,\mathcal{E}}_{\text{disc}}(\pi) I^{\mathcal{E}}_G(\pi,f). \end{aligned}$$

It follows easily from the definition (9.4) and the trivial case ($M = G$) of Lemma 12.1 that this last expression equals

$$\sum_{\tau\in\Pi(G'(\mathbf{A}),t)} a^{G'}_{\text{disc}}(\tau)\hat{I}_{G'}(\tau,f').$$

This in turn is just equal to $\hat{I}_{\text{disc},t}(f')$. Theorem B therefore provides an identity

$$I_{\text{disc},t}(f) = \hat{I}_{\text{disc},t}(f') \tag{17.7}$$

between the "discrete parts" of the trace formulas of G and G'.

Instead of using characters on $G(\mathbf{A})^1$ it is sometimes more convenient to deal with characters on $G(\mathbf{A})$ which are equivariant with respect to a subgroup of the center. For example, we could take

$$A_{G,\infty} = A_{G_{\mathbf{Q}}}(\mathbf{R})^0,$$

where $G_{\mathbf{Q}}$ is obtained from G by restricting scalars from F to $\mathbf{Q}$, and $A_{G_{\mathbf{Q}}}$ is the corresponding $\mathbf{Q}$-split component of the center. Then $A_{G,\infty}$ is a subgroup of $\prod_{v\in S_\infty} A_G(F_v)$. The map

$$H_G : A_{G,\infty} \to \mathfrak{a}_G$$

is an isomorphism, which we use to pull back the Haar measure on $\mathfrak{a}_G$ to a Haar measure on $A_{G,\infty}$. If μ belongs to $i\mathfrak{a}_G^*$, define

$$I_{\text{disc},t,\mu}(f) = \int_{A_{G,\infty}} I_{\text{disc},t}(f_a)e^{\mu(H_G(a))}da,$$

where

$$f_a(x) = f(ax), \qquad x \in G(\mathbf{A}),\ a \in A_{G,\infty}.$$

This serves to transform the characters on $G(\mathbf{A})^1$ which occur in $I_{\text{disc,t}}$ to μ-equivariant characters on $G(\mathbf{A})$. We can of course repeat the same construction for $G'(\mathbf{A})$. Since

$$(f_a)' = f'_{a'},$$

$$e^{\mu(H_G(a))} = e^{\mu'(H_{G'}(a'))},$$

and

$$da' = \ell da,$$

we obtain

$$\ell I_{\text{disc},t,\mu}(f) = I_{\text{disc},t,\mu'}(f') \tag{17.8}$$

from the identity (17.7).

CHAPTER 3

Base Change

1. Weak and strong base change: definitions

In this chapter E/F will denote a cyclic extension of degree l of number fields. We write v for the places of F, w for the places of E; other notations are as in Chapter I. In particular, $\mathbf{A} = \mathbf{A}_F$, $\mathbf{A}_E = \mathbf{A} \otimes E$, and G again stands for $GL(n)$.

If π is an automorphic representation of $G(\mathbf{A})$, we have $\pi = \bigotimes_v \pi_v$ where π_v is unramified for almost all v.

For any finite prime v unramified in E, we have the base change homomorphism $b : \mathcal{H}_{E_v} \to \mathcal{H}_{F_v}$ (cf. §I.5; we use the notation there). By duality, to an unramified representation π_v, we may associate an unramified representation $\Pi_v = \bigotimes_{w|v} \Pi_w$ of $G(E_v) = \prod_{w|v} G(E_w)$.

In terms of Hecke eigenvalues (cf. e.g. §6.3) the correspondence is described as follows: if f_v is the residual degree of E above an unramified v, then for any $w|v$:

$$(t_{\pi,v})^{f_v} = t_{\Pi,w}. \tag{1.1}$$

DEFINITION 1.1.: *Let π, Π denote automorphic representations of $G(\mathbf{A})$, $G(\mathbf{A}_E)$ respectively. We say that Π is a* weak base change lift *(to $G(\mathbf{A}_E)$) of π if the relation (1.1) is satisfied for almost all finite primes v, w.*

This definition may be strengthened using the theory of local base change:

DEFINITION 1.2.: *We say that Π is a* strong *base change lift of π if, for any (finite or infinite) $w|v$, the component Π_w is a base change lift of π_v.*

We will use Definition 1.2 only when the components of π and Π are generic, so that, according to §I.6, base change is expressed by character identities.

2. Some results of Jacquet and Shalika

To extract the lifting results from the identity of traces obtained in Chapter II, we will have to use deep facts about L-functions of pairs of representations proved by Jacquet and Shalika in [27(a),(b)]. We now review those results.

Let F be a number field, $\mathbf{A} = \mathbf{A}_F$. Assume π, σ are cuspidal automorphic representations of $GL(n, \mathbf{A})$ and $GL(m, \mathbf{A})$ respectively. We will assume that π, σ are *unitary*.

Let S be a finite set of primes such that π, σ are unramified outside S. We form the L-function

$$L^S(s, \pi \otimes \sigma) = \prod_{v \notin S} \det(1 - t_{\sigma,v} \otimes t_{\pi,v} q_v^{-s})^{-1}.$$

Here $t_{\pi,v}$ and $t_{\sigma,v}$ denote the Hecke matrices, considered as diagonal endomorphisms of $\mathbf{C}^n$, $\mathbf{C}^m$: their tensor product is an endomorphism of $\mathbf{C}^{nm}$.†

We will use the following properties of these L-functions:

(2.1) The Euler product L^S is absolutely convergent for $\mathrm{Re}\, s > 1$. (cf. [27(a), Thm. 5.3].)

(2.2) Let X be the set of s on the line $\mathrm{Re}\, s = 1$ such that $\pi \otimes |\ |^{s-1}$ is equivalent to $\tilde{\sigma}$, the contragredient of σ. (Thus X contains at most one point.) Then the function L^S extends continuously to the line $\mathrm{Re}\, s = 1$ with X removed. Moreover, it does not vanish there.

(2.3) If $s_0 \in X$, the limit

$$\lim_{\substack{s \to s_0 \\ \mathrm{Re}\, s \geq 1}} (s - s_0) L^S(s, \pi \otimes \sigma)$$

exists and is finite and non-zero.

(cf. [27(b), Prop. 3.6]. The non-vanishing part of these results is due to Shahidi [36(a)]).

More generally, suppose that π and σ are cuspidal automorphic representations of Levi components $M \subset GL(n)$ and $L \subset GL(m)$ of parabolic subgroups. For almost all v we again have the conjugacy classes $t_{\pi,v} \subset GL(n, \mathbf{C})$ and $t_{\sigma,v} \subset GL(m, \mathbf{C})$. We then have the following consequence of the facts above ([27(b), Theorem 4.4]).

(2.4) Suppose that $m = n$ and that $t_{\pi,v} = t_{\sigma,v}$ for almost all v. Then the pairs (M, π) and (L, σ) are conjugate in $GL(n)$.

†Jacquet-Shalika write $L^S(s, \pi \times \sigma)$. We use the older $\otimes$ because we will need the symbol $\times$ for something else.

3. Fibers of global base change

In this paragraph we prove a result which in essence describes the fibers of the global base change correspondence; we will need to know it, however, before proving the lifting results, and its statement has nothing to do with base change. (Note that it has been used already in §I.6.) At this point we do not assume l prime.

THEOREM 3.1.: *Let π, π' be cuspidal automorphic representations of $G(\mathbf{A})$. Assume that, for almost all v:*

$$(t_{\pi,v})^{f_v} = (t_{\pi',v})^{f_v}.$$

Then $\pi' = \pi \otimes \chi$, for some character χ of $F^* N(\mathbf{A}_E^*)\backslash \mathbf{A}^*$.

Proof. Let η be a character of $\mathbf{A}^*$ vanishing exactly on $F^* N(\mathbf{A}_E^*)$. We compare the products $\prod_{i=1}^{l} L^S(s, \pi \otimes \tilde{\pi}' \otimes \eta^i)$ and $\prod_{i=1}^{l} L^S(s, \pi \otimes \tilde{\pi} \otimes \eta^i)$ where $\tilde{\sigma}$ denotes the contragredient of σ.

If v is a finite place of F, the factor of the first product at v is equal to the inverse of

$$\prod_{i=1}^{l} \det(1 - t_v \otimes \tilde{t}'_v \zeta_v^i q_v^{-s})$$

with $\tilde{t}'_v$ denoting the adjoint of $t'_v = t_{\pi',v}$, and $\zeta_v = \eta(\tilde{\omega}_v)$. We take S so large that E/F is unramified for $v \notin S$. Then ζ_v is a root of unity of order f_v. Consequently this product is equal to

$$\det\left(1 - t_v^{f_v} \otimes (t'_v)^{f_v} q_v^{-f_v s}\right)^{l/f_v}$$

which, by assumption, is equal to the corresponding term in the second product. We have therefore

$$\prod_{i=1}^{l} L^S(s, \pi \otimes \tilde{\pi}' \otimes \eta^i) = \prod_{i=1}^{l} L^S(s, \pi \otimes \tilde{\pi} \otimes \eta^i).$$

We may assume π, π' unitary. By (2.2) and (2.3), the product on the right has a pole at $s = 1$; so the product on the left must have one also, and since its terms do not vanish on the line $\mathrm{Re}\,(s) = 1$, we see using again the results in §2 that $\pi' = \pi \otimes \eta^i$ for some i. ∎

4. Weak lifting

In this section we will prove the results concerning weak lifting. We must first restrict the class of automorphic representations that we consider: we have to do so because of our ignorance of the residual spectrum of $GL(n)$. Assuming the conjectural description of the discrete spectrum given in [24(d)], our results could be extended to all the automorphic forms appearing in the decomposition of $L^2(G(F)\backslash G(\mathbf{A}))$.

DEFINITION 4.1.: *We will say that the automorphic representation π of $G(\mathbf{A})$ is* induced from cuspidal *if there is a cuspidal unitary representation σ of $M(\mathbf{A})$, where $P = MN$ is an F-parabolic subgroup of G, such that*

$$\pi = \operatorname{ind}_{M(\mathbf{A})N(\mathbf{A})}^{G(\mathbf{A})} (\sigma \otimes 1).$$

Note that π is then unitary irreducible ([4]).

We now state in one theorem the main results concerning base change for cyclic representations *of prime degree.*

We will denote by π a representation of $G(\mathbf{A})$, by Π a representation of $G(\mathbf{A}_E)$. Let η be a character of $\mathbf{A}^*$ vanishing exactly on $F^* N(\mathbf{A}_E^*)$. Assume $n = ab$ is a decomposition of n. If Π_i $(i = 1, \dots b)$ is an automorphic representation of $GL(a, \mathbf{A})$, we denote by $\Pi_1 \times \cdots \times \Pi_b$ the representation of $G(\mathbf{A})$ induced from the representation $\Pi_1 \otimes \cdots \otimes \Pi_b \otimes 1$ of the parabolic subgroup of type $(a, \dots a)$. We will write "Π lifts π" to say that Π is a weak lifting of π in the sense of Definition 1.1.

THEOREM 4.2: (WEAK LIFTING). *All representations are induced from cuspidal; E/F is cyclic of prime degree l.*

(a) *Assume π is cuspidal, $\pi \not\cong \pi \otimes \eta$. Then there is a unique σ-stable representation Π of $G(\mathbf{A}_E)$ lifting π; Π is cuspidal.*

(b) *Assume $\pi \cong \pi \otimes \eta$, π cuspidal. Then there is a cuspidal representation Π_1 of $GL(n/l, \mathbf{A}_E)$, with $\Pi_1 \not\cong \Pi_1^\sigma$, such that $\Pi = \Pi_1 \times \cdots \times \Pi_1^{\sigma^{l-1}}$ is the only lift of π.*

(c) *Assume π is induced from cuspidal. Then there exists Π, induced from cuspidal, unique, lifting π.*

(d) *Assume Π is cuspidal, $\Pi \cong \Pi \circ \sigma$. Then there is π cuspidal lifting to Π; all such π are conjugate by tensor product by a power of η; they satisfy $\pi \not\cong \pi \otimes \eta$.*

(e) *Write $n = lm$; let Π_1 be a cuspidal representation of $GL(m, \mathbf{A}_E)$, $\Pi_1 \not\cong \Pi_1^\sigma$. Then $\Pi_1 \times \Pi_1^\sigma \times \cdots \times \Pi_1^{\sigma^{l-1}} = \Pi$ is σ-stable and lifts some cuspidal representation π; π is unique and $\pi \cong \pi \otimes \eta$.*

(f) *Assume* Π *is induced from cuspidal and* σ*-stable. Then* Π *lifts at least one* π*;* π *is then induced from cuspidal.*

Before starting the proof, we recall that at the end of Chapter II we obtained an identity (17.8) of the discrete parts of the trace formulas for $G(\mathbf{A}_F)$ and $G(\mathbf{A}_E) \rtimes \sigma$. Write F^*_∞ for the subgroup of the Archimedean idèles

$$\prod_{v \in S_\infty} F^*_v$$

obtained by taking the diagonal image of the group of positive real numbers. We shall regard F^*_∞ as a subgroup of the center of $G(\mathbf{A}_F)$. Let μ be a unitary character of F^*_∞. Then $\mu_E = \mu \circ N$ is a unitary character of E^*_∞. In the present context, (17.8) may be stated as the identity of the expressions

(4.1)
$$\sum_M |W_0^M| \, |W_0^G|^{-1} \sum_{s \in W(\mathfrak{a}_M)_{\text{reg}}} |\det(s-1)_{\mathfrak{a}_M^G}|^{-1} \operatorname{tr}(M(s,0)\rho_{Q,t,\mu}(0,f)),$$

and

(4.2)
$$l \sum_M |W_0^M| \, |W_0^G|^{-1} \sum_{s \in W(\mathfrak{a}_M)_{\text{reg}}} |\det(s-1)_{\mathfrak{a}_M^G}|^{-1} \operatorname{tr}(M(s,0)\sigma\rho_{Q,t,\mu_E}(0,\phi)),$$

in which $t \geq 0$, and ϕ and f are functions on $G(\mathbf{A}_E)$ and $G(\mathbf{A}_F)$ which are associated in the sense of §I.3. Here

$$\rho_{Q,t,\mu}(0) = \bigoplus_{\pi} \rho_{Q,\pi}(0)$$

is the representation of $G(\mathbf{A})$ induced from the subspace of μ-equivariant automorphic forms on $M(\mathbf{A})$ which decomposes as a direct sum of irreducible representations π such that the imaginary part of the Archimedean infinitesimal character of π has norm t, while

$$\rho_{Q,t,\mu_E}(0) = \bigoplus_{\pi_E} \rho_{Q,\pi_E}(0)$$

is the analogous representation for $G(\mathbf{A}_E)$.

We now begin the proof of Theorem 4.2. We start with (a). Assume that $\pi \not\cong \pi \otimes \eta$ is cuspidal. We must find Π lifting π. We assume all statements of Theorem 4.2 known up to $n-1$. We consider the identity of (4.1) with (4.2). Note that since the imaginary infinitesimal characters have fixed norm, the expressions (4.1) and (4.2) each contain a fixed finite number of terms as soon as the K_F and K_E types of f and ϕ have been fixed.

Let S be a finite set of places containing all ramified places of π. Then, taking f_v unramified for $v \notin S$, we may write (4.1) as

$$\operatorname{tr} \pi_S(f_S) \prod_{v \notin S} f_v^\vee(t_{\pi,v}) + \ldots$$

where (if the ramification of f is fixed) the remainder is a finite combination of independent characters of $\mathcal{H}^S = \bigotimes_{v \notin S} \mathcal{H}_v$. By (2.4), these characters are independent of the character $f \to \prod f_v^\vee(t_{\pi,v})$ determined by π. (One needs to recall that any contribution to (4.1) from the noncuspidal discrete spectrum of M is obtained by induction from a cuspidal automorphic representation of a proper Levi subgroup of M.)

Let $\mathcal{H}_E^S = \bigotimes_{v \notin S} \bigotimes_{w|v} \mathcal{H}_w$. We have the base change homomorphism b : $\mathcal{H}_E^S \to \mathcal{H}^S$ (taking S so large that E/F is unramified outside S). If a representation π' of $G(\mathbf{A})$ yields the same character of $\mathcal{H}_E^S$ as π, we must have, by Theorem 3.1, $\pi' = \pi \otimes \eta^i$ for some i. Thus, if $f^S = b\phi^S$, (4.1) equals

$$\left[\sum_i \operatorname{trace}(\pi_S \otimes \eta_S^i)(f_S)\right] \prod_{v \notin S} b\phi_v^\vee(t_{\pi,v}) + \ldots,$$

the terms in the remainder being independent homomorphisms of $b(\mathcal{H}_E^S)$. The term in square brackets is of the form $\sum n_i \operatorname{trace} \pi_i(f_S)$, with $n_i > 0$, and can therefore be made $\neq 0$ for some f_S. The identity then shows that there is a representation Π of $G(\mathbf{A}_E)$, occurring in (4.2), which is a weak lift of π. We want to show that Π is cuspidal. We will need the following lemma.

LEMMA 4.3.: *Assume* Π_i is a weak lift of π_i *(*$i = 1, 2$*)* *(*Π_i, π_i *automorphic). Then, for large* S*:*

$$L^S(s, \Pi_1 \otimes \Pi_2) = \prod_{i=1}^{l} L^S(s, \pi_1 \otimes \pi_2 \otimes \eta^i).$$

Of course the product on the left must be taken on places w above $v \notin S$. The proof is an easy computation, left to the reader. ∎

By the theory of Eisenstein series, we may write Π as a subquotient of a representation $\Pi_1 \times \Pi_2 \times \cdots \times \Pi_r$, where $n = n_1 + \cdots + n_r$ and Π_i is a cuspidal representation of $GL(n_i, \mathbf{A}_E)$. Then Π^σ is a subquotient of $\Pi_1^\sigma \times \cdots \times \Pi_r^\sigma$. Using Theorem 4.4 of [27(b)], we see that since $\Pi \cong \Pi^\sigma$, Π

is a subquotient of a product

$$(\Pi_1 \times \Pi_1^\sigma \times \dots \Pi_1^{\sigma^{l-1}} \times \dots \times (\Pi_u \times \Pi_u^\sigma \times \dots \times \Pi_u^{\sigma^{l-1}})$$
$$\times \Pi_{u+1} \times \Pi_{u+2} \times \dots \times \Pi_t;$$

the factors on the first line satisfy $\Pi_i \not\cong \Pi_i^\sigma$, the other ones are σ-stable. We want to show that $u = 0$ and $t = 1$.

Let us denote by $\lambda(\Pi_i) \in \mathbf{R}$ the parameter λ defined by $|\omega_{\Pi_i}| = |\ |_{\mathbf{A}_E}^{\lambda n_i}$, where $|\ |_{\mathbf{A}_E}$ is the adèle norm on $\mathbf{A}_E^*$, ω_{Π_i} is the central character of Π_i and $|\omega_{\Pi_i}|$ its (complex) absolute value. (The same notation applies over F.)

Up to a reordering, we may assume that $\lambda(\Pi_u)$ or $\lambda(\Pi_t)$ is minimal amongst the $\lambda(\Pi_i)$. Assume first that $\lambda(\Pi_t)$ is minimal. If $n_t = n$, our proof is complete. So, we may assume $n_t < n$. Since $L^S(\Pi \otimes \Pi') = \prod_i L^S(\Pi_i \otimes \Pi')$, for any automorphic Π', we see, using the properties in §2 and the minimality of $\lambda(\Pi_t)$, that

$$L^S(s, \Pi \otimes \tilde{\Pi}_t) = L^S(s, \Pi_t \otimes \tilde{\Pi}_t) \prod_{i \neq t} L^S(s + \lambda_i - \lambda_t, \Pi_i^0 \otimes \tilde{\Pi}_t^0)$$

(where $\lambda_i = \lambda(\Pi_i)$ and $\Pi_i^0 = \Pi_i \otimes |\ |^{-\lambda_i}$ is unitary) has a pole at $s = 1$. However, since $n_t < n$, we may apply Theorem 4.2(d) to Π_t: if Π_t lifts the representation $\pi_t \otimes \eta^j$ $(j = 1, \dots l)$ we have by Lemma 4.3:

$$L^S(s, \Pi \otimes \tilde{\Pi}_t) = \prod_j L^S(s, \pi \otimes \tilde{\pi}_t \otimes \eta^j).$$

Therefore one of the factors on the right should have a pole at $s = 1$: this is impossible (§2) since $n_t \neq n$, $L^S(s, \pi \otimes \tilde{\pi}_t \otimes \eta^j) = L^S(s - \lambda_t, \pi \otimes (\tilde{\pi}_t)^0 \otimes \eta^j)$ and $\lambda_t \leq 0$ by minimality.

So we are reduced to the case that $(u > 0)$ and $\lambda(\Pi_u)$ is minimal. By similar arguments, we see that $L^S(s, \Pi \otimes (\tilde{\Pi}_u \times \dots \times \tilde{\Pi}_u^{\sigma^{l-1}}))$ has a pole of order at least l at $s = 1$. Assume first that $n_u < n/l$. Using Theorem 4.2(e) inductively we have a representation π_u of $GL(ln_u, \mathbf{A})$ such that $\Pi_u \times \dots \times \pi_u^{\sigma^{l-1}}$ lifts π_u; π_u is cuspidal and $\pi_u \cong \pi_u \otimes \eta$. By Lemma 4.3, $\prod_j L(s, \pi \otimes \tilde{\pi}_u \otimes \eta^j)$ has a pole of order at least l at $s = 1$; moreover, we have again $\lambda(\pi_u) \leq 0$. Since $ln_u < n$, we obtain, as above, a contradiction.

We are left with the case when $n_u = n/l$. Then Π is a subquotient of $\Pi_u \times \dots \times \Pi_u^{\sigma^{l-1}}$: since this is induced in the unitary range, and therefore irreducible, we have

$$\Pi \cong \Pi_u \times \dots \times \Pi_u^{\sigma^{l-1}}.$$

But then $L(s, \Pi \otimes \tilde{\Pi})$ has a pole of order l at $s = 1$; since $L(s, \Pi \otimes \tilde{\Pi}) = \prod_j L(s, \pi \otimes \tilde{\pi} \otimes \eta^j)$, this contradicts the fact that $\pi \not\cong \pi \otimes \eta^j$ for $j \neq l$. We have proved part (a) of the Theorem—the uniqueness of Π is obvious by (2.4).

We begin the proof of (b) in the same manner and obtain likewise $\Pi_1 \times \cdots \times \Pi_r$ lifting π. We have assumed $\pi \cong \pi \otimes \eta$ and we must now show that $t = 0$ and $u = 1$. Assume first $\lambda(\Pi_t)$ minimal. If $n_t < n$, the argument given above still holds. Assume $n_t = n$. We would then have a cuspidal representation $\Pi = \Pi_t$ of $GL(n, \mathbf{A}_E)$ lifting π. In the identity

$$L(s, \Pi \otimes \tilde{\Pi}) = \prod_j L(s, \pi \otimes \tilde{\pi} \otimes \eta^j),$$

the left-hand side has a pole of order 1 and the right-hand side a pole of order l at $s = 1$, whence a contradiction.

So we see that $\lambda(\Pi_u)$ must be minimal. If $ln_u < n$, we use Theorem 4.2(e) inductively to obtain π_u lifted by $\Pi_u \times \cdots \times \Pi_u^{\sigma^{l-1}}$. Proceeding as for (a), we obtain a contradiction by comparing L-functions. Thus $n_u = n/l$, and by irreducibility of the induced representation we must have $\Pi = \Pi_u \times \cdots \times \Pi_u^{\sigma^{l-1}}$. This yields (b); again uniqueness follows from (2.4).

To prove (c), we just use (a) or (b) as the case may be, and then induce. Again, the representation Π we obtain is induced from cuspidal. Since unitary induction produces irreducible representations, the uniqueness of Π comes from (2.4).

We now prove the "going down" part of the theorem, starting with (d). Assume $\Pi \cong \Pi \circ \sigma$ given. The identity of (4.1) with (4.2) shows the existence of at least one representation π of $G(\mathbf{A})$ lifted by Π. We first show that π is cuspidal.

We may write π as a subquotient of $\pi_1 \times \cdots \times \pi_r$, where π_i is a cuspidal representation of $GL(n_i, \mathbf{A}_F)$. We assume $\lambda(\pi_r)$ minimal. We must show that $r = 1$. Assume not. Using inductively Theorem 4.2(a) or (b) we may lift π_r to Π_r or $\Pi_r \times \Pi_r^{\sigma} \times \cdots \times \Pi_r^{\sigma^{l-1}}$. In the first case, the identity

$$L^S(s, \Pi \otimes \tilde{\Pi}_r) = \prod_j L^S(s, \pi \otimes \tilde{\pi}_r \otimes \eta^j)$$

yields a pole for $L^S(s, \Pi \otimes \tilde{\Pi}_r)$ at $s = 1$. Since $n \neq n_r$ and $\lambda(\Pi_r) \leq 0$, this is impossible by §2. In the second case, we obtain likewise a pole for

$$\prod_{i=0}^{l-1} L^S(s, \Pi \otimes \Pi_r^{\sigma^i})$$

at $s = 1$; this is impossible for the same reasons. So $r = 1$ and π is cuspidal. Moreover, the identity

$$L^S(s, \Pi \times \tilde{\Pi}) = \prod_j L^S(s, \pi \otimes \tilde{\pi} \otimes \eta^j)$$

shows that the product on the right has only a simple pole, whence $\pi \not\cong \pi \otimes \eta$. Finally, Theorem 3.1 now shows that the $\pi \otimes \eta^i$ are the only representations lifted by Π. (Note that we have already shown that any such is cuspidal, so we can use Theorem 3.1).

For (e), assume $\Pi_1 \not\cong \Pi_1^\sigma$ is a cuspidal representation of $GL(m, \mathbf{A}_E)$. Then the representation $\Pi = \Pi_1 \times \cdots \times \Pi_1^{\sigma^{l-1}}$ of $GL(n, \mathbf{A}_E)$ is unitary irreducible, and σ-stable since $\Pi^\sigma = \mathrm{ind}(\Pi_1^\sigma \otimes \cdots \otimes \Pi_1^{\sigma^l} \otimes 1)$ is isomorphic to Π by the standard intertwining operator. Therefore, for a suitable choice of ϕ, the term

$$\mathrm{trace}\,(M(s,0)\sigma\rho_{Q,\pi_E}(0,\phi))$$

in formula (4.2), gives a non-zero contribution to the twisted trace formula.

We get a number of terms with the same family of Hecke eigenvalues from formula (4.2): they will appear for each Levi subgroup $M \supset M_0$ conjugate to the standard Levi subgroup of type $(m, m, \ldots m)$; and, M being fixed, for the $l!$ representations of $M(\mathbf{A}_E)$ obtained by permutation of the components of $\Pi_1 \otimes \Pi_1^\sigma \otimes \cdots \otimes \Pi_1^{\sigma^{l-1}}$.

Because of the obvious symmetries (cf. the remarks in [1(c), p. 1293–94]), the contributions from the different M are equal; if we write, as usual, W_A^M for the Weyl group $N_M(A)/Z_M(A)$ of a special torus A in the Levi group M, the number of relevant Levi subgroups is equal to

$$\frac{|W_0^G|}{|W_0^M|\,|W_A^G|},$$

A being the split component of M.

Consider now the $l!$ terms associated to a fixed—say, the standard—M. In the case of $GL(n)$, Shahidi [36(c)] has shown that the intertwining operators $M(s,\lambda)$ could be normalized in the way predicted by Langlands [30(d), Appendix II]. In particular, let $\pi_E = \Pi_1 \otimes \cdots \otimes \Pi_1^{\sigma^{l-1}}$; for $\tau \in \mathfrak{S}_l$, denote by $\tau\pi_E$ the representation of $M(\mathbf{A}_E)$ obtained by permuting the indices by τ. If s is the Coxeter element in $W(\mathfrak{a}) = W_A^G$ such that $s\sigma\pi_E = \pi_E$—so s is the permutation sending $(1, \ldots l)$ onto $(l, 1, \ldots, l-1)$—the Coxeter element s' such that $s'\sigma(\tau\pi_E) = \tau\pi_E$ is clearly $\tau s \tau^{-1}$. Let us write $M(s, 0, \pi'_E)$ when we want to specify the representation π'_E we consider. By

Shahidi's result we may write:

$$M(s,0,\pi_E) = N(s,0,\pi_E)\cdot m(s,0,\pi_E)$$

where $N(s,0,\pi_E)$ is the normalized intertwining operator (denoted by $\mathbf{a}(0,\pi_E,s)$ in [36(c)]) and

$$m(s,0,\pi_E) = \prod_{j=1}^{l} \frac{L(0,\Pi_1\otimes\tilde{\Pi}_1^{\sigma^j})}{L(1,\Pi_1\otimes\tilde{\Pi}_1^{\sigma^j})\varepsilon(0,\Pi_1\otimes\tilde{\Pi}_1^{\sigma^j},\psi)}. \tag{4.3}$$

It is easy to infer from [36(c)] that we have also

$$M(\tau s\tau^{-1},0,\tau\pi_E) = N(\tau s\tau^{-1},0,\tau\pi_E)m(s,0,\pi_E):$$

the normalizing constant is the same. The normalized operators satisfy the product relation

$$N(s_1s_2,0,\pi) = N(s_1,0,s_2\pi)N(s_2,0,\pi)$$

([36(c), Thm. 3.1]), from which we obtain

$$N(\tau s\tau^{-1},\tau\pi_E) = N(\tau,s\pi_E)N(s,\pi_E)N(\tau^{-1},\tau\pi_E). \tag{4.4}$$

(We have dropped the mention of "0" from the notation.)

Moreover, the operators $M(s,\lambda,\pi)$ are obtained by integration over subgroups which can be taken over F; that operation commutes with the action of $\mathrm{Gal}\,(E/F)$, and therefore

$$M(\tau^{-1},\tau\pi_E)\sigma = \sigma M(\tau^{-1},\tau\sigma^{-1}\pi_E).$$

Since the normalizing factors m are invariant if we conjugate representations by σ—assuming we take the character ψ in (4.3) invariant by $\mathrm{Gal}\,(E/F)$—we also have

$$N(\tau^{-1},\tau\pi_E)\sigma = \sigma N(\tau^{-1},\tau\sigma^{-1}\pi_E).$$

Combining this equality with (4.4) yields

$$N(\tau s\tau^{-1},\tau\pi_E)\sigma = N(\tau,s\pi_E)[N(s,\pi_E)\sigma]N(\tau^{-1},\tau\sigma^{-1}\pi_E).$$

Using the fact that $s\pi_E = \sigma^{-1}\pi_E$, and the product relation, we obtain

$$N(\tau,s\pi_E)^{-1} = N(\tau^{-1},\tau\sigma^{-1}\pi_E).$$

Since the normalizing constants are equal, we have then:

$$\begin{aligned}&\operatorname{trace}(M(\tau s\tau^{-1},\tau\pi_E)\sigma\rho_{Q,\tau\pi_E}(\phi))\\ =&\operatorname{trace}(N(\tau,s\pi_E)M(s,\pi_E)\sigma N(\tau,s\pi_E)^{-1}\rho_{Q,\tau\pi_E}(\phi))\\ =&\operatorname{trace}(N(\tau,s\pi_E)M(s,\pi_E)\sigma\rho_{Q,\pi_E}(\phi)N(\tau,s\pi_E)^{-1})\\ =&\operatorname{trace}(M(s,\pi_E)\sigma\rho_{Q,\pi_E}(\phi)),\end{aligned}$$

using the basic property of intertwining operators. We have shown that all contributions associated to elements of $\mathfrak{S}_l$ are equal; remembering the counting involved, we see that the sum of all terms of the type considered in (4.3) is equal to

$$|\det(s-1)|^{-1}\operatorname{trace}(M(s,0)\sigma\rho_{Q,\pi_E}(0,\phi)).$$

It is easy to check that, for the Coxeter element, s, $|\det(s-1)|=l$. Now the identity of (4.1) and (4.2) implies the existence of some representation π of $G(\mathbf{A})$ lifted by Π. Remark that, if (π_i), $i\in I$, is the set of all such representations, we have, by the identity of traces:

$$\sum_i \operatorname{trace}\pi_i(f)=\operatorname{trace}(M(s,0)\sigma\Pi(\phi))$$

for associated f and ϕ. Since the operator $M(s,0)$ is unitary, the local theory already implies that there must be only one representation on the left. We will also obtain this fact from the consideration of L-functions.

We now proceed to show that π is cuspidal.

Assume π is a subquotient of $\pi_1\times\cdots\times\pi_r$, a product of cuspidal representations; we take $\lambda(\pi_r)$ minimal. If $r\neq l$, we may, by induction, lift π_r to Π_r or $\Pi_r\times\Pi_r^\sigma\times\cdots\times\Pi_r^{\sigma^{l-1}}$.

In the first case, we have

$$L^S(s,(\Pi_1\times\Pi_1^\sigma\times\cdots\times\Pi_1^{\sigma^{l-1}})\otimes\tilde{\Pi}_r)=\prod_j L^S(s,\pi\otimes\tilde{\pi}_r\otimes\eta^j).$$

The right hand side has a pole at $s=1$; thus the left-hand side has one also; since Π_1 is unitary and $\lambda(\Pi_r)\leq 0$, this implies (using again §2) that $L^S(s,\Pi_1^{\sigma^i}\otimes\tilde{\Pi}_r)$ has a pole at 1, for some i. Therefore $m=n_r$ and $\Pi_r\cong\Pi_1^{\sigma^i}$, which is impossible since Π_r is σ-stable and Π_1 is not.

In the second case, we have

$$\prod_{i,j=0}^{l=1} L^S(s,\Pi_1^{\sigma^i}\otimes(\tilde{\Pi}_r)^{\sigma^j})=\prod_j L^S(s,\pi\otimes\tilde{\pi}_r\otimes\eta^j). \tag{4.5}$$

The right-hand side has a pole at $s = 1$; thus we must have a pole at $s = 1$ for some $L^S(s, \Pi_1^{\sigma^i} \otimes (\tilde{\Pi}_r)^{\sigma^j})$. This implies again that $n_r/l = m$ and $\Pi_r = \Pi_1^{\sigma^k}$ for some k.

Thus we see that $r = 1$ and π is cuspidal. Moreover, in (4.5), the left-hand side has a pole of order l exactly at 1. Thus the same must be true for the left-hand side $\prod_j L^S(s, \pi \otimes \tilde{\pi} \otimes \eta^j)$. Therefore $\pi \cong \pi \otimes \eta$.

This proves (e).

Finally, (f) is then proved by induction. Of course there are a finite number of π lifted by Π, and their number can be determined using the cuspidal representation defining Π and parts (d) and (e) of the theorem. ∎

In the next proposition we collect some further properties of base change; the local analogues have been proved earlier. We denote, as earlier, by ω_π the central character of π.

PROPOSITION 4.4.: (i) *The notion of base change lifting is independent of the choice of the generator σ of Σ.*

(ii) *If Π is a base change lift of π,*

$$\omega_\Pi = \omega_\pi \circ N_{E/F}.$$

(iii) *Assume*

$$\begin{array}{c} E \\ | \\ F \\ | \\ L \end{array}$$

is a diagram of Galois extensions, with E/F cyclic as above. Let $\tau \in \mathrm{Gal}(E/L)$. Then, for π, Π representations of $G(\mathbf{A})$ and $G(\mathbf{A}_E)$, π^τ and Π^τ are defined.

Then, if Π lifts π, Π^τ lifts π^τ.

Proof. (i) is clear since the notion of weak base change defined by (1.1) is independent of σ.

(ii) is also an easy consequence of (1.1), since the central characters are determined by their values almost everywhere.

For (iii) we may use the proof of the local analogue at the end of §I.6.2. If Π lifts π, then at almost all finite primes, Π_v lifts π_v—in the sense of Hecke eigenvalues and also, by the considerations in §I.6.2, in the sense of character identities. By the proof there, Π_v^τ lifts π_v^τ in the character sense. But for two unramified representations, one checks easily that this identity of characters is equivalent to the identity of Hecke eigenvalues (1.1). Thus Π^τ is a (weak) lift of π^τ. ∎

Remark. When we have proved Theorem 5.1, Proposition 4.4 will immediately hold for *strong* lifting.

5. Strong lifting

With the notions defined in §1, we will now prove the following (the assumptions are as in Theorem 4.2):

THEOREM 5.1: (STRONG LIFTING). *Assume π, Π are representations induced from cuspidal of $G(\mathbf{A})$, $G(\mathbf{A}_E)$ respectively.*

If Π is a weak lifting of π, then Π is in fact a strong lifting of π.

Proof. Since weak and strong lifting survive induction, we may assume that π is cuspidal. We distinguish two cases, corresponding to parts (a) and (b) of Theorem 4.2.

Assume first that $\pi \not\cong \pi \otimes \eta$. Then, separating strings of Hecke eigenvalues in the identity (4.1) as in the proof of Theorem 4.1, we obtain the equality

$$\sum_{i=1}^{l} \operatorname{trace}(\pi \otimes \eta^i)(f) = l \operatorname{trace}(\Pi(\phi) I_\sigma),$$

whenever f and ϕ are associated. Note that by the vanishing conditions on the orbital integrals of f (Proposition I.3.1), this is in fact equivalent to

$$\operatorname{trace} \pi(f) = \operatorname{trace}(\Pi(\phi) I_\sigma).$$

Let v be a place of F. This global identity obviously implies that, $I_{\sigma,v}$ being the normalized intertwining operator at v between $\Pi_v = \bigotimes_{w|v} \Pi_w$ and $\Pi_v \circ \sigma$, we have, for f_v and $\phi_v = \bigotimes_{w|v} \phi_w$ associated:

$$\operatorname{trace} \pi_v(f_v) = c \operatorname{trace}(\Pi_v(\phi_v) I_{\sigma,v}),$$

c being some non-zero constant. By Weyl's integration formulas, this implies

$$\Theta_{\Pi_v,\sigma} = c\Theta_{\pi_v} \circ \mathcal{N} \tag{5.1}$$

where $\Theta_{\Pi_v,\sigma}$ denotes the character twisted by $I_{\sigma,v}$.

Now π_v, as a local component of a cuspidal π, is a generic representation. By an easy extension of the results of §I.6 (cf. [11(a)] in the real case), there exists a generalized principal series representation Π_v^0 of $G(E_v)$ such that, for the normalized operator I_σ of §I.2, extended to the non-unitary parameters in the obvious way:

$$\Theta_{\Pi_v^0,\sigma} = \Theta_{\pi_v} \circ \mathcal{N} \tag{5.2}$$

where $\Theta_{\Pi_v^0,\sigma}$ is defined by I_σ. From (5.1) and (5.2) we obtain the equality $c\Theta_{\Pi_v^0,\sigma} = \Theta_{\Pi_v,\sigma}$. Of course, Π_v^0 might so far be reducible. This is a linear relation between standard (= generalized principal series) twisted characters. However, the same argument as in the non-twisted case (cf. [11(d), Prop. 2]) shows that these twisted characters are independent—here we have to use the Langlands classification for σ-stable representations, cf. before Proposition I.6.9, and the independence of the twisted characters (Lemma I.6.3). Therefore we must have $\Pi_v^0 \cong \Pi_v$, $c = 1$ and

$$\Theta_{\Pi_v,\sigma} = \Theta_{\pi_v} \circ \mathcal{N}.$$

This finishes the proof in this case.

If $\pi \cong \pi \otimes \eta$, the identity (4.1) reads—using the counting arguments in the proof of Theorem 4.2(e)

$$\operatorname{trace} \pi(f) = \frac{1}{l!} \sum_w \operatorname{trace}(M(s_w, 0)\sigma \rho_{w\pi_E}(0, \phi)).$$

Here $\pi_E = \Pi_1 \otimes \Pi_1^\sigma \otimes \cdots \otimes \Pi_1^{\sigma^{l-1}}$, where Π_1 is cuspidal and $\Pi_1 \not\cong \Pi_1^\sigma$: thus π_E is a representation of $M(\mathbf{A}_E)^1$. The subscript w runs over $\mathfrak{S}_l$; $w\pi_E$ is the obvious permutation of π_E and s_w is the only cycle of length l in $\mathfrak{S}_l$ such that $s_w \cdot (w\pi_E) = \sigma(w\pi_E)$. All terms on the right side are proportional, with the same constant, to trace$(\Pi(\phi)I_\sigma)$ where I_σ is now normalized as in §I.2, and $\Pi = \Pi \times \cdots \times \Pi_1^{\sigma^{l-1}}$ is generic.

Thus we obtain an identity

$$\operatorname{trace} \pi(f) = c \operatorname{trace}(\Pi(\phi)I_\sigma).$$

From then on the argument is the same (note that the local components of Π are induced from generic representations and irreducible, and hence generic). This finishes the proof. ∎

6. Base change lift of automorphic forms in cyclic extensions

In this section we will show that the base change results we obtained allow one to "induce" cusp forms on linear groups from one global field to another. This is expected as a part of the general Langlands conjectures and is best explained in terms of the (conjectural!) Tannaka group. We refer to Langlands [30(c)] and to [26(b)] for more information.

Let F be a number field. From what is known so far about the properties of automorphic forms, it seems natural to surmise that automorphic representations of $GL(n\mathbf{A}_F)$ (in fact isobaric representations [30(c)] should correspond bijectively to completely reducible representations of degree n of some conjectural group, denoted by $G_{\Pi(F)}$.

Of course the local analogue is the so-called "local conjecture" of Langlands, describing representations of *local* groups $GL(n,F)$ by representations of degree n of the modified Weil group $W_F \times SL(2,\mathbf{C}) = W'_F$. Local and global conjectures should be compatible, i.e., there should be, for each completion F_v of the global field F, natural homomorphisms

$$W'_{F_v} \xrightarrow[\iota_v]{} G_{\Pi(F)}$$

such that, if the automorphic representation π is associated to $\phi : G_{\Pi(F)} \to GL(n,\mathbf{C})$, π_v should be associated to the representation $\phi \circ \iota_v$ of W'_{F_v}.

Now assume E/F is an extension of global fields, that we take to be cyclic. It seems, again, natural to expect an exact sequence, analogous to the one relating Weil groups:

$$1 \to G_{\Pi(E)} \to G_{\Pi(F)} \to \mathrm{Gal}(E/F) \to 1.$$

In particular, a representation of $G_{\Pi(E)}$, of degree n, should induce to yield a representation of degree nl of $G_{\Pi(F)}$. Consequently, one should be able to associate, to an automorphic representation of $GL(n,\mathbf{A}_E)$, an automorphic representation of $GL(nl,\mathbf{A}_F)$. It is easy now to describe directly (without the Tannaka group) what this correspondence should be. Let π_E be the automorphic representation of $GL(n,\mathbf{A}_E)$.

For almost all places w of E, the representation $\pi_{E,w}$ is unramified, and is thus naturally associated to a representation of degree n of W_E which is a sum of unramified Abelian characters. We define a (hypothetical) representation π_F of $GL(nl,\mathbf{A}_F)$ by defining its Hecke eigenvalues almost everywhere. Let v be a place of F, unramified in E and such that $\pi_{E,w}$ is unramified for any $w|v$. We define the local representations of W_{F_v} as follows:

(i) If v is inert, we have an exact sequence

$$1 \to W_{E_w} \to W_{F_v} \to \mathrm{Gal}(E/F) \to 1.$$

We obtain a representation of W_{F_v}, of degree nl, by inducing that of W_{E_w}. (Of course w is the only place of E above v.)

(ii) If v splits in E, and $w_1, \ldots w_l$ are the places of E above v, we obtain a representation of $W_{F_v} \cong W_{E_{w_i}}$ by taking the direct sum of the representations of $W_{E_{w_i}}$ $(i = 1, \ldots l)$.

(iii) The composite case may be left to the reader.

DEFINITION 6.1.: (E/F cyclic). *If π_E, π_F are automorphic representations of $GL(n, \mathbf{A}_E)$, $GL(nl, \mathbf{A}_F)$ respectively, we say that π_F is* automorphically induced *from π_E if their Hecke eigenvalues are associated as in* (i)–(iii) *above.*

THEOREM 6.2.: *Let E/F be a cyclic extension of global fields of degree l (prime or not).*

Then, if π_E is a representation of $GL(n, \mathbf{A}_E)$ induced from cuspidal, there exists one, and only one, representation π_F of $GL(n, \mathbf{A}_F)$ automorphically induced from π_E. Moreover π_F is induced from cuspidal.

Notice first that it suffices to prove the theorm for π_E cuspidal; the general case follows by induction. We will see in the proof that π_F is induced from cuspidal, and therefore determined by the knowledge of its Hecke eigenvalues at almost all primes. Thus the uniqueness of π_F is obvious.

To prove the existence of a representation π_F with the correct Hecke eigenvalues at almost all primes, we first notice that automorphic induction, as defined in Definition 6.1, satisfies the usual property of "induction by stages". Namely, if $E/E_1/F$ is a diagram of extensions, all cyclic, if π_{E_1} is automorphically induced from π_E and π_F from π_{E_1}, then π_F is automorphically induced from π_E. (The verification in terms of Hecke eigenvalues—or Weil group representations—using Definition 6.1, is left to the reader).

Hence Theorem 6.2 can be proved by considering a sequence of cyclic extensions of prime order. In this case, assuming E/F cyclic of order l, we may reformulate Definition 6.1 as follows. Assume E_w/F_w is an unramified field extension, with χ_w an unramified character of E_w. Then

$$\mathrm{ind}_{W_{E_w}}^{W_{F_v}} \chi_w = \sum_{\chi_v \circ N = \chi_w} \chi_v;$$

it is the sum of the l characters of F_v^* that compose with the norm to yield χ_w. Using this fact, it is easy to see that Definition 6.1 is equivalent to the following formulas at almost all places. Let ζ be a primitive l^{th} root of unity.

Then π_F is automorphically induced from π_E if:

$$t_{\pi_F,v} = t_{\pi_E,w_1} \oplus t_{\pi_E,w_2} \oplus \cdots \oplus t_{\pi_E,w_l} \tag{6.1}$$

$$(v \text{ split into } w_1, \ldots w_l)$$

$$t_{\pi_F,v} = t^{1/l}_{\pi_E,w} \oplus \zeta t^{1/l}_{\pi_E,w} \oplus \cdots \oplus \zeta^{l-1} t^{1/l}_{\pi_E,w} \tag{6.2}$$

$$(v \text{ inert}, w|v).$$

In (6.2), $t^{1/l}$ denots any l^{th} root of the diagonal matrix t; note that $t_{\pi_F,v}$ is unambiguously defined (up to permutation of entries), since we then add all the products by powers of ζ.

To prove the existence of π_F, we rely on Theorem 4.2(d) and (e). Assume first that $\pi_E \cong \pi_E^\sigma$, where σ generates $\mathrm{Gal}(E/F)$. By Theorem 4.2(d), π_E lifts exactly l representations $\pi_n, \pi_n \otimes \eta, \ldots \pi_n \oplus \eta^{l-1}$, where π_n is a cupidal representation of $G((n, \mathbf{A}_F)$. Set

$$\pi = \pi_F = \pi_n \times (\pi_n \otimes \eta) \times \cdots \times (\pi_n \otimes \eta^{l-1}).$$

We check that the $t_{\pi,v}$ are given by (6.1) and (6.2). Note that we have

$$t_{\pi_n,v} = t_{\pi_E,w}, \qquad v \text{ split}, w|v, \tag{6.3}$$

$$(t_{\pi_n,v})^l = t_{\pi_E,w}, \qquad v \text{ inert} \tag{6.4}$$

In particular, $t_{\pi_E,w_i} = t_{\pi_E,w_j}$ if $w_i \neq w_j$, w_j above v. It is then obvious that (6.1) is satisfied.

Assume v is inert: thus $\eta(\bar{\omega}_v) = \zeta$, a primitive l^{th} root. Then

$$t_{\pi,v} = t_{\pi_n,v} \oplus \zeta t_{\pi_n,v} \oplus \cdots \oplus \zeta^{l-1} t_{\pi_n,v}.$$

By (6.4), $t_{\pi_n,v}$ is an lth root of $t_{\pi_E,w}$. So $t_{\pi,v}$ has the value specified by (6.2).

Assume now that $\pi_E \not\cong \pi_E^\sigma$. By Theorem 4.2(e), $\pi_E \times \pi_E^\sigma \times \cdots \times \pi_E^{\sigma^{l-1}}$ defines a unique representation π of $GL(nl, \mathbf{A}_F)$; moreover, $\pi \otimes \eta \cong \pi$. Again, we check that (6.1) and (6.2) are satisfied by π. By construction we have

$$t_{\pi,v} = t_{\pi_E,w_1} \oplus \cdots \oplus t_{\pi_E,w_l}, \qquad v \text{ split}, \tag{6.5}$$

$$t^l_{\pi,v} = t_{\pi_E,w} \oplus \cdots \oplus t_{\pi_E,w}, \qquad v \text{ inert}, \tag{6.6}$$

(It is easy to check (6.5) by composing Galois action and parabolic induction at a split place.) Thus (6.1) is satisfied. As for (6.2), note that $\pi \cong \pi \otimes \eta$; if ζ is identified with a diagonal matrix, this implies that $\zeta t_{\pi,v} = s \cdot t_{\pi,v}$ where s is an element of $\mathfrak{S}_{nl}$. Setting $t = t_{\pi,v}$, we see that, up to a reordering of the indices, we can write

$$t = (t_1, \zeta t_1, \dots \zeta^{l-1} t_1, t_2, \dots \zeta^{l-1} t_2, \dots \zeta^{l-1} t_n).$$

Set $T = t_{\pi_E, w} = (T_1, \dots T_n)$. We have the equality (6.6), which is true of course modulo permutation. It implies that, up to reordering,

$$(\underbrace{t_1^l, t_1^l, \dots t_1^l}_{l \text{ terms}}, \dots \underbrace{t_n^l, t_n^l, \dots t_n^l}_{l \text{ terms}})$$

is equal to l times the segment $(T_1, T_2, \dots T_n)$.

Obviously this means that, upon reordering, we may assume $t_i = T_i^{1/l}$ where the lth roots are arbitrary; this implies that—always $\bmod \mathfrak{S}_{nl}$—t is equal to $T^{1/l} \oplus \zeta T^{1/l} \oplus \cdots \oplus \zeta^{l-1} T^{1/l}$. That is the equality (6.2). Theorem 6.2 is proved. ∎

We now observe that more information may be obtained on the representation π_F. We will need two lemmas:

LEMMA 6.3.: *Assume π_E is cuspidal and $\pi_E \cong \pi_E^\sigma$, where σ is a generator of* Gal(E/F). *Then there are exactly l representations of $GL(n, \mathbf{A}_F)$ lifted by π_E. They are of the form $\pi_F, \pi_F \otimes \eta_{E/F}, \dots, \pi_F \otimes \eta_{E/F}^{l-1}$, where π_F is one of them; π_F is cuspidal, $\pi_F \cong \pi_F \otimes \eta_{E/F}$.*
Here of course $\eta_{E/F}$ is a generating character of $F^* N \mathbf{A}_E^* \backslash \mathbf{A}_F^*$.

Proof. Of course if l is prime this is part of Theorem 4.2. We reduce to the prime case. Let $E/E_1/F$ be a composite extension, with $[E : E_1]$ prime. Given π_E, we obtain π_{E_1} lifted by π_E (Theorem 4.2), then π_F lifted by π_{E_1} using Lemma 6.3 inductively. By transitivity of the lifting identities (1.1) we see that π_E lifts π_F and $\pi_F \otimes \eta_{E/F}^i$ for any i. Moreover the identity

$$L^S(s, \pi_E \otimes \tilde{\pi}_E) = \prod_i L^S(s, \pi_F \otimes \tilde{\pi}_F \otimes \eta_{E/F}^i)$$

shows that the $\pi_F \otimes \eta^i$ are distinct; one has only to compare the orders of poles at $s = 1$ on each side. ∎

LEMMA 6.4.: *Assume π_E is cuspidal and $\pi_E \not\cong \pi_E^{\sigma^i}$ for any $i < l$. Then there is a unique representation π_F of $GL(nl, \mathbf{A}_F)$ lifted by $\pi_E \times \pi_E^\sigma \times \cdots \times \pi_E^{\sigma^{l-1}}$. It is cuspidal and $\pi_F \cong \pi_F \otimes \eta_{E/F}$.*

Proof. Again, assume $l = l_1 k_1$ is composite, $[E : E_1] = l_1$, $[E_1 : F] = k_1$. Thus E_1 is the field fixed by $\tau = \sigma^{k_l}$. Since $\pi_E \cong \pi_E^\tau$, we obtain a representation of $GL(nl_1, \mathbf{A}_{E_1})$, say π_{E_1}—this is part of Theorem 4.2. It is cuspidal. Moreover, assume $\pi_{E_1} \cong \pi_{E_1}^{\sigma^i}$ for some $i < k_1$. By Proposition 4.4(iii), we get $\pi_E \cong \pi_E^{\sigma^i}$, contrary to our assumption. Thus $\pi_{E_1} \not\cong \pi_{E_1}^{\sigma^i}$, and applying Lemma 6.4 inductively we get π_F, a cuspidal representation of $GL(nl, \mathbf{A}_F)$. Again, the identity of L-functions shows that $\pi_F \cong \pi_F \otimes \eta_{E/F}$. ∎

We now give a more explicit description of the "induced" representation obtained in Theorem 6.2. Assume that $\pi_E \cong \pi_E^\tau$, where $\tau = \sigma^a$ for $1 < a < l$, and a is minimal: $\pi_E \not\cong \pi_E^{\sigma^i}$ if $i < a$. Let L be the fixed field of τ. By Lemma 6.3, π_E lifts the representations $\pi_L, \pi_L \otimes \eta_{E/L}, \ldots, \pi_L \otimes \eta_{E/L}^{b-1}$ of $GL(n, \mathbf{A}_L)$. Now $\pi_L \not\cong \pi_L^{\sigma^i}$ for $i < a$, since $\pi_E \not\cong \pi_E^{\sigma^i}$ (again Proposition 4.4(iii)); the same applies to all the twists up to $\pi_L \otimes \eta_{E/L}^{b-1}$. Threefore there is a unique cuspidal representation π_F^0 of $GL(na, \mathbf{A}_F)$ lifted by $\pi_L \times \cdots \times \pi_L^{\sigma^{a-1}}$. By class field theory, we may write $\eta_{E/L} = \eta_{E/F} \circ N_{L/F}$. Just by using Theorem 3.1, we see that the representation of $GL(nl, \mathbf{A}_F)$ automorphically induced from Π_E (which exists by Theorem 6.2) is $\pi_F^0 \times \pi_F^0 \otimes \eta_{E/F} \times \cdots \times \pi_F^0 \otimes \eta_{E/F}^{b-1}$: the theorem implies that one is the twist of the other by a power of $\eta_{E/F}$, and both are stable by such a twist. We have $\eta_{E/F}^b = 1$ on $N_{L/F}\mathbf{A}_L^*$, and $\eta_{E/F}^b$ is a generator of the character group of $F^* N\mathbf{A}_L^* \backslash \mathbf{A}_F^*$. Thus by Lemma 6.4 applied to L/F, we see that $\pi_F^0 \otimes \eta_{E/F}^b \cong \pi_F^0$. We record this in

COROLLARY 6.5.: *Under the assumptions of Theorem 6.2, assume π_E cuspidal. Let $\tau = \sigma^a$, a minimal, be a generator of the stabilizer of π_E in* Gal(E/F). *Let L be the fixed field of τ. Then the representation of $GL(nl, \mathbf{A}_F)$ automorphically induced from π_E is of the form*

$$\pi_F = (\pi_F^0 \otimes \eta_1) \times \cdots \times (\pi_F^0 \otimes \eta_b)$$

where π_F^0 is a cuspidal representation of $GL(na, \mathbf{A}_F)$ and η_i ranges over the characters of $F^* N\mathbf{A}_D^* \backslash \mathbf{A}_F^*$ modulo those vanishing on $N\mathbf{A}_L^*$. Moreover, for any character η of $F^* N\mathbf{A}_L^* \backslash \mathbf{A}_F^*$, $\pi_F^0 \otimes \eta \cong \pi_F^0$.

We end up this section with a lemma which has been used in §I.6. Notations are as above.

LEMMA 6.6.: *Assume π_F is cuspidal and satisfies $\pi_F \cong \pi_F \otimes \eta_{E/F}^b$ with b minimal. Then, if Π is the lift of π_F to $GL(n, \mathbf{A}_E)$, we have $\Pi = \Pi_1 \times \Pi_1^\sigma \times \cdots \times \Pi_1^{\sigma^{a-1}}$, Π_1 being a cuspidal representation of $GL(b, \mathbf{A}_E)$ such that $\Pi_1 \not\cong \Pi_1^{\sigma^i}$ for $i < a = \frac{n}{b}$.*

Proof. Using Theorem 4.2 inductively we may find Π; it is induced from cuspidal. Since $\Pi \cong \Pi^\sigma$, we see that Π is a product (for the operation denoted by $\times$) of blocks of type $\Pi_1 \times \cdots \times \Pi_1^{\sigma^{a-1}}$ with Π_1 cuspidal and $\Pi_1^{\sigma^a} \cong \Pi_1$, and a minimal. Using the "going down" Lemma 6.4, and the considerations of §2, one easily sees that there must be only one block if π_F is cuspidal. Finally, one checks that $a = \frac{n}{b}$ by using the identity of L-functions (Lemma 4.3) at $s = 1$.

7. The strong Artin conjecture for nilpotent groups

Let E/F be a Galois extension of number fields, with Galois group Γ. If r is a complex representation of Γ, we denote by $L(s,r)$ the Artin L-function associated to r ([2(a), (b)]). It is a product of local factors $L_v(s,r)$.

We will call "strong Artin conjecture" the following assertion: if r is irreducible, there should exist a cuspidal representation π of $\mathrm{GL}(n,\mathbf{A}_F)$—where n is the degree of r—such that, for some finite set of places of F containing all ramified primes for r and π:

$$L_v(s,r) = L_v(s,\pi). \qquad (v \notin S)$$

This conjecture was made by Langlands [30(a)]. Since the Frobenius eigenvalues of r at the other places are unitary, it is easy to check that it would imply the holomorphicity of $L(s,r)$.

THEOREM 7.1.: *Assume that E/F is a Galois extension of number fields with nilpotent Galois group Γ. Then, if r is any irreducible complex representation of Γ, the strong Artin conjecture is true for r.*

Of course this says nothing new about the Artin conjecture itself, which is true in this case since irreducible characters are monomial.

Remark. The method used to prove Proposition I.6.9 implies the stronger result that $L(s,r) = L(s,\pi)$, i.e., the L-functions coincide at all places.

We will in fact deduce Theorem 7.1 from a stronger result.

Let us call a representation r of $\mathrm{Gal}(E/F) = \Gamma$ *automorphic* [27(b)] if there is an automorphic representation π of $\mathrm{GL}(n,\mathbf{A}_F)$, where n is the degree of r, such that $L_v(s,r) = L(s,\pi_v)$, $(v \notin S)$ for large S.

PROPOSITION 7.2.: *Let E/F be a* solvable *Galois extension of number fields, with Galois group Γ. Assume that r is an irreducible representation of Γ, and that its character belongs to the subgroup of the Grothendieck group of characters of Γ spanned over $\mathbf{Z}$ by characters of the form*

$$\mathrm{ind}_{\Gamma_0}^{\Gamma}(\chi), \tag{7.1}$$

where Γ_0 is a subgroup of Γ admitting a subinvariant series

$$\Gamma_0 \lhd \Gamma_1 \lhd \Gamma_2 \cdots \lhd \Gamma_n = \Gamma$$

with all factors cyclic, and χ is an Abelian character of Γ_0.

Then r is automorphic, associated to a cuspidal representation. In particular, the strong Artin conjecture holds for r.

Since nilpotent groups are monomial, and any subgroup Γ_0 of a nilpotent Γ has the property stated in Proposition 7.2, this implies Theorem 7.1.

Remark. E. C. Dade [16] has shown that the characters verifying the assumptions of Proposition 7.2 are monomial. Therefore the Artin conjecture was already known for r. We don't know for which finite groups the group of characters is spanned by characters of type (7.1).

Proof of Proposition 7.2. We will rely on a result of Jacquet-Shalika [27(b)]. Consider first a representation $r = \mathrm{ind}(\chi)$ of Γ as in (7.1). (Thus r is not, in general, irreducible). If L_0 is the fixed field of Γ_0, the character χ of $\mathrm{Gal}(E/L_0)$ is associated, by Abelian class field theory, to a character of $L_0^*\backslash\mathbf{A}_{L_0}^*$ that we denote also by χ. Using Theorem 6.2 repeatedly in the tower of fields associated to the normal series $\Gamma_0 \subset \Gamma_1 \ldots \subset \Gamma$, we otain an automorphic representation π of $\mathrm{GL}(n, \mathbf{A}_F)$, where $n = \deg r = [\Gamma : \Gamma_0]$. The identities (i–iii) in Definition 6.1 show that π has (for almost all primes) Hecke eigenvalues corresponding to the Frobenius eigenvalues of r. Therefore, r is automorphic.

If now r is an irreducible representation of degree n of Γ satisfying the assumptions of Proposition 7.2, we may write it, in the Grothendieck group, as $r = r' - r''$, where r' and r'' are sums of representations of the form $\mathrm{ind}(\chi)$, as in (7.1). By the previous paragraph, r' and r'' are automorphic.

By Theorem 4.7 of [27(b)], we conclude that r is also automorphic, associated to an automorphic representation π of $\mathrm{GL}(n, \mathbf{A}_F)$, and π is cuspidal. This concludes the proof. ∎

The proof of Theorem 7.1 implies that the automorphic representations of $GL(n, \mathbf{A}_F)$ associated to representation of $\mathrm{Gal}(E/F)$ can be *multiplied* by arbitrary automorphic representations of $GL(m, \mathbf{A}_F)$. Recall that if π, τ are cuspidal representations of $GL(n, \mathbf{A}_F)$ and $GL(m, \mathbf{A}_F)$ respectively, Langlands' principle of functoriality implies that there should exist an automorphic representation $\Pi = \pi \boxtimes \tau$ of $GL(mn, \mathbf{A}_F)$ whose Hecke eigenvalues satisfy (up to permutations, and at almost all primes):

$$t_{\Pi,v} = t_{\pi,v} \otimes t_{\tau,v} \tag{7.2}$$

(see [8(b), 30(c)]. The representation Π will not, in general, be cuspidal but should be *induced from cuspidal*, in the sense of this chapter, for unitary π, τ.

THEOREM 7.3.: *Let E/F be a finite nilpotent extension of F, with Galois group Γ. Let r be an irreducible representation of Γ of degree n, and π the associated cuspidal representation of $GL(m, \mathbf{A}_F)$, there exists a unique automorphic representation $\pi \boxtimes \tau$ of $GL(mn, \mathbf{A}_F)$ verifying (7.2); it is induced from cuspidal.*

Proof. Write $r = \mathrm{ind}_{\Gamma_0}^{\Gamma}(\chi)$, $\Gamma_0 = \mathrm{Gal}(E/F_0)$ being a subgroup of Γ and χ a one-dimensional character of Γ_0. We can find a tower of fields $F = F_r \subset F_{r-1} \subset \cdots \subset F_0$ associated to subgroups $\Gamma_0 \subset \Gamma_1 \subset \cdots \subset \Gamma_r = \Gamma$ with cyclic quotients (of prime order). The representation π associated to r is then obtained by automorphic induction (Thm. 6.2) from $GL(1, \mathbf{A}_{F_0})$ to $GL(n, \mathbf{A}_F)$, where $n = \deg(r)$, from the one-dimensional character of $GL(1, \mathbf{A}_{F_0})$ associated to χ. We denote it by $\iota_{F_0}^F(\chi)$, ι standing for *automorphic induction.*

In the same situation, we may also apply automorphic induction to any representation π_{F_0} of $G_n(\mathbf{A}_{F_0})$. Consider the representation τ_{F_0} of $G_m(\mathbf{A}_{F_0})$ obtained from τ by repeated base change in the tower $(F, F_{r-1}, \ldots, F_0)$. We denote it by $\rho_{F\to F_0}(\tau)$; it is obtained by *automorphic restriction* from τ (for Galois representations, this would translate restriction). We will prove that

$$\iota_{F_0}^F(\chi \otimes \rho_{F\to F_0}\tau) \cong \iota_{F_0}^F\chi \boxtimes \tau, \tag{7.3}$$

this being taken to mean that the representation on the left (which exists by the results of this chapter) has Hecke eigenvalues equal almost everywhere to those of the (conjectured) right-hand side. This implies the existence of $\pi\boxtimes\tau$; its uniqueness follows from the fact that the operations involved in $\iota_{F_0}^F$ and $\rho_{F\to F_0}$ preserve the category of representations induced from cuspidal: these are determined by their Hecke eigenvalues almost everywhere.

One should notice that, in the case that τ is associated to a representation r' of $\mathrm{Gal}(E/F)$ (maybe for a larger, non-nilpotent E/F), (7.3) just translates on the automorphic side the standard isomorphism

$$\mathrm{ind}_{\Gamma_0}^{\Gamma}(\chi \otimes \mathrm{res}_{\Gamma\to\Gamma_0} r') \cong \mathrm{ind}_{\Gamma_0}^{\Gamma}\, \chi \otimes r' \tag{7.4}$$

between representations of Γ.

Notice also that (7.3) can be obtained by repeated cyclic lifting. Indeed, replacing $F = F_r$ by F_{r-1} in (7.3), and τ by $\rho_{F\to F_{r-1}}\tau$, assume we have proved

$$\iota_{F_0}^{F_{r-1}}(\chi \otimes \rho_{F\to F_0}\tau) \cong \iota_{F_0}^{F_{r-1}}\chi \boxtimes \rho_{F\to F_{r-1}}\tau \tag{7.5}$$

in the sense indicated above (note that $\rho_{F_{r-1}\to F_0} \circ \rho_{F\to F_{r-1}} = \rho_{F\to F_0}$). Then, applying (7.3) in the cyclic extension of prime degree F_{r-1}/F_r, we obtain:

$$\begin{aligned}\iota_{F_0}^{F}(\chi \otimes \rho_{F\to F_0}\tau) &= \iota_{F_{r-1}}^{F}\iota_{F_0}^{F_{r-1}}(\chi \otimes \rho_{F\to F_0}\tau)\\ &= \iota_{F_{r-1}}^{F}(\iota_{F_0}^{F_{r-1}}\chi \boxtimes \rho_{F\to F_{r-1}}\tau)\\ &= \iota_{F_0}^{F}\chi \boxtimes \tau.\end{aligned}$$

The first equality uses only the obvious transitivity of ι; the second is (7.5); the third uses (7.3) in the prime cyclic extension F_{r-1}/F; however, we see that we must extend (7.3) so as to deal with the non-Abelian representation $\iota_{F_0}^{F_{r-1}}\chi$. Therefore the proof will be completed by the following lemma:

LEMMA 7.4.: *Let E/F be a cyclic extension of prime degree, Π a representation of $G_n(\mathbf{A}_E)$ induced from cuspidal, τ a cuspidal representation of $G_m(\mathbf{A}_F)$. Assume $\Pi \boxtimes \rho_{F\to E}\tau$ exists and is induced from cuspidal. Then*

$$\iota_E^F(\Pi \boxtimes \rho_{F\to E}\tau) \cong \iota_E^F\Pi \boxtimes \tau.$$

Of course *exists* means that there is an (induced from cuspidal) representation with the correct eigenvalues almost everywhere, and the last equality has the meaning explained after (7.3).

The proof is an easy computation, relying on formulas (6.1) and (6.2) in §6, and is left to the reader (alternately, one can use, in the inert case at least, the interpretation of the functors ι and ρ in terms of Weil group representations (Def. 6.1) and an obvious extension of (7.4) to this case). ∎

The proof of Theorem 7.3 is now complete. ∎

We conclude with a tantalizing remark. Consider the regular representation r of a solvable Galois group Γ on the space $\mathbf{C}[\Gamma]$ of functions on Γ:

$$r = \bigoplus_{\rho\in\widehat{\Gamma}} \deg(\rho)\cdot\rho$$

where $\widehat{\Gamma}$ is the dual of Γ. By solvable base change, as in the proof of Proposition 7.2, we know that there exists an automorphic representation π of $\mathrm{GL}(n, \mathbf{A}_F)$, where $n = [E : F]$, such that

$$L(s,\pi) = L(s,r) = \prod_{\rho}(L(s,\rho)^{\deg\rho}.$$

(As remarked after Theorem 7.1, we may even get the identity of L-functions at all places.) Write π, as in [25b], as a formal sum of cuspidal representations π_i of $\mathrm{GL}(n_i, \mathbf{A}_F)$; $\pi = d_1\pi_1 + d_2\pi_2 + \cdots + d_k\pi_k$, π_i non-isomorphic,

$d_1 n_1 + d_2 n_2 + \cdots + d_k n_k = n$. The consideration of the poles of $L(s, \pi \otimes \pi)$ and $L(s, r \otimes r)$ at $s = 1$ shows that

$$d_1^2 + d_2^2 + \cdots + d_k^2 = \sum_\rho \deg(\rho)^2 = n.$$

Showing that each $\rho \in \widehat{\Gamma}$ is associated to a cuspidal π_i, however, seems to be difficult.

Bibliography

1. Arthur, J.,

(a) The characters of discrete series as orbital integrals, Inv. Math. **32** (1976), 205–261.

(b) The trace formula in invariant form, Ann. of Math. **114** (1981), 1–74.

(c) On a family of distributions obtained from Eisenstein series II: Explicit formulas, Amer. J. Math. **104** (1982), 1289–1336.

(d) On a family of distributions obtained from orbits, Canad. J. Math. **38** (1986), 179–214.

(e) The local behaviour of weighted orbital integrals, Duke Math. J. **56** (1988), 223–293.

(f) Intertwining operators and residues I. Weighted characters, J. Funct. Anal. (to appear).

(g) The invariant trace formula I. Local theory, Journal of the Amer. Math. Soc. **1** (1988), 323–383.

(h) The invariant trace formula II. Global theory, Journal of the Amer. Math. Soc. **1** (1988), 501–554.

(i) The characters of supercuspidal representations as weighted orbital integrals, Proc. Indian Acad. Sci. (Math. Sci.) **97** (1987), 3–19.

2. Artin, E.

(a) Über eine neue Art von L-Reihen, Hamb. Abh. **1** (1923).

(b) Zur Theorie der L-Reihen mit allgemeinen Gruppencharakteren, Hamb. Abh. **8** (1930), 292–306.

3. Artin, E., and Tate, J., *Class Field Theory*, Benjamin, New York, 1967.

4. Bernstein, J., P-invariant distributions on GL(N) and the classification of unitary representations of GL(N) (non-archimedean case), *in* Lie Group Representations II, Herb ed., Springer LN 1041 (1984), 50–102.

5. Bernstein, J., and Deligne, P., Le "centre" de Bernstein, in *Représentations des Groupes Réductifs sur un Corps Local*, Hermann, Paris, 1984, 1–32.

6. Bernstein, J., Deligne, P., Kazhdan, D., Trace Paley-Wiener theorem for reductive groups, preprint.
7. Bernstein, J., Zelevinsky, A. V.,
(a) Representations of the group GL(n,F), where F is a nonarchimedean local field, Uspekhi Mat. Nauk **31** (3) (1976), 5–70 = Russian Math. Surveys **31** (3) (1976), 1–68.
(b) Induced representations of reductive p-adic groups I, Ann. Sc. E.N.S., 4e série, **10** (1977), 441–472.
8. Borel, A.,
(a) *Linear Algebraic Groups*, Benjamin, New York, 1967.
(b) Automorphic L-functions, *in* Proc. Symp. Pure Math. XXXIII, Part II, A.M.S., Providence (1979), 27–62.
9. Borel, A., and Wallach, N., *Continuous Cohomoglogy, Discrete Subgroups, and Representations of Reductive Groups*, Annals of Math. Studies, Princeton U. Press, 1980.
10. Casselman, W.,
(a) *Introduction to the theory of admissible representations of p-adic reductive groups*, mimeographed notes.
(b) The Steinberg character as a true character, Proc. Symp. Pure Math. XXVI, A.M.S., Providence (1974), 413–417.
(c) Characters and Jacquet modules, Math. Ann. **230** (1977), 101–105.
(d) GL_n, *in* Algebraic Number Fields, Frölich ed., Academic Press (1977), 663–704.
11. Clozel, L.,
(a) Changement de base pour le représentations tempérees des groupes réductifs réels, Ann. Sc. E.N.S., 4e série, **15** (1982), 45–115.
(b) Théorème d'Atiyah-Bott pour les variétés p-adiques et caractères des groupes réductifs, Mém. Soc. Math. France **15**, 39–64.
(c) Sur une conjecture de Howe—I, Compositio Math. **56** (1985), 87–110.
(d) On limit multiplicities of discrete series representations in spaces of automorphic forms, Invent. Math. **83** (1986), 265–284.
(e) Characters of non-connected, reductive p-adic groups, Can. J. Math. **39** (1987), 149–167.
12. Clozel, L., and Delorme, P.,
(a) Le Théorème de Paley-Wiener invariant pour les groupes de Lie réductifs, Inv. Math. **77** (1984), 427–453.
(b) Sur le Théorème de Paley-Wiener invariant pour les groupes de Lie réductifs réels, C.R. Acad. Sc. Paris, t.300, série I, **11** (1985), 331–333.

13. Clozel, L., Labesse, J. P., and Langlands R. P., Morning Seminar on the Trace Formula, Lecture Notes, Institute for Advanced Study, Princeton.
14. Deligne, P., Les constantes des équations fonctionelles des fonctions L, Antwerp II, Springer LN 349 (1973), 501–595.
15. Deligne, P., Kazhdan, D., Vignéras, M.-F., Représentations des algèbres centrales simples p-adiques, in *Représentations des Groupes Réductifs sur un Corps Local*, Hermann, Paris, 1984, 33–117.
16. E. C. Dade, Accessible characters are monomial, J. Alg. **117** (1988), 256–266.
17. Doi, K., and Naganuma, H.,
(a) On the algebraic curves uniformized by arithmetical automorphic functions, Ann. of Math. **86** (1967), 449–460.
(b) On the functional equation of certain Dirichlet series, Invent. Math. **9** (1969), 1–14.
18. D. Flath, A comparison of the automorphic representations of GL(3) and its twisted forms, Pacific J. Math. **97** (2) (1981), 373–403.
19. Y. Flicker, *The Trace Formula and Base Change for GL(3)*, Springer LN 927 (1982).
20. Harish-Chandra
(a) Fourier transforms on a semisimple Lie algebra I, Amer. J. Math. **79** (1957), 193–257.
(b) *Automorphic Forms on Semi-simple Lie Groups*, Notes by J.G.M. Mars, Springer LN 68 (1968).
(c) *Harmonic Analysis on Reductive p-adic Groups*, Springer LN 162 (1970).
(d) Harmonic analysis on reductive p-adic groups, Proc. Symp. in Pure Math. XXVI (1974), 167–192.
(e) Harmonic analysis on real reductive groups III. The Maass-Selberg relations and the Plancherel formula, Ann. of Math. **104** (1976), 117–201.
(f) The Plancherel formula for reductive p-adic groups, Collected Papers, Vol. IV, Springer-Verlag, 353–367.
(g) Admissible invariant distributions on reductive p-adic groups, Queen's Papers in Pure Appl. Math. **48** (1978), 281–347.
21. Henniart, G., La conjecture de Langlands locale pour GL(3), Mém. S.M.F., n° 11–12 (1984).
22. Howe, R., The Fourier transform and germs of characters, Math. Ann. **208** (1974), 305–322.
23. Howe, R., and Moy, A., Harish-Chandra homomorphisms for p-adic groups, preprint.

24. Jacquet, H.,
(a) Automorphic forms on GL(2) II, Springer LN 278 (1972).
(b) Generic representations, Non-commutative harmonic analysis, Springer LN 587 (1977), 91–101.
(c) Principal L-functions of the linear group, *in* Proc. Symp. Pure Math. XXXIII, Part II, A.M.S. Providence (1979), 63–86.
(d) On the residual spectrum of GL(n), Lie Group Representations II, Herb ed., Springer LN 1041, 1984.
25. Jacquet, H., and Langlands, R. P., *Automorphic Forms on GL(2)*, Springer LN 114, (1970).
26. Jacquet, H. Piatetskii-Shapiro, I.I., Shalika, J.A.,
(a) Facteurs L et ϵ du groupe linéaire: théorie archimédienne, C. R. Acad. Sc. Paris, t. 293 (1981) 13–18.
(b) Rankin-Selberg convolutions, Am. J. Math. **105** (1983), 367–464.
27. Jacquet, H., and Shalika, J.A.,
(a) On Euler products and the classification of automorphic representations I, Am. J. Math. **103** (1981), 499–558.
(b) On Euler products and the classification of automorphic representations II, Am. J. Math. **103** (1981), 777–815.
28. Kazhdan, D., Cuspidal geometry of p-adic groups, to appear in Israel J. Math.
29. Kottwitz, R.,
(a) Orbital integrals on GL(3), Am. J. Math. **102** (1980), 327–384.
(b) Sign changes in harmonic analysis on reductive groups, Transactions A.M.S. **278** (1983), 289–297.
(c) Base change for unit elements of Hecke algebras, preprint.
30. Langlands, R. P.,
(a) Problems in the theory of automorphic forms, Lectures in Modern Analysis and Applications, Springer LN 170 (1970), 18–86.
(b) On the classification of irreducible representations of real reductive groups, mimeographed notes, Institute for Advanced Study, Princeton 1973.
(c) Automorphic representations, Shimura varieties, and motives. Ein Märchen, *in* Proc. Symp. Pure Math. XXXIII, Part II, A.M.S. Providence (1979), 205–246.
(d) *On the Functional equations satisfied by Eisenstein series*, Springer LN 544 (1976).
(e) *Base change for GL(2)*, Princeton U. Press, 1980.

31. Piatetskii-Shapiro, I.I., Multiplicity one theorems, *in* Proc. Symp. Pure Math. XXXIII, Part II, A.M.S., Providence (1979), 209–212.

32. Repka, J., Base change for tempered irreducible representations of $GL(n, \mathbf{R})$, Pacific J. Math. **93** (1981), 193–200.

33. Rogawski, J.,
(a) Applications of the building to orbital integrals, thesis, Princeton University, 1980.
(b) Representations of GL(n) and division algebras over a p-adic field, Duke Math. J. **50** (1983), 161–169.
(c) The trace Paley-Wiener theorem in the twisted case, preprint.

34. Saito, H., Automorphic Forms and Algebraic Extensions of Number Fields, Lectures in Math., Kyoto Univ. 1975.

35. Serre, J.-P., *Corps Locaux*, Hermann, Paris 1968.

36. Shahidi, F.,
(a) On nonvanishing of L-functions, Bull. A.M.S. (new series) **2** (1980), 442–464.
(b) On certain L-functions, Am. J. Math. 103 (1981), 297–295.
(c) Local coefficients and normalization of intertwining operators for GL(n), Compositio Math. **48** (1983), 271–295.
(d) Fourier transforms of intertwining operators and Plancherel measures for GL(n), Am. J. Math. **106** (1984), 67–111.
(e) Local coefficients as Artin factors for real groups, preprint.

37. Shalika, J., The multiplicity one theorem for GL(n), Ann. of Math. (2) **100** (1974), 171–193.

38. Shelstad, D.,
(a) Characters and inner forms of a quasi-split group over **R**, Compositio Math. **39** (1979), 11–45.
(b) Base change and a matching theorem for real groups, Non Commutative Harmonic Analysis and Lie Groups, Springer LN 880 (1981), 425–482.

39. Shintani, T., On liftings of holomorphic cusp forms, *in* Proc. Symp. Pure math. XXXIII, Part II, Providence (1979), 79–110.

40. Silberger, A.,
(a) The Langlands quotient theorem for p-adic groups, Math. Ann. **236** (1978), 95–104.
(b) *Introduction to Harmonic Analysis on Reductive p-adic Groups*, Princeton U. Press (1979).

41. Vignéras, M. F.,
(a) Caractérisation des intégrales orbitales sur un groups réductif p-adique, J. Fac. Sc. Univ. Tokyo, Sec. IA, **29** (1981), 945–962.
(b) Correspondances entre représentations automorphes de GL(2) sur une extension quadratique et de GSp(4) sur **Q**, conjecture locale de Langlands pour GSp(4), preprint.
42. Zelevinsky, A. V., Induced representations of reductive p-adic groups II. On irreducible representations of GL(n), Ann. Sc. E.N.S. (Ser. 4) **13** (1980), 165–210.